Silkworm Biofactory

Enoch Y. Park

Silkworm Biofactory

A Breakthrough Platform for High-Yield Protein Production

Springer

Enoch Y. Park
Shizuoka University
Shizuoka, Japan

ISBN 978-981-92-0055-9 ISBN 978-981-92-0056-6 (eBook)
https://doi.org/10.1007/978-981-92-0056-6

This Springer imprint is published by the registered company Springer Nature Singapore Pte Ltd.
The registered company address is: 152 Beach Road, #21-01/04 Gateway East, Singapore 189721, Singapore

Preface

Insects are among the most populous organisms on Earth, with an estimated 3 to 50 million species. It is no exaggeration to say they cover the planet. However, a significant number of these species remain unknown, suggesting that many valuable resources remain unexplored. Each insect possesses remarkable survival strategies, making them ecologically and economically significant to human society.

The silkworm, classified within the phylum Arthropoda and class Insecta, is a member of this diverse group. Notably, the silkworm has played a crucial role in sericulture, producing silk for more than 5,000 years alongside human civilization. The historic Silk Road, extending more than 6,400 km, facilitated economic, cultural, and technological exchanges across Eurasia, significantly influencing the development of Eastern and Western civilizations. However, the invention of synthetic fibers, such as nylon, in the 1930s gradually shifted production from natural to artificial fibers, leading to a global decline in sericulture.

Silkworms produce exceptional amounts of protein, spitting out approximately 1.2 to 1.5 km of silk fiber per individual. Genetic engineering techniques developed in the 1970s have enabled gene manipulation in silkworms, facilitating further advances. Thus, repurposing silkworms as hosts for protein production beyond traditional sericulture is a logical progression. This book presents research findings on the silkworm primarily as a host for protein production, based on extensive studies.

This book is organized into eight chapters.

Chapter 1 shows the history of the silkworm and sericulture. Archaeological evidence of silk production dating back 5,000 years reveals the development of sericulture and the emergence of the Silk Road, a 6,400 km route that transformed trade between East and West. The Silk Road facilitated the exchange of goods such as spices, tea, and ceramics from the East, and horses, honey, wine, and gold from the West, laying the foundation for global trade and cultural exchange. The chapter also details Japan's sericulture, which became a major industry in the 1800s and, by the 1900s, made Japan the world's leading silk producer. However, the industry declined due to World War II and the invention of synthetic fibers, which shifted the focus from natural to artificial fibers and from agriculture to industry. Recently, new technologies have been developed to expand sericulture beyond silk production.

Chapter 2 introduces the origins of silkworm biotechnology, with a focus on silkworm disease. The development of electron microscopy in the 1950s enabled the identification of the causative agent, leading to extensive research on baculoviruses. The chapter outlines the life cycle and infection mechanisms of baculoviruses, genome sequencing, and the use of silkworms as hosts for recombinant protein expression.

Chapter 3 presents key achievements essential to the development of silkworm gene expression systems. Traditional homologous recombination enabled the successful expression of interferon in silkworms. The chapter also introduces novel recombinant protein expression methods that overcome limitations of conventional techniques, including the development of the bacmid expression system, which combines properties of baculoviruses and plasmids, allowing gene replication in both silkworms and *Escherichia coli*. Transgenic expression systems that fuse target proteins to silk proteins, such as sericin, are also described, along with the resulting data.

Chapter 4 addresses the tendency of recombinant proteins expressed in silkworms to degrade. To prevent liquefaction in silkworms, bacmids lacking protease and chitinase were developed. Additional improvements, such as promoter modifications, are discussed, resulting in more robust bacmid expression systems.

Chapter 5 examines the quality of recombinant proteins expressed in silkworms. Posttranslational modifications are essential for maintaining protein function. The chapter focuses on modifications such as disulfide bond formation, phosphorylation, and glycosylation in silkworm-produced recombinant proteins, demonstrating that silkworms can achieve posttranslational modifications comparable to those in mammalian cells.

Chapter 6 presents the practical recombinant protein production using the silkworm bacmid expression system. Recombinant proteins are grouped into three types: secretory proteins, membrane proteins, and virus-like particles (VLPs). The chapter outlines expression methods for each category and further distinguishes between membrane-enveloped VLPs (eVLPs) and non-enveloped VLPs (neVLPs), outlining their expression characteristics. Given the importance of VLPs as vaccine candidates, purification methods are also introduced and discussed.

Chapter 7 focuses on engineering VLPs. Efficient antigen display on VLP surfaces enables strong immune induction. Displaying multiple antigens enables immunity against various pathogens, making efficient surface display a major research challenge. For eVLPs, antigen fusion or co-expression with VLPs is used, whereas for neVLPs, covalent coupling of antigen molecules to VLP monomers is the primary method. The chapter introduces the creation of bivalent malaria VLPs and tetravalent dengue VLPs using these techniques, along with various published VLP modifying techniques.

Chapter 8 discusses the industrial applications and future outlook of the silkworm biofactory for recombinant protein production. It traces the commercialization history, starting with the first successful veterinary interferon in the 1990s, and highlights recent advancements in scalable production and downstream processing. The chapter reviews various commercialized silkworm-derived products, emphasizing the platform's advantages, such as cost-effectiveness and appropriate posttranslational modifications. Looking ahead, it explores the integration of transgenic technologies and artificial intelligence to optimize these systems.

Ultimately, the chapter envisions the historic Silk Road transforming into a modern "Bio-Road" (From Silk Road to Bio-Road) championing a paradigm shift from traditional cell-based bioreactors to a robust, highly efficient silkworm biofactory.

Shizuoka, Japan — Enoch Y. Park

Acknowledgments

I am deeply appreciative of my students and fellow researchers for their dedication and collaborative spirit within my laboratory. The prospect of no longer working alongside distinguished colleagues from around the world on pioneering endeavors is one I will greatly miss. Over the past two decades, this research has been steadily advanced at Shizuoka University, Japan, with the overarching goal of establishing a silkworm-based biofactory.

The chapters of this volume are thoughtfully structured to provide a comprehensive exploration of the silkworm expression system. It is my hope that both students and researchers will find in these pages a deeper understanding of the field, including the inherent challenges and promising opportunities that each methodological approach presents.

I wish to express my special appreciation to Mei Hann Lee, senior editor, whose insightful suggestions and steadfast encouragement were pivotal in realizing this book. My sincere thanks also go to the team at Springer Nature—particularly Nandhini Natchatra Rajangam, Momoko Asawa, and Sean Eickhoff—for their exemplary professional guidance and support throughout the publication process.

Above all, I am profoundly grateful to my wife, whose steadfast support in every conceivable way has sustained my work, "From Silk Road to Bio Road," throughout my extended academic journey.

The inception of this book dates to October 2025, with its completion in July 2026, and publication anticipated in September 2026.

Shizuoka University
Shizuoka, Japan

Enoch Y. Park

Competing Interests The author has no competing interests to declare that are relevant to the content of this manuscript.

Contents

Chapter 1
History of Silkworm

Abstract This chapter provides a historical overview of sericulture, tracing its origins to ancient China during the Yangshao period (5000–3000 BC). Archaeological and phylogenetic evidence confirms that the domesticated silkworm (*Bombyx mori*) evolved from the wild silk moth (*Bombyx mandarina*) in mainland China. The narrative highlights the profound economic and cultural impact of the Silk Road, a vast trade network that eventually broke the Chinese monopoly and facilitated the global spread of silk textiles and sericulture technology across Asia, the Middle East, and Europe. Furthermore, the chapter details the evolution of the silk industry in Japan. Introduced around the third century, Japanese sericulture underwent rapid modernization during the Meiji era with the establishment of mechanized, state-run facilities like the Tomioka Silk Mill. This industrialization propelled Japan to become the world's leading raw silk producer by the early twentieth century. By chronicling these pivotal historical developments, the chapter illustrates how millennia of technological innovation and global trade in traditional silk production established the foundational infrastructure and biological expertise that support modern silkworm biotechnology.

Keywords Silkworm • *Bombyx mori* • Sericulture • Silk Road • Tomioka Silk Mill

1.1 Beginning of Sericulture

It is very difficult to determine the definitive origins, location, and timing of sericulture. We believe that sericulture started during the Yangshao period (5000–3000 BC) in China (Barber 1992). Archaeological artifacts suggest that sericulture originated in mainland China, where flat silk pieces, silk belts, and silk ropes dating back to around 2750 BC (estimated) have been found in the ruins in Zhejiang Province (Zhejiang SW 1960; Zhou 1980). In 1926, halved cocoon shells were discovered at a site in Xiyin village, Shanxi Province, China, and dated to the Saito pottery period (2500–2000 BC). Since these cocoon shells are smaller than those of today's

E. Y. Park, *Silkworm Biofactory*, https://doi.org/10.1007/978-981-92-0056-6_1

domestic or mulberry silkworms, it has not been confirmed that they are silkworm cocoons. Still, it was shown that they closely resemble the cocoons of the silkworm *Rondotia menciana*, which is now found near the Yangtze River basin (Inoue 2010). Silk products were also in use at least as early as the Shang dynasty, also known as the Yin dynasty. The presence of characters such as "silkworm," "mulberry," "thread," and "woven silk" in oracle bone scripts excavated from Shang ruins, along with the analysis of cloth attached to bronze axes and jars showing that they were silk fabrics, indicates that fiber-manipulation skills were developed during the Shang dynasty (1700–1100 BC) (Keightley 1978). In China, it was forbidden to take sericultural technology out of the country, especially after the unification of China by the Qin dynasty (Liu 1952), which is believed to have strengthened the industry's control. For almost 3000 years, sericultural techniques remained confined to the Chinese mainland due to problems in transporting eggs, which hatch in a few weeks, and in procuring mulberry trees to feed the silkworms. According to some, the first silkworms were brought out of China in the first century AD by a woman who married the king of the ancient Khotan kingdom in Tibet, who carried mulberries and silkworm eggs in a hat (Hill 2003). Sericulture in China, which dates back to ancient times, gradually spread to the surrounding regions. It is said that sericulture reached the Lelang region of the Korean Peninsula in the first century BC (Kameyama 2009) or the first century AD (Inoue 2010), when a large number of people migrated from China and settled there during the Han dynasty. According to Western historians, mulberry-tree cultivation spread to India through Tibet during the first century BC. The cultivation of mulberry trees, along with the rearing of silkworms, began in the areas flanking the Brahmaputra and Ganges rivers (Savithri et al. 2013). In the sixth to seventh century AD, the smuggling of silkworm eggs into the Byzantine Empire facilitated its establishment in the Mediterranean, where it remained a monopoly for centuries (Byzantine Silk).

It is believed that sericulture began in Japan from mainland China at the end of the Yayoi period (second–third century AD). As China unified under Qin Shi Huang (221 BC) and tightened control, silkworm eggs are thought to have been shipped during the Han dynasty. The earliest record of sericulture in Japan is a piece of plain silk excavated from the Arita ruins in the Sawara district of Fukuoka City, dating back to around 200 BC (Kameyama 2009). Because Japan was well-suited to mulberry cultivation, sericulture is believed to have been introduced there relatively early. In 195 AD, silkworm eggs were introduced from Korea, and sericulture and silk-weaving techniques were temporarily introduced from China in 283 AD. During the Nara period (710–794 AD), sericulture became widespread throughout the country, such as the Tohoku region and Hokkaido. From an archaeological perspective, the beginnings of sericulture in the surrounding areas of China are shown in Table 1.1.

There are two historical views regarding the origin of the silkworm (*Bombyx mori*). One view is that the wild species *Bombyx mandarina*, which belongs to the same family as the silkworm, feeds on mulberry leaves and boxwood leaves, and hibernates in its egg stage, is considered to be the ancestor of the silkworm. The other theory posits that various wild silkworms, or the wild Indian silkworm

Table 1.1 Beginning of sericulture in various regions

Time / Region		25–20 BC	17–11 BC	1 BC	1 AD	Early 3 AD	Late 3 AD	Late 4 AD	Early 5 AD	Mid 6 AD	Late 6 AD
China	Shanxi Province	△									
	Henan Province		△ ○								
Lelang (North-western Korean peninsula)				△Kameyama (2009)							
Khotan (Xinjiang Uygur Region)					△ ◎						
Eastern Ye (North-eastern Korean peninsula)							○				
Mahan (South-western Korean peninsula)							○				
Japan	Kyushu					○ (243–250 AD)					
	Kinai (South-western Japan)							△ ○			
Yigo (Kumul Xinjiang Uygur Region)									○		
Gaochang (Xinjiang Uygur Region)										○	
Eastern Roman (Byzantine)											◎ (500–562 AD)
Persia (Northern)											○

This table was adapted from permission (Inoue 2010). Copyright 2010 Sericultural Technology Research Institute of The Dainipponn Silk Foundation
△: Existence of archaeological data on sericulture. ○: Existence of sericulture. ◎: Occurrence of sericulture

(*Thophila huttoni*), native to the Himalayan region of India, evolved into domesticated silkworms. While these two hypotheses have been proposed, advances in genetics have led to the conclusion that the latter possibility is highly unlikely. Specifically, the silkworm has $n = 28$ chromosomes and an XY male sex chromosome, whereas the Indian silkworm has $n = 31$ chromosomes. That is, silkworm chromosomes are $n = 28$, and the sex chromosomes are XY male type, whereas in the case of Indian silkworms, $n = 31$, and even in the case of Eri silkworm (*Philasamia cynthiaricini*) of the family *Saturniidae*, which is relatively easy to breed indoors and inhabits in the Assam region, $2n = 28$, and the sex chromosomes are either XY female type or XO female type (Kiyokawa 2004). In contrast, *B. mandarina* is widely distributed throughout East Asia, primarily in China, and is known to be an XY female type with $n = 28$ (China and Far East Asia) on $= 27$ (Japan, Korea, etc.). In both cases, hybridization with silkworms is possible, and the first generation (F1) and second generation (F2) of the hybrid are said to have high fertility. Archaeological evidence and cytogenetic information suggest that *B. mori* was first domesticated from *B. mandarina* in China, a finding supported by molecular phylogenetic studies (Li et al. 2005; Yoshitake 1968; Arunkumar et al. 2006).

1.2 Silk Road

Silk textiles produced by silkworms eventually formed a significant economic sphere, becoming a key trade route connecting Asia and Europe. Its route is 6,400 km long and played a crucial role in promoting economic, cultural, political, and religious interactions between the eastern and western worlds. From the second century BC to the eighteenth century AD, silk textiles, as well as tea, dyes, perfumes, and porcelain, from the East, and horses, camels, honey, wine, and gold, from the West, flowed along this route, along with cultural influences. The term "Silk Road" was coined in the nineteenth century and later became widely recognized worldwide. Recognizing its historical significance, UNESCO designated the routes of the Chang'an-Tianshan Corridor of the Silk Road (Christian 2000) (Fig. 1.1) as a World Heritage Site in 2014 under the name "The Silk Roads."

The Silk Road consisted of several routes. The main route begins in Chang'an and splits at Anxi County and then heads to Turpan and Dunhuang. At Dunhuang, the path splits, heading to Aksu and Hotan, and rejoins at Kashgar. Both routes, Turpan-Yining and Kashgar, converge at Kokand, connecting to the ancient city of Merv in Turkmenistan. From Merz, the route was nearly a straight line west through Qumis, Iran, Mesopotamia, and the northern tip of Dura-Europos in the Syrian Desert, where Mediterranean trading ships regularly plied their routes to Italy. The land route goes either north or south to North Africa.

This maritime route was used for trade with Indian dynasties. Spices and fine textiles were transported from India to Rome (Tsang 2000). Dynasties in South India and Ceylon, in addition to trading with Europe, crossed the Bay of Bengal,

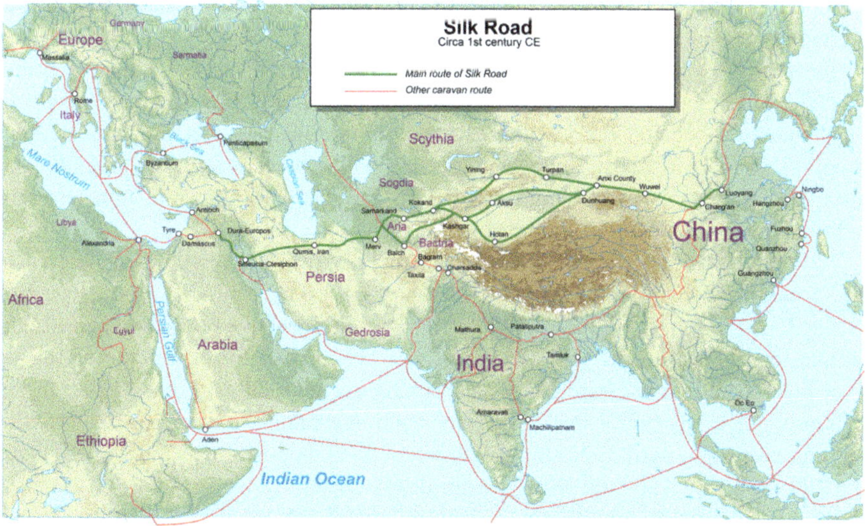

Fig. 1.1 Silk road route (https://commons.wikimedia.org/wiki/File:Silk_Road_in_the_I_century_AD_-_en.svg)

traversed the Strait of Malacca, and opened a route to the South China Sea, thereby conducting trade with Southeast Asian nations and China (Fig. 1.1).

1.3 Sericulture of Japan

Sericulture in Japan began around the third century (Inoue 2010). At 701, enacting the Taiho Code, the mulberry trees were mandated to be planted in every household, and silk was contributed as tribute. As a result, sericulture spread from the Kinki region to the Kanto and Tohoku regions during the Nara period (710–794). By the Heian period (794–1185), it had expanded to cover almost the entire nation. Ancient sericulture and silk-reeling techniques were harmonized and stabilized, becoming methods suited to Japan's climate and soil.

With the decline of the manorial system in the Middle Ages, sericulture remained a supplementary activity for farmers, and no technological improvements were made. In the early modern period, a natural economy centered on agriculture and self-sufficiency based on land formed the foundation. However, as surplus production became possible through self-sufficiency, agriculture shifted from a focus on rice and wheat to a specialization in local products. Technological improvements were also made, and demand for silk textiles increased. This commercialization led to the independent development of businesses, including silkworm egg production, sericulture, silk reeling, and weaving, and helped the establishment of sericultural technology.

In 1859, Yokohama opened as a port, and as silkworm disease, called pebrine, spread across Europe, reducing silk production, Japanese raw silk and silkworm eggs suddenly became the leading export commodities. Meanwhile, the wave of modernization in Japanese science and technology reached sericulture and silk production. Through measures such as inviting European technicians, Japanese silk manufacturing showed rapid progress.

Following the Sino-Japanese War, the export of raw silk continued to increase, and foreign currency earnings grew steadily. By 1934, raw silk production had reached 45,243 tons (Fig. 1.2). However, it suffered a decisive blow from World War II, and by 1947, after the war ended, the raw silk production had declined to just over 7186 tons. In 1957, raw silk production recovered to 20,014 tons; however, the collapse in raw silk prices in 1958 caused cocoon production to stagnate, and it did not increase again until 1974. Japan had maintained its position as the world's leading silk producer since the Meiji Restoration in 1868; however, in 1963, it ceded that position to China and became a net importer of silk cocoons (Fig. 1.2).

Japan's sericulture industry has progressed in response to the demands of each era throughout its evolution. Compared to 1900–1904, 60 years later in 1960–1964, raw silk production per 10 acres increased fivefold from 2.3 kg to 11.7 kg, and labor hours per kilogram of cocoons were reduced to one-fifth (Inoue 2010). This demonstrates how wisdom and technological innovation were applied to address the challenges demanded by each era.

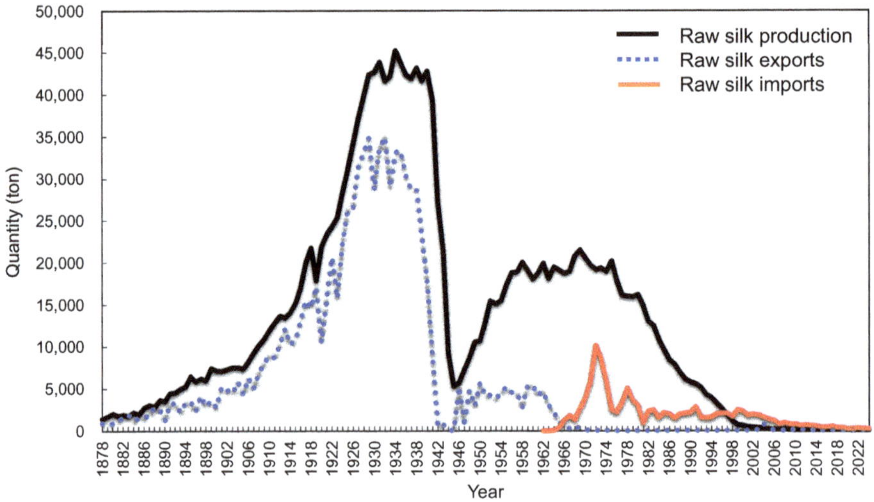

Fig. 1.2 Raw silk production in Japan. (Source: Surveys by the Ministry of Agriculture, Forestry and Fisheries, Production Bureau, 1878–2008; Surveys by the Sericultural Technology Research Institute of The Dainipponn Silk Foundation, 2009–2024)

Based on these technical developments, the Meiji government made urgent efforts to improve the quality of raw silk and increase production. In 1870, the Meiji government decided to establish a state-run model machine silk mill using domestic capital. The fundamental principles of this model factory were the introduction of Western-style silk reeling technology, the employment of foreigners as instructors, the recruitment of female workers from across the nation, and the completion of their training, which would then enable them to return to their hometowns and become instructors of machine silk reeling. Based on these principles, the Frenchman Paul Brunat, who had been working as a raw silk inspector in Yokohama, was invited to construct a large-scale, state-built machine-based silk mill, the Tomioka Silk Mill, in 1871. Its winding room, approximately 140 m long, housed 300 winding machines, making it the world's largest silk mill of its time. The main buildings were completed the following year, 1872, and operations commenced. In the reeling room where raw silk was extracted from cocoons, apprentice spinners from across the nation gathered, marking the beginning of full-scale machine silk production. After the foreign instructors left in 1876, the factory was run entirely by Japanese staff. Although the factory did not consistently operate at a profit during this government-run period, the high-quality raw silk produced was highly valued in international markets. By the time the initial objectives of promoting machine silk reeling and training technicians had been achieved, the factory was sold off to multiple companies as part of a government initiative to privatize state-run factories. This transition led to the mass production of high-quality raw silk and improvements in silkworm strains. In 1938, it became independent and was renamed Tomioka Silk Mill Co., Ltd. However, in 1939, it merged with Katakura Silk Spinning Co., Ltd. (now Katakura Industries Co., Ltd.), which was then Japan's largest silk manufacturing company. After World War II, automatic reeling machines were introduced, and the mill continued to operate for many years. However, due to the decline of Japan's silk industry, it ultimately ceased operations in March 1987. Even after operations ceased, most of the buildings were carefully preserved by Katakura Industries Co., Ltd. In July 2005, the site was designated a National Historic Site. In July 2006, the main buildings were designated Important Cultural Properties. In June 2014, it was inscribed on the World Heritage List. Furthermore, in December of the same year, three buildings—the West Cocoon Storage House, the East Cocoon Storage House (Fig. 1.3), and the Reeling House (Fig. 1.4)—were designated as National Treasures. The sericulture industry cultivated cutting-edge technology in the past, and it awaits the day when it will be revived in advanced industries beyond silk production, thereby contributing to human welfare.

Fig. 1.3 East silk cocoon storage facility of Tomioka Silk Mill (National Treasure). (Source: Web page of SILKOOL Tomioka (http://www.tomioka-silk.jp/))

Fig. 1.4 Silk reeling facility of Tomioka Silk Mill (National Treasure). (Source: Web page of SILKOOL Tomioka (http://www.tomioka-silk.jp/))

References

Arunkumar KP, Metta M, Nagaraju J (2006) Molecular phylogeny of silkmoths reveals the origin of domesticated silkmoth, *Bombyx mori* from Chinese *Bombyx mandarina* and paternal inheritance of *Antheraea proylei* mitochondrial DNA. Mol Phylogenet Evol 40:419–427. https://doi.org/10.1016/j.ympev.2006.02.023

Barber EJW (1992) Prehistoric textiles: the development of cloth in the Neolithic and Bronze Ages with special reference to the Aegean (reprint, illustrated ed.). Princeton Unversity Press

Christian D (2000) Silk roads or steppe roads? The silk roads in world history, realms of the silk roads: ancient and modern, pp 67–94

Hill JE (2003) Annotated translation of the chapter on the Western Regions according to the Hou Hanshu, 2nd Draft Edition. Appendix A

Inoue M (2010) Yousan (Sericulture). Sericultural Technology Research Institute of the Dainipponn Silk Foundation, Ibaraki Prefecture

Kameyama M (2009) Atsumizokutojohuku yayoijidaiwotukuriagetahitotati (The Atsumi clan and Jofuku: the people who created the Yayoi period). Ryufu Shobo, Nagano City

Keightley DN (1978) Sources of Shang history. The Oracle-bone Inscriptions of Bronze Age China. University of California Press, Berkeley

Kiyokawa Y (2004) Tayounarusekainosannsigyou—Takasankarakuwagomade— (The diverse world of sericulture— from domesticated silkworms to wild silkworms—). The Institute of Economic Research, Hitotsubashi University, Tokyo

Li A, Zhao Q, Tang S, Zhang Z, Pan S, Shen G (2005) Molecular phylogeny of the domesticated silkworm, *Bombyx mori*, based on the sequences of mitochondrial cytochrome b genes. J Genet 84:137–142. https://doi.org/10.1007/BF02715839

Liu KKC (1952) The silkworm and Chinese culture. Osiris 10:129–194. https://doi.org/10.1086/368551

Savithri G, Sujathamma P, Neeraja P (2013) History of sericulture—a glance. ZENITH Int J Multidisc Res 3:173–180

Tsang C-H (2000) Recent advances in the iron age archaeology of Taiwan. Bull Indo-Pac Prehist Assoc 20:153–158. https://doi.org/10.7152/bippa.v20i0.11751

Yoshitake N (1968) Phylogenetic aspects on the origin of Japanese race of the silkworm, *Bombyx mori* L. J Sericol Sci Japan 37:83–87

Zhejiang SW (1960) Wuxing Qianshanyang yizhi dierci fajue baogao (Report on the first and second excavations at the site of Qianshanyang in Wuxing County). Kaogu Xuebao, 2:73–92

Zhou K (1980) Qianshanyang canjuanpian chutu de qishi, (illuminating remarks on the fragments of silk excavated at Qianshanyang). Wenwu 1:74–77

Chapter 2
Beginning of Silkworm Biotechnology

Abstract This chapter explores the biological and virological foundations of silkworm biotechnology, beginning with an overview of the silkworm (*Bombyx mori*) life cycle and the historical impact of infectious diseases on the global silk industry. It provides a detailed examination of baculoviruses, particularly *Autographa californica* multiple nucleopolyhedrovirus (AcMNPV) and *Bombyx mori* nucleopolyhedrovirus (BmNPV). The chapter delineates the biphasic baculovirus life cycle, highlighting the distinct roles of budded viruses (BVs) for systemic propagation within the host and occlusion-derived viruses (ODVs) for robust inter-host transmission. Furthermore, it analyzes the genomic organization of these large, double-stranded DNA viruses. Despite sharing over 90% sequence homology, AcMNPV and BmNPV exhibit strict host specificities, which are primarily governed by the viral DNA helicase gene (p143). By elucidating the molecular mechanisms of viral replication, host-range determination, and viral-induced host liquefaction (mediated by enzymes like chitinase and cathepsin), this chapter establishes the critical virological principles that enabled the transformation of pathogenic baculoviruses into powerful expression vectors for modern biotechnology.

Keywords *Autographa californica* multiple nucleopolyhedrovirus (AcMNPV) • *Bombyx mori* nucleopolyhedrovirus (BmNPV) • Baculovirus • Baculovirus genome • Budded virus • Occlusion-derived virus • DNA helicase

2.1 Silkworm Life Cycle

Silkworm larvae hatch from eggs and go through four molts. During the larval stage, they voraciously feed on mulberry leaves, growing to 10,000 times their original weight in about 25 days (Fig. 2.1). Larvae that reach the fifth instar produce silk (sericultural silk) to spin a cocoon around themselves, pupating inside. These cocoons are then shipped, dried, and processed into raw silk. Silkworms bred for reproduction are kept without drying their cocoons. They emerge as adults after

E. Y. Park, *Silkworm Biofactory*, https://doi.org/10.1007/978-981-92-0056-6_2

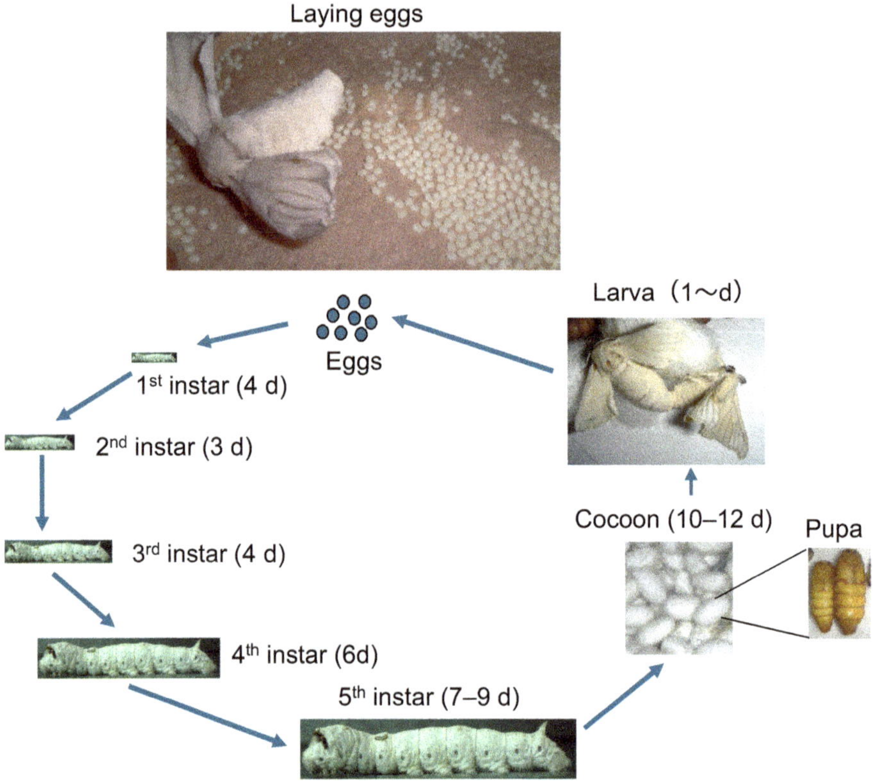

Fig. 2.1 Silkworm life cycle. Silkworm moths lay eggs, which take about 2 weeks to hatch. Some eggs are nondormant and hatch immediately, while others are dormant and do not hatch until the next spring. The hatched larvae are also called "hairy silkworms." They grow by eating mulberry leaves and molting repeatedly, increasing in size from the first to the fifth instar. The stage just before molting, when the larva becomes immobile, is called "dormant." They pupate approximately 22–27 days after hatching. They spin silk inside the cocoon, forming a cocoon about 6 cm long. Inside the cocoon, they pupate and undergo metamorphosis into adults over about 2 weeks. The adult moth (silkworm moth) emerges by breaking out of the cocoon. Its lifespan is short, lasting only a few days to about a week. It does not feed during this time, focusing solely on reproduction (mating and egg-laying). It lays eggs within a few days, passing its life on to the next generation

approximately 2 weeks. The adults quickly mate and lay around 500 eggs. The adult moths do not eat, barely move, and live for about a week before dying.

2.2 Silkworm Infectious Disease

Sericulture has spread worldwide, and silk production has become a global industry. Additionally, supplying silkworm eggs to farmers became a significant business. However, by the nineteenth century, European sericulture was frequently plagued by silkworm epidemics. Consequently, they began purchasing silkworms as far afield as Syria and Egypt, sourcing and selling healthy silkworm eggs. However, silkworm plagues spread across overseas, Africa, and India, and the silkworm merchants who had painstakingly purchased after long journeys were nearly all struck down by disease. Many merchants who had built fortunes on silkworm eggs could no longer continue their trade. Before the cause of the plague could be scientifically understood, the entire world became captive to an inexplicable curse.

Meanwhile, Japan adopted a policy of national isolation from 1639 to 1854, prohibiting trade with foreign countries for 210 years. Therefore, some silkworm egg merchants secretly entered Yokohama Port, obtained silkworm eggs, wrapped them in silk cloth, placed them in wooden boxes, and sealed them. They then boarded ships bound for the continent, reached the Russian coast, and set their course as far north as possible. This was intended to hinder the eggs' life cycle (Fig. 2.1) through exposure to cold, thereby prolonging the time before hatching. They traversed 4000 km of the Siberian plains, crossed the Ural Mountains, and arrived in Saint Petersburg. There, they bought chunks of ice for gold, packed them around the eggs, and loaded them into the holds of merchant ships bound for Hamburg. Then, the wooden boxes containing the silkworm eggs boarded southbound trains to reach their destination (Baricco 1996). However, despite all this effort, many silkworms died during the journey.

The French government sent Louis Pasteur to investigate the cause of the silkworm epidemic. During that time, a microscope was available to observe objects smaller than what is invisible to the naked eye. Under the observation of an optical microscope, dark brown spots were visible on the bodies of both larval and adult individuals. This is the origin of the name "pebre disease" (la pébrine), and according to Pasteur, the name was given by Armand de Catoufage. To the people of southern France in the nineteenth century, when the disease was rampant, the small, dark brown spots scattered across the bodies of the silkworm larvae resembled pepper sprinkled on them. This disease caused the mass death of domestic silkworms in Europe in the nineteenth century. However, under the guidance of bacteriologist Louis Pasteur, the disease was brought under control through improvements in breeding methods (Smith 1999).

By the 1950s, the true nature of viruses had been recognized. It became clear that a disease-causing factor could pass through unglazed filters with mesh sizes that bacteria could not penetrate. This factor, significantly smaller than bacteria, could cause disease in healthy individuals even in tiny quantities and demonstrated the existence of viruses—entities capable of transmitting pathogenicity across generations, meaning they could replicate. By the early 1900s, it was also hypothesized that silkworm diseases, such as pebre disease, were similarly caused by viruses and were intensively researched. Insect viruses often form crystalline structures containing numerous virus particles, known as inclusion bodies. Consequently, when silkworms become infected, small black dots appear on the surface of their bodies. The occlusion bodies can be observed under an optical microscope, allowing for easy detection of viral infection.

2.3 Baculovirus

Baculoviruses belong to insect viruses. Many insect viruses synthesize protein-based crystalline structures called occlusion bodies that encapsulate viral particles (Table 2.1).

In other animal and plant viruses, structures called occlusion bodies, which are specific to viral infection, may sometimes be observed; however, they are often products of the host's response to the infection. In the case of insect viruses, all occlusion bodies are crystalline structures composed of proteins encoded by the virus's own genes actively synthesized by the virus itself. With the advent of the light microscope, a characteristic of silkworm infectious disease was revealed by

Table 2.1 Insect viruses

Family/Subfamily	Genus	Genome		Occlusion body formation
		Size (kb)	Gene numbers	
Double-stranded DNA viruses				
Poxviridae Entomopoxvirinae	*Alphaentomopoxvirus*	246	49	Yes
	Betaentomopoxvirus	228–307	49	Yes
Iridoviridae	*Iridovirus*	200	150	No
	Chloriridovirus	191	126	No
Baculoviridae	*Nucleopolyhedrovirus*	100–180	100–150	Yes
	Granulovirus	100–180	100–150	Yes
Nimaviridae	*Whispovirus*	280	117	No
Single-stranded DNA viruses				
Parvoviridae/ Densovirinae	*Densovirus*	4–6	2	Yes
	Iteravirus	5	3	Yes
	Brevidensovirus	4	2	No
Double-stranded RNA viruses				
Reoviridae	*Idnoreovirus*	20–30	10	Yes
	Cypovirus	25	10–12	Yes
Birnaviridae	*Entomobirnavirus*	6–10	5	Yes
Negative single-stranded RNA viruses				
Rhabdoviridae	*Vesiculovirus*	11–16	5	Yes
	Cytorhabdovirus	10–20	5–7	Yes
	Nucleorhabdovirus	11–15	5–6	Yes
Orthomyxoviridae	*Thogotovirus*	10	7–9	No
Bunyaviridae	*Orthobunyavirus*	11–15	6	Yes/No
	Nairovirus	11–20	6	Yes/No
	Tospovirus	15–20	6	Yes
Positive single-stranded RNA viruses				
Flaviviridae	*Flavivirus*	10–12	7	No
	Hepacivirus	10–11	3–5	Yes
Togaviridae	*Alphavirus*	11–12	3	No

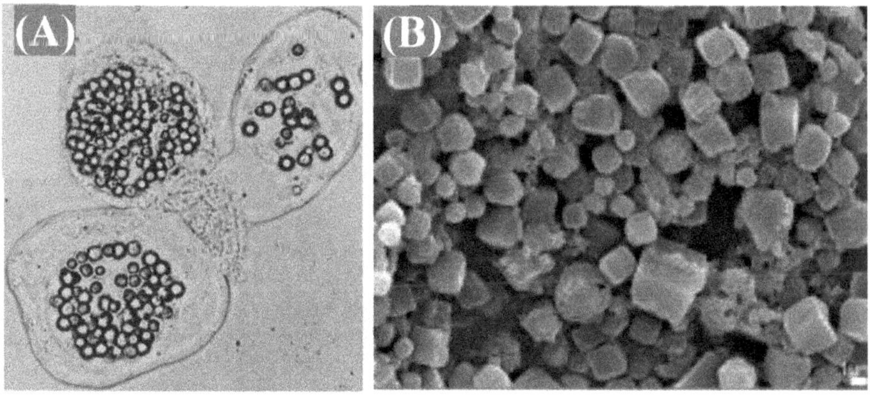

Fig. 2.2 Occlusion bodies and polyhedra images. (**a**) *Trichoplusia ni* ovarian cells containing polyhedra (×8000). (Adapted from permission (Volkman and Summers 1975). Copyright 1975 American Society for Microbiology). (**b**) Scanning electron micrograph image of the purified recombinant occlusion bodies of native AcMNPV. (Adapted from permission (Silva et al. 2022). Copyright 2022 Springer Nature)

the presence of lightly refractile occlusion bodies (Fig. 2.2a), which were symptomatic of the affected insects. This structure was renamed as a polyhedra. With advances in microscopy, occlusion bodies, including infectious particles, were visualized by electron microscopy (Fig. 2.2b).

Many insect viruses form occlusion bodies composed of a single type of protein, maintaining beautiful shapes such as regular hexahedra or icosahedra. These occlusion bodies are remarkably durable and withstand typical environmental stresses, enabling the virions they contain to remain infectious for extended periods. They can survive passage through the gastrointestinal tract of birds, which facilitates their widespread spread (Entwistle et al. 1978). Viruses that form occlusion bodies can be categorized into four types: nuclear polyhedrosis viruses (NPVs) (Fig. 2.3a), cytoplasmic polyhedrosis viruses (CPVs) (Fig. 2.3b), granulosis viruses (GVs) (Fig. 2.3c), and Entomopoxviruses (Fig. 2.3d).

Two main types of polyhedrosis diseases have been identified: one, called nuclear polyhedrosis virus (NPV), forms polyhedra inside the cell nucleus, while the other, known as cytoplasmic polyhedrosis virus (CPV), produces occlusion bodies in the cytoplasm. The NPVs contain DNA with a rod-shaped morphology (Fig. 2.3e). In contrast, CPVs were observed to have icosahedral capsids and belong to the Reoviridae family, a group of viruses characterized by segmented, double-stranded RNA genomes. The other group of baculoviruses, characterized by small, granular, ellipsoidal occlusion bodies (Steinhaus et al. 1949), was reported and named granulosis viruses (GVs) based on the occlusion body morphology. Entomopoxviruses are characterized by forming a spherical occlusion body during the later stages of infection, within which mature virus particles become occluded.

This contrasts with baculoviruses that use the trachea as a conduit to disseminate infection. The NPVs and GVs were unified into the Baculoviridae in honor of

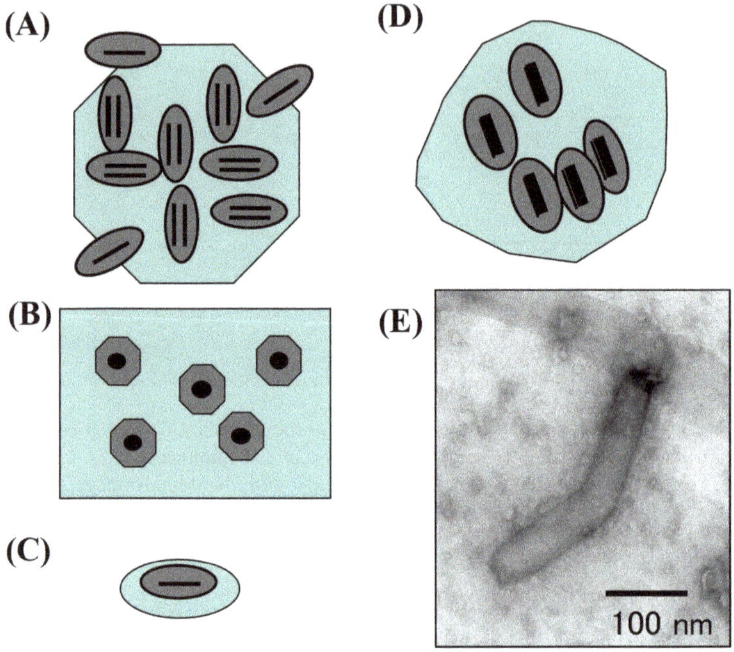

Fig. 2.3 Illustration of occlusion bodies formed by insects. Nuclear polyhedrosis viruses (**a**), cytoplasmic polyhedrosis viruses (**b**), granulosis viruses (**c**), Entomopoxviruses (**d**), and baculoviruses observed by a transmission electron microscope (**e**)

various historical individuals who had done early research (Vago et al. 1974). The name "baculovirus" was proposed by Mauro Martignoni because of the rod-shaped morphology of their virions, which is derived from the Latin baculum, meaning "cane," "walking stick," or "staff."

The polyhedra contained up to 200 baculoviruses, which are rod-shaped nucleocapsids that are 230–385 nm in length and 40–60 nm in diameter (Ackermann and Smirnoff 1983). Baculoviruses are a highly diverse group of viruses that package double-stranded, circular, supercoiled genomes ranging in size from approximately 80–180 kb and encode 90–180 genes. Typically, there are two forms of virions produced: occlusion-derived virions (ODVs) and budded virions (BVs). While both types share similar nucleocapsid structures, they differ in the sources and compositions of their envelopes and play distinct roles in the viral life cycle.

2.4 Host Specificity of Baculovirus

Over the years, baculoviruses have been studied in a variety of invertebrate species. Well-studied hosts are insects of the orders Lepidoptera, Diptera, and Hymenoptera. *Autographa californica* multinucleocapsid nucleopolyhedrovirus (AcMNPV) can infect 39 species of Lepidopteran larvae belonging to 13 families (Gershburg et al. 1997). Advancements in tissue culture techniques and molecular research have enabled the development and isolation of genetically modified AcMNPVs exhibiting unique traits. The *Bombyx mori* nucleopolyhedrovirus (BmNPV) is a significant pathogen affecting the mulberry silkworm, *B. mori*, and is responsible for substantial losses in silk production. Comparative genomic analyses have shown that BmNPV and AcMNPV share more than 90% sequence similarity (Gomi et al. 1999).

The two closely related baculoviruses, AcMNPV and BmNPV, exhibit a high degree of DNA sequence similarity but differ in host specificity (Maeda et al. 1993). AcMNPV replicates in *Spodoptera frugiperda* cells and *Trichoplusia ni* cells but does not in *B. mori* cells. In contrast, BmNPV replicates in *B. mori* cells but does not in either *S. frugiperda* cells or *T. ni* cells (Kondo and Maeda 1991). Coinfection of *B. mori* cells with AcMNPV and BmNPV prevents the production of both AcMNPV and BmNPV; however, coinfection of *S. frugiperda* cells with AcMNPV and BmNPV leads to the replication of both BmNPV and AcMNPV, generating host-range expanded viruses that can replicate in both types of cells (Kamita and Maeda 1993). Genomic analysis of these host-range expanded viruses revealed that a baculovirus DNA helicase gene, p143, is responsible for the host-range expansion (Kamita and Maeda 1997; Maeda et al. 1993). AcMNPV-P143 (Ac-P143) and BmNPV-P143 (Bm-P143) share about 96% identity in their amino acid sequences (Kamita and Maeda 1997); however, recombinant AcMNPVs with short sequence substitutions in Bm-p143 that correspond to Ac-p143 sequences exhibit the host-range expanded phenotype (Kamita and Maeda 1997; Maeda et al. 1993). By fine-mapping the host-range specificity region using silkworm larvae, it has been demonstrated that the substitution of only two amino acids, S564N and F577L, is sufficient to alter the host-range specificity of AcMNPV (Argaud et al. 1998).

2.5 Life Cycle of Baculovirus

The NPVs form occlusion bodies called polyhedra within the nucleus, which can contain numerous viral particles and serve as a reservoir for the virus. GVs also form occlusion bodies that contain only one virus particle in each granule. Viruses that do not form occlusion bodies have no safe place to store themselves, so they must constantly infect hosts in the wild. Such viruses are found in tropical insects.

Infection by NPVs involves a complex process. In nature, when a larva ingests occlusion bodies, they dissolve in the midgut due to the alkalinity of digestive fluids and the action of proteases. From within, rod-shaped viral particles with membrane structures release onto and invade intestinal cells (Fig. 2.4a). After initial replication there, they replicate in most tissues throughout the body, including blood cells and fat cells. At this stage, two distinct types of viral particles are formed. One type is budded viruses (BVs), which are released by budding and enveloped by the plasma membrane of the cell compartment that facilitates infection. These BVs are not occluded and spread the infection from cell to cell within the insect, primarily

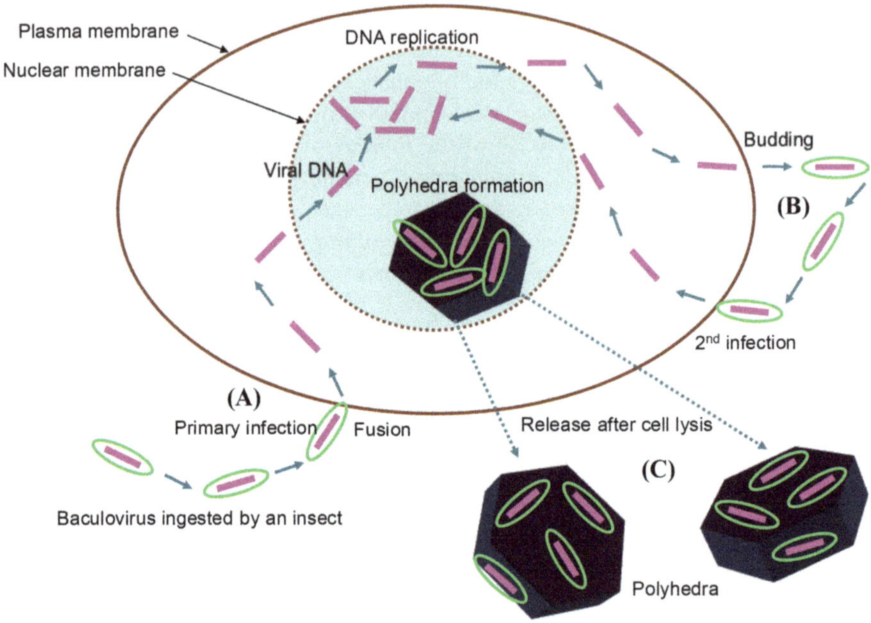

Fig. 2.4 Illustration of the life cycle of a baculovirus. When an insect ingests occlusion bodies, these structures dissolve in the midgut, releasing occlusion-derived viruses that subsequently infect the midgut epithelial cells. Virions then bud from the host cell, initiating the primary infection (**a**). During the early phase, additional budded viruses are generated, propagating the infection throughout the host organism (**b**). In the late stage of infection, polyhedra form; the infected cell ultimately lyses, releasing occlusion bodies (**c**)

produced during the early and mid-stage of infection, enabling the spread throughout the insect (Fig. 2.4b). While highly capable of infecting other cells within the body, they rarely cause infection when ingested by the larva. The other type is occlusion-derived viruses (ODVs), which are occluded within the nucleus during the late stage of infection and are adapted for use outside the host insect, specifically for infecting midgut cells. In the late stage of infection, these types of viral particles were produced, encapsulated, and formed occlusion bodies. When the cell dies, it releases the occlusion bodies (Fig. 2.4c).

Some baculoviruses express enzymes such as chitinase or protease that facilitate the disintegration of infected larvae late in infection. The viral chitinase is clustered with several lepidopteran chitinases; for example, the AcMNPV (Ac126) and BmNPV (Bm103) proteins show over 60% amino acid sequence identity to *B. mori* chitinase. Although similar in amino acid similarity, they exhibit different properties; the AcMNPV chitinase stays in the endoplasmic reticulum and functions under alkaline conditions, whereas the host enzyme is secreted and exhibits reduced activity at higher pH levels (Daimon et al. 2006; Thomas et al. 1998). Furthermore, chitinase, in conjunction with a viral proteinase (cathepsin, Ac127), contributes to the liquefaction of insects. Experimental deletion of either the chitinase or cathepsin gene demonstrated that the remaining functional protein maintained its activity for several days postmortem. Retention of viral chitinase within the endoplasmic reticulum may delay liquefaction in infected insects, thereby enabling the virus to continue its replication cycle (Hawtin et al. 1997). Their infection efficiency differs for different tissues; ODVs infect midgut epithelial cells up to 10,000 times more efficiently than BVs, whereas conversely, BVs are up to 1000 times more effective at infecting cultured cells than ODVs (Volkman and Summers 1977; Volkman et al. 1976). Both types are essential for replication and transmission in nature.

NPV genes are classified based on their expression timing: genes expressed before the initiation of genomic DNA replication are called early genes, while genes whose expression begins at or after the initiation of genomic DNA replication are called late genes. Early genes include those encoding proteins involved in viral DNA replication and proteins that regulate the expression of other early and late genes. The promoters of early genes are recognized by host factors and are transcribed by the host RNA polymerase II, similar to host genes. Late genes encode viral structural proteins and proteins involved in late gene expression, including p10 that participates in nuclear envelope breakdown and cell lysis, and polyhedrin, the major component of occlusion bodies.

Once the viral genes are transferred to the nucleus and begin to replicate, the initial viral genes are expressed. The action of the resulting gene products gradually suppresses the expression of host genes, leading to the selective transcription of viral genes. In the final stage, large amounts of polyhedrin proteins are synthesized, ultimately leading to the cell's death. The viral genome consists of circular double-stranded DNA, making it a large DNA virus with a total size of approximately 130 kb. Apart from approximately 100 bp repetitive sequences found at five locations across the genome, most of it consists of gene translation regions, containing around 100 genes throughout the entire virus. For such studies at the gene level of NPVs, AcMNPV is commonly used. Among the approximately 100 or more genes, only a very limited number are immediately expressed upon infection. These promoters have structures similar to those of the host, enabling the host's transcription system to produce viral mRNA. Such gene products then act on the virus's specific promoter sequences, causing only viral genes to be expressed. In contrast, host gene expression gradually declines. By the late stages of infection, synthesis declines even for viral proteins initially expressed, and only polyhedrin proteins are selectively synthesized. Polyhedrin proteins can reach as much as 20–30% of all proteins.

2.6 Baculovirus Genome

The complete nucleotide sequences of nucleopolyhedrovirus (NPV) genomes have been determined. Among these, the AcMNPV and BmNPV genomes are the most extensively studied. The AcMNPV genome comprises 134 kb with an A + T content of 59%. It contains 154 methionine-initiated open reading frames (ORFs) of at least 150 nucleotides in length (Ayres et al. 1994). These ORFs are distributed uniformly across both strands of the AcMNPV genome, arranged as non-overlapping reading frames and separated by short intergenic regions. On the other hand, the BmNPV genome was 128 kb long, with a G + C content of 40%, and contained 136 ORFs encoding proteins longer than 60 amino acids (Gomi et al. 1999). Although BmNPV and AcMNPV differ phenotypically, their genome organization is highly similar. Over 90% of the BmNPV genome shares identity with approximately three-quarters of the AcMNPV genome's open reading frames (ORFs). The annotation of AcMNPV and their homologs in BmNPV, including gene products, is summarized (Table 2.2). The BmNPV genome lacked homologs of seven or eight AcMNPV ORFs and contained five ORFs related to Ac2. Such genomic information will likely be widely used not only for fundamental investigations of gene function but also for applications such as gene expression in insect cells or silkworms, as well as protein production.

Table 2.2 AcMNPV genes, their homologs in BmNPV, and their gene products (Reproduced with permission (Rohrmann 2019))

AcMNPV gene	Homologs in BmNPV	Gene product	References
Ac1 (168aa:19.3 kDa)	Bm130 (168aa:19.3 kDa)	Protein tyrosine phosphatase (PTP)	Sheng and Charbonneau (1993)
Ac2 (328aa:38.8 kDa)	Bm131(349aa:40.1 kDa), Bm22 (317aa:35.4 kDa), Bm80 (239:27.5 kDa), Bm81 (318aa:35.9 kDa), Bm132 (241aa:27.8 kDa)	Baculovirus repeated orf (BRO)	Shrestha et al. (2018) and Zemskov et al. (2000)
Ac3 (53aa:5.6 kDa)		Conotoxin-like (Ctl)	Olivera et al. (1994)
Ac6 (210aa:23.9 kDa)	Bm135 (210aa:23.8 kDa)	Late expression factor (LEF)-2	Wu et al. (2010)
Ac8 (245aa:28.6 kDa)	Bm1 (245aa:28.8 kDa)	Polyhedrin	Ji et al. (2010)
Ac9 (543aa:60.7 kDa)	Bm2 (542aa:60.9 kDa)	Pp78/83	Kitts and Possee (1993)
Ac10 (272aa:32 kDa)	Bm3 (275aa:32.4 kDa)	Protein kinase-1 (PK-1)	Reilly and Guarino (1994)
Ac14 (266aa:30.8 kDa)	Bm6 (270aa:31.1 kDa)	Lef-1, DNA primase	Lu and Miller (1995b)
Ac15 (506aa:57 kDa)	Bm7 (506aa:57.0 kDa)	Ecdysteroid UDP-glucosyl transferase (EGT)	O'reilly and Miller (1989)
Ac16 (225aa:25.9 kDa)	Bm8 (229aa:26.2 kDa)	DA26	Braunagel et al. (2003)
Ac20/21 (417aa:47.7 kDa)	Bm12 (440aa:50.0 kDa)	Actin-rearrangement-inducing factors (arif-1)	Roncarati and Knebel-Morsdorf (1997)

(continued)

Table 2.2 (continued)

AcMNPV gene	Homologs in BmNPV	Gene product	References
Ac22 (382aa:43.8 kDa)	Bm13 (382aa:43.8 kDa)	Per os infectivity factor (PIF)	Pijlman et al. (2002)
Ac23 (690aa:79.9 kDa)	Bm14 (673aa:78.0 kDa)	Fusion protein (F)	Xu et al. (2019)
Ac24 (169aa:19.2 kDa)	Bm15 (169aa:19.4 kDa)	Protein kinase interacting protein (PKIP)	Fan et al. (1998)
Ac25 (316aa:36.6 kDa)	Bm16 (317aa:36.7 kDa)	DNA binding protein (DBP)	Mikhailov et al. (2008)
Ac27 (286aa:33.3 kDa)	Bm18 (292aa:34.0 kDa)	Inhibitor of apoptosis (Iap-1)	Clem (2015)
Ac28 (173aa:20.4 kDa)	Bm19 (173aa:20.3 kDa)	Lef-6	Passarelli and Miller (1994)
Ac31 (151aa:16.2 kDa)	Bm23 (151aa:16.3 kDa)	Super oxide dismutase (SOD)	Tomalski et al. (1991)
Ac32 (181aa:20.6 kDa)	Bm24 (182aa:20.8 kDa)	Fibroblast growth factor (FGF)	Detvisitsakun et al. (2005)
Ac33 (182aa:20.8 kDa)		Polynucleotide kinase (PNK)	Wang et al. (2007a)
Ac35 (77aa:8.7 kDa)	Bm26 (77aa:8.7 kDa)	Ubiquitin-like protein	Ma et al. (2015)
Ac36 (275aa:31.3 kDa)	Bm27 (277aa:31.5 kDa)	Pp31	Guarino and Summers (1986)
Ac 37 (112aa:13.1 kDa)	Bm28 (112aa:13.1 kDa)	Lef-11	Lu and Miller (1995b)
Ac38 (216aa:25.3 kDa)	Bm29 (217aa:25.5 kDa)	Nucleotide diphosphate X (Nudix)	Mildvan et al. (2005)
Ac40 (401aa:47.5 kDa)	Bm31 (399aa:47.3 kDa)	P47	Li et al. (1999)
Ac41 (181aa:21.1 kDa)	Bm32 (183aa:21.1 kDa)	Lef-12	Todd et al. (1995)
Ac42 (506aa:59.1 kDa)	Bm33 (506aa:59.2 kDa)	Global transactivator (Gta)	Katsuma et al. (2008)
Ac46 (704aa:79.1 kDa)	Bm37 (702aa:79.2 kDa)	ODV-E66	Hong et al. (1994)
Ac47 (88aa:10.5 kDa)	(Bm38:89aa:10.5 kDa)	Translin-associated factor X (Trax)-like	
Ac48 (113aa:12.9 kDa)		ETM (the mid-sized orf in the EcoRI T fragment)	
Ac49 (285aa:32.1 kDa)		Proliferating cell nuclear antigen (PCNA)	Johnson and O'donnell (2005)

(continued)

Table 2.2 (continued)

AcMNPV gene	Homologs in BmNPV	Gene product	References
Ac50 (876aa:101.8 kDa)	Bm39 (877aa:101.8 kDa)	Lef-8	Passarelli et al. (1994)
Ac51 (318aa:37.5 kDa)	Bm40 (319aa:37.8 kDa)	DNA domain protein	Langer et al. (1992)
Ac53a (78aa:8.6 kDa)	Bm42a (78aa:8.6 kDa)	Lef-10	Lu and Miller (1994)
Ac54 (365aa:42.1 kDa)	Bm43 (365aa:42.0 kDa)	V1054 capsid protein	Olszewski and Miller (1997)
Ac58, Ac58/59, (172aa:20.3 kDa)	Bm47 (171aa:37.8 kDa)	ChaB homolog	Harrison and Bonning (2003)
Ac60 (87aa:10.1 kDa)	Bm48 (83aa:9.7 kDa)	ChaB homolog	Arunkumar et al. (2006)
Ac61 (214aa:25.2 kDa)	Bm49 (214aa:25.3 kDa)	Few polyhedra (FP, FP-25 K)	Harrison et al. (1996)
Ac62 (516aa:59.3 kDa)	Bm50 (490aa:56.4 kDa)	Lef-9	Guarino et al. (1998)
Ac64 (302aa:34.8 kDa)	Bm52 (294aa:33.8 kDa)	GP37, Fusolin	Vialard et al. (1990)
Ac65 (984aa:114.3 kDa)	Bm53 (986aa:114.4 kDa)	DNA polymerase	Tomalski et al. (1988)
Ac66 (808aa:94 kDa)	Bm54 (805aa:93.3 kDa)	Desmoplakin-like	Braunagel et al. (2003)
Ac67 (385aa:44.6 kDa)	Bm55 (385aa:44.9 kDa)	Single-stranded DNA binding protein (SSB), Lef-3	Lu and Miller (1995b)
Ac68 (192aa:22.3 kDa)	Bm56 (134aa:15.8 kDa)	PIF-6	Li et al. (1999)
Ac69 (262aa:30.4 kDa)	Bm57 (262aa:30.4 kDa)	RNA Cap (Nucleoside-2′-O)-Methyltransferase	Wu and Guarino (2003)
Ac70 (290aa:34.4 kDa)		Host cell-specific factor-1 (Hcf-1)	Lu and Miller (1995a)
Ac71 (249aa:28.6 kDa)	Bm58 (249aa:28.7 kDa)	Inhibitor of apoptosis-2 (Iap-2)	Clem (2015)
Ac77 (379aa:44.4 kDa)	Bm63 (379aa:44.3.0 kDa)	Very late factor-1 (Vlf-1)	Yang and Miller (1999)
Ac80 (409aa:45.4 kDa)	Bm66 (403aa:44.9 kDa)	GP41	Whitford and Faulkner (1992)
Ac82 (180aa:19.8 kDa)	Bm68 (181aa:20.1 kDa)	Telokin-like protein (TLP)	Holden et al. (1996)
Ac83 (847aa:96.2 kDa)	Bm69 (839aa:95.8 kDa)	VP91, PIF-8	Russell and Rohrmann (1997)

(continued)

Table 2.2 (continued)

AcMNPV gene	Homologs in BmNPV	Gene product	References
Ac86 (684aa:80.8 kDa)		Polynucleotide 5′-kinase (PNK)/ Polynucleotide 3′-phosphatase (PNL)	Martins and Shuman (2004)
Ac87 (126aa:15 kDa)	Bm70 (126aa:15.1 kDa)	P15	Lu and Iatrou (1997)
Ac88 (264aa:30.1 kDa)	Bm71 (267aa:30.7 kDa)	CG30	Thiem and Miller (1989a)
Ac89 (347aa:39 kDa)	Bm72 (350aa:39.3 kDa)	VP39	Thiem and Miller (1989b)
Ac90 (464aa:53.9 kDa)	Bm73 (465aa:54.0 kDa)	LEF-4	Guarino et al. (1998)
Ac92 (259aa:30.9 kDa)	Bm75 (259aa:30.9 kDa)	P33, Sulfhydryloxidase (SOX)	Long et al. (2009)
Ac93 (161aa:18.4 kDa)	Bm76 (161aa:18.4 kDa)	P18	Yuan et al. (2011)
Ac94 (228aa:25.5 kDa)	Bm77 (228aa:25.6 kDa)	ODV-E25 (p25, 25 k, e25)	Wu and Passarelli (2010)
Ac95 (1221aa:143 kDa)	Bm78 (1222aa:143.6 kDa)	DNA helicase (p143)	Gordon and Carstens (1984)
Ac96 (173aa:19.8 kDa)	Bm79 (182aa:21.0 kDa)	PIF-4	Fang et al. (2009)
Ac98 (320aa:38 kDa)	Bm82 (320aa:38.0 kDa)	38 K	Olszewski and Miller (1997)
Ac99 (265aa:31 kDa)	Bm83 (265aa:31.1 kDa)	LEF-5	Passarelli and Miller (1993)
Ac100 (55aa:6.9 kDa)	Bm84 (65aa: 8.1 kDa)	P6.9 (arginine/serine/ threonine-rich DNA binding protein)	Wilson et al. (1987)
Ac101 (361aa:41.5 kDa)	Bm85 (362aa:41.6 kDa)	BV/ODV C42	Braunagel et al. (2003)
Ac102 (122aa:13.3 kDa)	Bm86 (123aa:13.5 kDa)	P12	Braunagel et al. (2003)
Ac103 (387aa:45.3 kDa)	Bm87 (387aa:45.4 kDa)	P45	Garavaglia et al. (2012)
Ac104 (691aa:79.9 kDa)	Bm88 (692aa:79.9 kDa)	VP80,VP87	Braunagel et al. (2003)
Ac105 (553aa:65.6 kDa)	Bm89 (289aa:34.3 kDa)	He65 (RNA editing ligase)	Ho and Shuman (2002)
Ac110 (56aa:6.8 kDa)	Bm92a	PIF-7	Jiantao et al. (2016)
Ac112/113 (258aa:30.9 kDa)		Apoptosis suppressor (Apsup)	Yamada et al. (2011)

(continued)

Table 2.2 (continued)

AcMNPV gene	Homologs in BmNPV	Gene product	References
Ac114 (424aa:49.3 kDa)	Bm94 (424aa:49.4 kDa)	Poly(ADP-ribose) glycohydrolase (PARG)	Zimmermann et al. (2018)
Ac115 (204aa:23 kDa)	Bm95 (204aa:23.0 kDa)	Pif-3	Wang et al. (2007b)
Ac119 (530aa:59.8 kDa)	Bm97 (527aa:59.8 kDa)	Pif-1	Wang et al. (2007b)
Ac123 (215aa:25 kDa)	Bm100 (225aa:26.0 kDa)	Protein kinase 2 (PK2)	Li and Miller (1995)
Ac125 (226aa:26.6 kDa)	Bm102 (227aa:26.6 kDa)	LEF-7	Harrison et al. (2018)
Ac126 (551aa:61.4 kDa)	Bm103 (552aa:61.8 kDa)	Chitinase	Daimon et al. (2006)
Ac127 (323aa:36.9 kDa)	Bm104 (323aa:36.9 kDa)	Cathepsin	Ishimwe et al. (2015)
Ac128 (530aa:60.6 kDa)	Bm105 (530aa:60.6 kDa)	GP64, GP67	Ardisson-Araujo et al. (2016)
Ac129 (198aa:22.1 kDa)	Bm106 (195aa:21.8 kDa)	P24-capsid	Wang et al. (2010)
Ac130 (106aa:12.1 kDa)	Bm107 (106aa:12.1 kDa)	GP16	Yang et al. (2014)
Ac131 (322aa:36.4 kDa)	Bm108 (315aa:35.4 kDa)	Calyx/polyhedron envelope (PE) protein, PP	Sajjan and Hinchigeri (2016)
Ac133 (419aa:48.3 kDa)	Bm110 (420aa:48.5 kDa)	Alkaline nuclease (AN)	Mikhailov et al. (2004)
Ac134	Bm111 (803aa:94.5 kDa)	P94	Clem et al. (1994)
Ac135 (299aa:34.8 kDa)	Bm112 (299aa:34.5 kDa)	P35	Escasa et al. (2006)
Ac136 (240aa:27.3 kDa)	Bm113 (240aa:27.3 kDa)	P26	Goenka and Weaver (2008)
Ac137 (94aa:10.3 kDa)	Bm114 (70aa:7.5 kDa)	P10	Alaoui-Ismaili and Richardson (1998)
Ac138 (645aa:73.9 kDa)	Bm115 (645aa:74.0 kDa)	P74-PIF	Zhou et al. (2005)
Ac 139 (449aa:52.6 kDa)	Bm116 (451aa:52.6 kDa)	ME53 (major early gene)	De Jong et al. (2009)
Ac141 (261aa:30.1 kDa)	Bm117 (261aa:30.1 kDa)	EXON0	Fang et al. (2007)
Ac142 (477aa:55.4 kDa)	Bm118 (476aa:55.5 kDa)	P49	Vanarsdall et al. (2007)
Ac143 (90aa:9.7 kDa)	Bm119 (101aa:10.4 kDa)	ODV-E18	Braunagel et al. (1996)

(continued)

Table 2.2 (continued)

AcMNPV gene	Homologs in BmNPV	Gene product	References
Ac144 (290aa:33.5 kDa)	Bm120 (290aa:33.5 kDa)	ODV-EC27	Belyavskyi et al. (1998)
Ac145 (97.0aa:8.9 kDa)	Bm121 (95aa:11.0 kDa)	P11	Lapointe et al. (2004)
Ac147 (582aa:66.9 kDa)	Bm123 (584aa:66.9 kDa)	Immediate early gene-1 (IE-1)	Guarino and Summers (1986)
Ac147–0 (636aa:72 kDa)		IE-0	Sokal et al. (2014)
Ac148 (376aa:40.9 kDa)	Bm124 (375aa:41.3 kDa)	ODV-E56, PIF-5	Wang et al. (2007b)
Ac151 (408aa:47 kDa)	Bm127 (422aa:48.8 kDa)	IE-2	Imai et al. (2005)
Ac153 (321aa:37.4 kDa)	Bm128 (309aa:36.1 kDa)	PE38	Krappa and Knebel-Morsdorf (1991)

2.7 Summary

This chapter discusses the biology of baculoviruses, with particular focus on AcMNPV and BmNPV, covering their host specificity, life cycle, genetic organization, and the molecular determinants of their host range. It explains the infection process in insect hosts, the roles of budded and occlusion-derived viruses, and the enzymatic mechanisms that facilitate viral spread and larval disintegration. The genome structures of AcMNPV and BmNPV, highlighting their sequence similarities and differences, as well as the functional implications of specific genes for viral replication and biotechnological applications, were revealed. This genetic information led to the development of silkworm biotechnology.

References

Ackermann HW, Smirnoff WA (1983) A morphological investigation of 23 baculoviruses. J Invertebr Pathol 41:269–280. https://doi.org/10.1016/0022-2011(83)90244-6

Alaoui-Ismaili MH, Richardson CD (1998) Insect virus proteins (FALPE and p10) self-associate to form filaments in infected cells. J Virol 72:2213–2223. https://doi.org/10.1128/JVI.72.3.2213-2223.1998

Ardisson-Araujo DM, Melo FL, Clem RJ, Wolff JL, Ribeiro BM (2016) A betabaculovirus encoded gp64 homolog codes for a functional envelope fusion protein. J Virol 90:1668–1672. https://doi.org/10.1128/JVI.02491-15

Argaud O, Croizier L, Lopez-Ferber M, Croizier G (1998) Two key mutations in the host-range specificity domain of the p143 gene of *Autographa californica* nucleopolyhedrovirus are required to kill Bombyx mori larvae. J Gen Virol 79(Pt 4):931–935. https://doi.org/10.1099/0022-1317-79-4-931

Arunkumar KP, Metta M, Nagaraju J (2006) Molecular phylogeny of silkmoths reveals the origin of domesticated silkmoth, *Bombyx mori* from Chinese *Bombyx mandarina* and paternal inheritance of *Antheraea proylei* mitochondrial DNA. Mol Phylogenet Evol 40:419–427. https://doi.org/10.1016/j.ympev.2006.02.023

Ayres MD, Howard SC, Kuzio J, Lopez-Ferber M, Possee RD (1994) The complete DNA sequence of *Autographa californica* nuclear polyhedrosis virus. Virology 202:586–605. https://doi.org/10.1006/viro.1994.1380

Baricco A (1996) Seta. Rizzoli editore, Italy

Belyavskyi M, Braunagel SC, Summers MD (1998) The structural protein ODV-EC27 of *Autographa californica* nucleopolyhedrovirus is a multifunctional viral cyclin. Proc Natl Acad Sci USA 95:11205–11210. https://doi.org/10.1073/pnas.95.19.11205

Braunagel SC, He H, Ramamurthy P, Summers MD (1996) Transcription, translation, and cellular localization of three *Autographa californica* nuclear polyhedrosis virus structural proteins: ODV-E18, ODV-E35, and ODV-EC27. Virology 222:100–114. https://doi.org/10.1006/viro.1996.0401

Braunagel SC, Russell WK, Rosas-Acosta G, Russell DH, Summers MD (2003) Determination of the protein composition of the occlusion-derived virus of *Autographa californica* nucleopolyhedrovirus. Proc Natl Acad Sci USA 100:9797–9802. https://doi.org/10.1073/pnas.1733972100

Clem RJ (2015) Viral IAPs, then and now. Semin Cell Dev Biol 39:72–79. https://doi.org/10.1016/j.semcdb.2015.01.011

Clem RJ, Robson M, Miller LK (1994) Influence of infection route on the infectivity of baculovirus mutants lacking the apoptosis-inhibiting gene p35 and the adjacent gene p94. J Virol 68:6759–6762. https://doi.org/10.1128/JVI.68.10.6759-6762.1994

Daimon T, Katsuma S, Kang W, Shimada T (2006) Comparative studies of *Bombyx mori* nucleopolyhedrovirus chitinase and its host ortholog, BmChi-h. Biochem Biophys Res Commun 345:825–833. https://doi.org/10.1016/j.bbrc.2006.04.112

De Jong J, Arif BM, Theilmann DA, Krell PJ (2009) *Autographa californica* multiple nucleopolyhedrovirus me53 (ac140) is a nonessential gene required for efficient budded-virus production. J Virol 83:7440–7448. https://doi.org/10.1128/JVI.02390-08

Detvisitsakun C, Berretta MF, Lehiy C, Passarelli AL (2005) Stimulation of cell motility by a viral fibroblast growth factor homolog: proposal for a role in viral pathogenesis. Virology 336:308–317. https://doi.org/10.1016/j.virol.2005.03.013

Entwistle PF, Adams PHW, Evans HF (1978) Epizootiology of a nuclear polyhedrosis virus in European spruce sawfly (*Gilpinia hercyniae*): the rate of passage of infective virus through the gut of birds during cage tests. J Invertebr Pathol 31:307–312. https://doi.org/10.1016/0022-2011(78)90221-5

Escasa SR, Lauzon HAM, Mathur AC, Krell PJ, Arif BM (2006) Sequence analysis of the *Choristoneura occidentalis* granulovirus genome. J Gen Virol 87:1917–1933. https://doi.org/10.1099/vir.0.81792-0

Fan X, Mclachlin JR, Weaver RF (1998) Identification and characterization of a protein kinase-interacting protein encoded by the *Autographa californica* nuclear polyhedrosis virus. Virology 240:175–182. https://doi.org/10.1006/viro.1997.8944

Fang M, Dai X, Theilmann DA (2007) *Autographa californica* multiple nucleopolyhedrovirus EXON0 (ORF141) is required for efficient egress of nucleocapsids from the nucleus. J Virol 81:9859–9869. https://doi.org/10.1128/JVI.00588-07

Fang M, Nie Y, Harris S, Erlandson MA, Theilmann DA (2009) *Autographa californica* multiple nucleopolyhedrovirus core gene ac96 encodes a per Os infectivity factor (PIF-4). J Virol 83:12569–12578. https://doi.org/10.1128/JVI.01141-09

Garavaglia MJ, Miele SA, Iserte JA, Belaich MN, Ghiringhelli PD (2012) The ac53, ac78, ac101, and ac103 genes are newly discovered core genes in the family Baculoviridae. J Virol 86:12069–12079. https://doi.org/10.1128/JVI.01873-12

Gershburg E, Rivkin H, Chejanovsky N (1997) Expression of the *Autographa californica* nuclear polyhedrosis virus apoptotic suppressor gene p35 in nonpermissive *Spodoptera littoralis* cells. J Virol 71:7593–7599. https://doi.org/10.1128/JVI.71.10.7593-7599.1997

Goenka S, Weaver RF (2008) The p26 gene of the *Autographa californica* nucleopolyhedrovirus: timing of transcription, and cellular localization and dimerization of product. Virus Res 131:136–144. https://doi.org/10.1016/j.virusres.2007.08.017

Gomi S, Majima K, Maeda S (1999) Sequence analysis of the genome of *Bombyx mori* nucleopolyhedrovirus. J Gen Virol 80(Pt 5):1323–1337. https://doi.org/10.1099/0022-1317-80-5-1323

Gordon JD, Carstens EB (1984) Phenotypic characterization and physical mapping of a temperature-sensitive mutant of *Autographa californica* nuclear polyhedrosis virus defective in DNA synthesis. Virology 138:69–81. https://doi.org/10.1016/0042-6822(84)90148-x

Guarino LA, Summers MD (1986) Functional mapping of a trans-activating gene required for expression of a baculovirus delayed-early gene. J Virol 57:563–571. https://doi.org/10.1128/JVI.57.2.563-571.1986

Guarino LA, Xu B, Jin J, Dong W (1998) A virus-encoded RNA polymerase purified from baculovirus-infected cells. J Virol 72:7985–7991. https://doi.org/10.1128/JVI.72.10.7985-7991.1998

Harrison RL, Bonning BC (2003) Comparative analysis of the genomes of Rachiplusia ou and *Autographa californica* multiple nucleopolyhedroviruses. J Gen Virol 84:1827–1842. https://doi.org/10.1099/vir.0.19146-0

Harrison RL, Jarvis DL, Summers MD (1996) The role of the AcMNPV 25K gene, "FP25," in baculovirus polh and p10 expression. Virology 226:34–46. https://doi.org/10.1006/viro.1996.0625

Harrison RL, Mowery JD, Rowley DL, Bauchan GR, Theilmann DA, Rohrmann GF, Erlandson MA (2018) The complete genome sequence of a third distinct baculovirus isolated from the true armyworm, *Mythimna unipuncta*, contains two copies of the lef-7 gene. Virus Genes 54:297–310. https://doi.org/10.1007/s11262-017-1525-0

Hawtin RE, Zarkowska T, Arnold K, Thomas CJ, Gooday GW, King LA, Kuzio JA, Possee RD (1997) Liquefaction of *Autographa californica* nucleopolyhedrovirus-infected insects is dependent on the integrity of virus-encoded chitinase and cathepsin genes. Virology 238:243–253. https://doi.org/10.1006/viro.1997.8816

Ho CK, Shuman S (2002) Bacteriophage T4 RNA ligase 2 (gp24.1) exemplifies a family of RNA ligases found in all phylogenetic domains. Proc Natl Acad Sci USA 99:12709–12714. https://doi.org/10.1073/pnas.192184699

Holden HM, Wesenberg G, Raynes DA, Hartshorne DJ, Guerriero V Jr, Rayment I (1996) Molecular structure of a proteolytic fragment of TLP20. Acta Crystallogr D Biol Crystallogr 52:1153–1160. https://doi.org/10.1107/S0907444996008128

Hong T, Braunagel SC, Summers MD (1994) Transcription, translation, and cellular localization of PDV-E66: a structural protein of the PDV envelope of *Autographa californica* nuclear polyhedrosis virus. Virology 204:210–222. https://doi.org/10.1006/viro.1994.1525

Imai N, Matsumoto S, Kang W (2005) Formation of *Bombyx mori* nucleopolyhedrovirus IE2 nuclear foci is regulated by the functional domains for oligomerization and ubiquitin ligase activity. J Gen Virol 86:637–644. https://doi.org/10.1099/vir.0.80523-0

Ishimwe E, Hodgson JJ, Clem RJ, Passarelli AL (2015) Reaching the melting point: degradative enzymes and protease inhibitors involved in baculovirus infection and dissemination. Virology 479–480:637–649. https://doi.org/10.1016/j.virol.2015.01.027

Ji X, Sutton G, Evans G, Axford D, Owen R, Stuart DI (2010) How baculovirus polyhedra fit square pegs into round holes to robustly package viruses. EMBO J 29:505–514. https://doi.org/10.1038/emboj.2009.352

Jiantao L, Zhu L, Zhang S, Deng Z, Huang Z, Yuan M, Wu W, Yang K (2016) The *Autographa californica* multiple nucleopolyhedrovirus ac110 gene encodes a new per os infectivity factor. Virus Res 221:30–37. https://doi.org/10.1016/j.virusres.2016.05.017

Johnson A, O'donnell M (2005) Cellular DNA replicases: components and dynamics at the replication fork. Annu Rev Biochem 74:283–315. https://doi.org/10.1146/annurev.biochem.73.011303.073859

Kamita SG, Maeda S (1993) Inhibition of *Bombyx mori* nuclear polyhedrosis virus (NPV) replication by the putative DNA helicase gene of *Autographa californica* NPV. J Virol 67:6239–6245

Kamita SG, Maeda S (1997) Sequencing of the putative DNA helicase-encoding gene of the *Bombyx mori* nuclear polyhedrosis virus and fine-mapping of a region involved in host range expansion. Gene 190:173–179

Katsuma S, Fujii T, Kawaoka S, Shimada T (2008) *Bombyx mori* nucleopolyhedrovirus SNF2 global transactivator homologue (Bm33) enhances viral pathogenicity in *B. mori* larvae. J Gen Virol 89:3039–3046. https://doi.org/10.1099/vir.0.2008/004887-0

Kitts PA, Possee RD (1993) A method for producing recombinant baculovirus expression vectors at high frequency. BioTechniques 14:810–817

Kondo A, Maeda S (1991) Host range expansion by recombination of the baculoviruses *Bombyx mori* nuclear polyhedrosis virus. J Virol 65:3625–3632

Krappa R, Knebel-Morsdorf D (1991) Identification of the very early transcribed baculovirus gene PE-38. J Virol 65:805–812. https://doi.org/10.1128/JVI.65.2.805-812.1991

Langer T, Lu C, Echols H, Flanagan J, Hayer MK, Hartl FU (1992) Successive action of DnaK, DnaJ and GroEL along the pathway of chaperone-mediated protein folding. Nature 356:683–689. https://doi.org/10.1038/356683a0

Lapointe R, Popham HJ, Straschil U, Goulding D, O'reilly DR, Olszewski JA (2004) Characterization of two *Autographa californica* nucleopolyhedrovirus proteins, Ac145 and Ac150, which affect oral infectivity in a host-dependent manner. J Virol 78:6439–6448. https://doi.org/10.1128/JVI.78.12.6439-6448.2004

Li Y, Miller LK (1995) Expression and functional analysis of a baculovirus gene encoding a truncated protein kinase homolog. Virology 206:314–323. https://doi.org/10.1016/s0042-6822(95)80047-6

Li L, Harwood SH, Rohrmann GF (1999) Identification of additional genes that influence baculovirus late gene expression. Virology 255:9–19. https://doi.org/10.1006/viro.1998.9546

Long CM, Rohrmann GF, Merrill GF (2009) The conserved baculovirus protein p33 (Ac92) is a flavin adenine dinucleotide-linked sulfhydryl oxidase. Virology 388:231–235. https://doi.org/10.1016/j.virol.2009.04.006

Lu M, Iatrou K (1997) Characterization of a domain of the genome of BmNPV containing a functional gene for a small capsid protein and harboring deletions eliminating three open reading frames that are present in AcNPV. Gene 185:69–75. https://doi.org/10.1016/s0378-1119(96)00632-4

Lu A, Miller LK (1994) Identification of three late expression factor genes within the 33.8- to 43.4-map-unit region of A*utographa californica* nuclear polyhedrosis virus. J Virol 68:6710–6718. https://doi.org/10.1128/JVI.68.10.6710-6718.1994

Lu A, Miller LK (1995a) Differential requirements for baculovirus late expression factor genes in two cell lines. J Virol 69:6265–6272. https://doi.org/10.1128/JVI.69.10.6265-6272.1995

Lu A, Miller LK (1995b) The roles of eighteen baculovirus late expression factor genes in transcription and DNA replication. J Virol 69:975–982. https://doi.org/10.1128/JVI.69.2.975-982.1995

Ma SS, Zhang Z, Xia HC, Chen L, Yang YH, Yao Q, Chen KP (2015) Evolutionary analysis of the ubiquitin gene of baculovirus and insect hosts. Genet Mol Res 14:9963–9973. https://doi.org/10.4238/2015.August.21.2

Maeda S, Kamita SG, Kondo A (1993) Host range expansion of *Autographa californica* nuclear polyhedrosis virus (NPV) following recombination of a 0.6-kilobase-pair DNA fragment originating from *Bombyx mori* NPV. J Virol 67:6234–6238

Martins A, Shuman S (2004) Characterization of a baculovirus enzyme with RNA ligase, polynucleotide 5′-kinase, and polynucleotide 3′-phosphatase activities. J Biol Chem 279:18220–18231. https://doi.org/10.1074/jbc.M313386200

Mikhailov VS, Okano K, Rohrmann GF (2004) Specificity of the endonuclease activity of the baculovirus alkaline nuclease for single-stranded DNA. J Biol Chem 279:14734–14745. https://doi.org/10.1074/jbc.M311658200
Mikhailov VS, Vanarsdall AL, Rohrmann GF (2008) Isolation and characterization of the DNA-binding protein (DBP) of the *Autographa californica* multiple nucleopolyhedrovirus. Virology 370:415–429. https://doi.org/10.1016/j.virol.2007.09.001
Mildvan AS, Xia Z, Azurmendi HF, Saraswat V, Legler PM, Massiah MA, Gabelli SB, Bianchet MA, Kang LW, Amzel LM (2005) Structures and mechanisms of Nudix hydrolases. Arch Biochem Biophys 433:129–143. https://doi.org/10.1016/j.abb.2004.08.017
O'reilly DR, Miller LK (1989) A baculovirus blocks insect molting by producing ecdysteroid UDP-glucosyl transferase. Science 245:1110–1112. https://doi.org/10.1126/science.2505387
Olivera BM, Miljanich GP, Ramachandran J, Adams ME (1994) Calcium channel diversity and neurotransmitter release: the omega-conotoxins and omega-agatoxins. Annu Rev Biochem 63:823–867. https://doi.org/10.1146/annurev.bi.63.070194.004135
Olszewski J, Miller LK (1997) Identification and characterization of a baculovirus structural protein, VP1054, required for nucleocapsid formation. J Virol 71:5040–5050. https://doi.org/10.1128/JVI.71.7.5040-5050.1997
Passarelli AL, Miller LK (1993) Identification of genes encoding late expression factors located between 56.0 and 65.4 map units of the *Autographa californica* nuclear polyhedrosis virus genome. Virology 197:704–714. https://doi.org/10.1006/viro.1993.1646
Passarelli AL, Miller LK (1994) Identification and transcriptional regulation of the baculovirus lef-6 gene. J Virol 68:4458–4467. https://doi.org/10.1128/JVI.68.7.4458-4467.1994
Passarelli AL, Todd JW, Miller LK (1994) A baculovirus gene involved in late gene expression predicts a large polypeptide with a conserved motif of RNA polymerases. J Virol 68:4673–4678. https://doi.org/10.1128/JVI.68.7.4673-4678.1994
Pijlman GP, Dortmans JC, Vermeesch AM, Yang K, Martens DE, Goldbach RW, Vlak JM (2002) Pivotal role of the non-hr origin of DNA replication in the genesis of defective interfering baculoviruses. J Virol 76:5605–5611. https://doi.org/10.1128/jvi.76.11.5605-5611.2002
Reilly LM, Guarino LA (1994) The pk-1 gene of *Autographa californica* multinucleocapsid nuclear polyhedrosis virus encodes a protein kinase. J Gen Virol 75(Pt 11):2999–3006. https://doi.org/10.1099/0022-1317-75-11-2999
Rohrmann GF (2019) Baculovirus molecular biology, 4th edn. National Center for Biotechnology Information (US), Bethesda
Roncarati R, Knebel-Morsdorf D (1997) Identification of the early actin-rearrangement-inducing factor gene, arif-1, from *Autographa californica* multicapsid nuclear polyhedrosis virus. J Virol 71:7933–7941. https://doi.org/10.1128/JVI.71.10.7933-7941.1997
Russell RL, Rohrmann GF (1997) Characterization of P91, a protein associated with virions of an *Orgyia pseudotsugata* baculovirus. Virology 233:210–223. https://doi.org/10.1006/viro.1997.8599
Sajjan DB, Hinchigeri SB (2016) Structural organization of baculovirus occlusion bodies and protective role of multilayered polyhedron envelope protein. Food Environ Virol 8:86–100. https://doi.org/10.1007/s12560-016-9227-7
Sheng Z, Charbonneau H (1993) The baculovirus *Autographa californica* encodes a protein tyrosine phosphatase. J Biol Chem 268:4728–4733
Shrestha A, Bao K, Chen YR, Chen W, Wang P, Fei Z, Blissard GW (2018) Global analysis of baculovirus *Autographa californica* multiple nucleopolyhedrovirus gene expression in the midgut of the lepidopteran host *Trichoplusia ni*. J Virol 92:10–1128. https://doi.org/10.1128/JVI.01277-18
Silva LA, Camargo BR, Araujo AC, Batista TL, Ribeiro BM, Ardisson-Araujo DMP (2022) Easily purified baculovirus/insect system expressed recombinant hepatitis B virus surface antigen fused to the N- or C-terminus of polyhedrin. Arch Virol 167:345–354. https://doi.org/10.1007/s00705-021-05305-6

Smith T (1999) Pasteur and insect pathogens. Nat Struct Biol 6:720–720. https://doi.org/10.1038/11483
Sokal N, Nie Y, Willis LG, Yamagishi J, Blissard GW, Rheault MR, Theilmann DA (2014) Defining the roles of the baculovirus regulatory proteins IE0 and IE1 in genome replication and early gene transactivation. Virology 468–470:160–171. https://doi.org/10.1016/j.virol.2014.07.044
Steinhaus EA, Hughes KM, Wasser HB (1949) Demonstration of tie Granulousis virus of the variegated cutworm. J Bacteriol 57:219–224
Thiem SM, Miller LK (1989a) A baculovirus gene with a novel transcription pattern encodes a polypeptide with a zinc finger and a leucine zipper. J Virol 63:4489–4497. https://doi.org/10.1128/JVI.63.11.4489-4497.1989
Thiem SM, Miller LK (1989b) Identification, sequence, and transcriptional mapping of the major capsid protein gene of the baculovirus *Autographa californica* nuclear polyhedrosis virus. J Virol 63:2008–2018. https://doi.org/10.1128/JVI.63.5.2008-2018.1989
Thomas CJ, Brown HL, Hawes CR, Lee BY, Min MK, King LA, Possee RD (1998) Localization of a baculovirus-induced chitinase in the insect cell endoplasmic reticulum. J Virol 72:10207–10212. https://doi.org/10.1128/JVI.72.12.10207-10212.1998
Todd JW, Passarelli AL, Miller LK (1995) Eighteen baculovirus genes, including lef-11, p35, 39K, and p47, support late gene expression. J Virol 69:968–974. https://doi.org/10.1128/JVI.69.2.968-974.1995
Tomalski MD, Wu JG, Miller LK (1988) The location, sequence, transcription, and regulation of a baculovirus DNA polymerase gene. Virology 167:591–600
Tomalski MD, Eldridge R, Miller LK (1991) A baculovirus homolog of a Cu/Zn superoxide dismutase gene. Virology 184:149–161. https://doi.org/10.1016/0042-6822(91)90831-u
Vago C, Aizawa K, Ignoffo C, Martignoni ME, Tarasevitch L, Tinsley TW (1974) Editorial: present status of the nomenclature and classification of invertebrate viruses. J Invertebr Pathol 23:133–134. https://doi.org/10.1016/0022-2011(74)90175-x
Vanarsdall AL, Pearson MN, Rohrmann GF (2007) Characterization of baculovirus constructs lacking either the Ac 101, Ac 142, or the Ac 144 open reading frame. Virology 367:187–195. https://doi.org/10.1016/j.virol.2007.05.003
Vialard JE, Yuen L, Richardson CD (1990) Identification and characterization of a baculovirus occlusion body glycoprotein which resembles spheroidin, an entomopoxvirus protein. J Virol 64:5804–5811. https://doi.org/10.1128/JVI.64.12.5804-5811.1990
Volkman LE, Summers MD (1975) Nuclear polyhedrosis virus detection: relative capabilities of clones developed from *Trichoplusia ni* ovarian cell line TN-368 to serve as indicator cells in a plaque assay. J Virol 16:1630–1637. https://doi.org/10.1128/JVI.16.6.1630-1637.1975
Volkman LE, Summers MD (1977) *Autographa californica* nuclear polyhedrosis virus: comparative infectivity of the occluded, alkali-liberated, and nonoccluded forms. J Invertebr Pathol 30:102–103. https://doi.org/10.1016/0022-2011(77)90045-3
Volkman LE, Summers MD, Hsieh CH (1976) Occluded and nonoccluded nuclear polyhedrosis virus grown in *Trichoplusia ni*: comparative neutralization comparative infectivity, and in vitro growth studies. J Virol 19:820–832. https://doi.org/10.1128/JVI.19.3.820-832.1976
Wang Y, Addess KJ, Chen J, Geer LY, He J, He S, Lu S, Madej T, Marchler-Bauer A, Thiessen PA, Zhang N, Bryant SH (2007a) MMDB: annotating protein sequences with Entrez's 3D-structure database. Nucleic Acids Res 35:D298–D300. https://doi.org/10.1093/nar/gkl952
Wang Y, Kleespies RG, Huger AM, Jehle JA (2007b) The genome of *Gryllus bimaculatus* nudivirus indicates an ancient diversification of baculovirus-related nonoccluded nudiviruses of insects. J Virol 81:5395–5406. https://doi.org/10.1128/JVI.02781-06
Wang R, Deng F, Hou D, Zhao Y, Guo L, Wang H, Hu Z (2010) Proteomics of the *Autographa californica* nucleopolyhedrovirus budded virions. J Virol 84:7233–7242. https://doi.org/10.1128/JVI.00040-10

Whitford M, Faulkner P (1992) Nucleotide sequence and transcriptional analysis of a gene encoding gp41, a structural glycoprotein of the baculovirus *Autographa californica* nuclear polyhedrosis virus. J Virol 66:4763–4768. https://doi.org/10.1128/JVI.66.8.4763-4768.1992

Wilson ME, Mainprize TH, Friesen PD, Miller LK (1987) Location, transcription, and sequence of a baculovirus gene encoding a small arginine-rich polypeptide. J Virol 61:661–666. https://doi.org/10.1128/JVI.61.3.661-666.1987

Wu X, Guarino LA (2003) *Autographa californica* nucleopolyhedrovirus orf69 encodes an RNA cap (nucleoside-2′-O)-methyltransferase. J Virol 77:3430–3440. https://doi.org/10.1128/jvi.77.6.3430-3440.2003

Wu W, Passarelli AL (2010) *Autographa californica* multiple nucleopolyhedrovirus Ac92 (ORF92, P33) is required for budded virus production and multiply enveloped occlusion-derived virus formation. J Virol 84:12351–12361. https://doi.org/10.1128/JVI.01598-10

Wu CP, Huang YJ, Wang JY, Wu YL, Lo HR, Wang JC, Chao YC (2010) *Autographa californica* multiple nucleopolyhedrovirus LEF-2 is a capsid protein required for amplification but not initiation of viral DNA replication. J Virol 84:5015–5024. https://doi.org/10.1128/JVI.02423-09

Xu W, Fan Y, Wang H, Feng M, Wu X (2019) *Bombyx mori* nucleopolyhedrovirus F-like protein Bm14 affects the morphogenesis and production of occlusion bodies and the embedding of ODVs. Virology 526:61–71. https://doi.org/10.1016/j.virol.2018.10.008

Yamada H, Shibuya M, Kobayashi M, Ikeda M (2011) Identification of a novel apoptosis suppressor gene from the baculovirus *Lymantria dispar* multicapsid nucleopolyhedrovirus. J Virol 85:5237–5242. https://doi.org/10.1128/JVI.00203-11

Yang S, Miller LK (1999) Activation of baculovirus very late promoters by interaction with very late factor 1. J Virol 73:3404–3409. https://doi.org/10.1128/JVI.73.4.3404-3409.1999

Yang M, Huang C, Qian DD, Li LL (2014) Functional characterization of *Autographa californica* multiple nucleopolyhedrovirus gp16 (ac130). Virology 464–465:341–352. https://doi.org/10.1016/j.virol.2014.07.030

Yuan M, Huang Z, Wei D, Hu Z, Yang K, Pang Y (2011) Identification of *Autographa californica* nucleopolyhedrovirus ac93 as a core gene and its requirement for intranuclear microvesicle formation and nuclear egress of nucleocapsids. J Virol 85:11664–11674. https://doi.org/10.1128/JVI.05275-11

Zemskov EA, Kang W, Maeda S (2000) Evidence for nucleic acid binding ability and nucleosome association of *Bombyx mori* nucleopolyhedrovirus BRO proteins. J Virol 74:6784–6789. https://doi.org/10.1128/jvi.74.15.6784-6789.2000

Zhou W, Yao L, Xu H, Yan F, Qi Y (2005) The function of envelope protein P74 from *Autographa californica* multiple nucleopolyhedrovirus in primary infection to host. Virus Genes 30:139–150. https://doi.org/10.1007/s11262-004-5623-4

Zimmermann L, Stephens A, Nam SZ, Rau D, Kubler J, Lozajic M, Gabler F, Soding J, Lupas AN, Alva V (2018) A completely reimplemented MPI bioinformatics toolkit with a new HHpred server at its core. J Mol Biol 430:2237–2243. https://doi.org/10.1016/j.jmb.2017.12.007

Chapter 3
Gene Expression in Silkworm

Abstract Silkworms (*Bombyx mori*) have emerged as highly efficient, cost-effective living bioreactors for large-scale recombinant protein production, offering superior cell-density capacity and accurate post-translational modifications compared to conventional bacterial and yeast expression systems. This chapter provides a comprehensive overview of silkworm-based gene expression technologies. It details the principles of the baculovirus expression system (BES), emphasizing the strategic use of strong viral promoters (e.g., *polh*, *p10*) and effective gene insertion techniques. To overcome the time-consuming purification processes of traditional BES, the BmNPV bacmid system is highlighted as a rapid, scalable, and biosafety-enhanced platform for transient protein expression in both larvae and pupae. Furthermore, the chapter explores the generation of transgenic silkworms utilizing the *piggyBac* transposon system. This approach enables stable, multigenerational integration of large foreign genes without genetic footprints. By targeting gene expression to specific silk-gland regions, recombinant proteins can be seamlessly incorporated into the silk cocoon's fibroin core or sericin layer, enabling simplified extraction. Together, these advanced expression systems position silkworm biotechnology as a highly promising and economically viable solution to meet the growing global demand for complex biopharmaceuticals and recombinant proteins.

Keywords BmNPV bacmid • Baculovirus expression system • Protein productivity • Polyhedrin promoter • Stable expression • *piggyBac*

3.1 Introduction

Since the development of gene expression technology in the 1970s, using *Escherichia coli*, a wide variety of methods and host organisms have been introduced. Starting with *E. coli*, protein expression systems have progressively expanded to encompass yeast, filamentous fungi, actinomycetes, insect cells, as well as both animal and plant cells. This diversification has enabled the successful production of numerous proteins across these platforms. In light of these recent advancements, it would be

E. Y. Park, *Silkworm Biofactory*, https://doi.org/10.1007/978-981-92-0056-6_3

prudent to examine the evolution of gene expression in silkworms and distinguish it from that in other expression systems. Embracing this silkworm-based method could greatly improve the efficiency and affordability of protein production in biotechnology. This chapter focuses on gene expression systems in silkworms and the related technologies involved.

3.1.1 Hosts for Recombinant Protein Production

Bacterial systems, especially *E. coli*, are the most well established (Cederbaum et al. 1984). Nevertheless, they lack the ability to carry out post-translational modifications and often encounter issues with protein solubility, which can result in the accumulation of inclusion bodies. Yeast strains such as *Pichia* can be suitable hosts for recombinant protein expression (Morikawa et al. 2007), as they tend to avoid inclusion bodies. Nevertheless, yeast may have lower production capacities and encounter glycosylation challenges (Morikawa et al. 2007).

Insect cell systems, such as *Spodoptera frugiperda* 9 (Sf9) and High Five cell lines, enable post-translational modifications and have a high growth rate (Cox 2012). However, they require equipped bioreactors with several control systems and expensive serum-free media (Krammer et al. 2010). Plant cells have become viable alternatives for protein production, offering a cost-effective and scalable production scale through transient expression, which eliminates lengthy transformation processes (Marsian and Lomonossoff 2016). Mammalian cells are ideal for producing complex proteins due to their advanced post-translational capabilities (Lalonde and Durocher 2017); however, they come with high production costs and generally lower productivity than other expression systems.

Silkworms emerge as a highly promising platform for large-scale recombinant protein production, offering significant potential to satisfy the increasing demands of contemporary biotechnology. With a history of over 5000 years of fulfilling human needs, the silk derived from these remarkable creatures has been a cornerstone of economic significance for centuries. Recently, silkworms have gained prominence in the biopharmaceutical field due to their capacity to produce eukaryotic recombinant proteins that undergo essential post-translational modifications.

The groundbreaking discovery in 1985, where silkworm larvae first demonstrated the capacity to produce recombinant proteins (Maeda et al. 1985), opened up new frontiers in protein expression technology. Since then, a variety of innovative techniques have been developed to harness the power of silkworms for recombinant protein production. The silkworm (*Bombyx mori*) larvae not only perform similar modifications to mammalian cells (Kato et al. 2010; Mitsudome et al. 2014; Pietrzyk et al. 2011), but they also provide a cost-effective strategy for high-yield protein production. The PCR-amplified gene is introduced into silkworms, enabling rapid protein collection within just 1 week (Motohashi et al. 2005). This innovative method outpaces traditional insect cell culture systems, which involve multiple steps to achieve the necessary cell concentration in bioreactors. Moreover, the infrastructure for silkworm rearing is well established across many Asian countries, making this approach both practical and accessible. The utilization of silkworms presents a cost-effective approach to recombinant protein production, requiring only a simple temperature-controlled environment and a supply of mulberry leaves for sustenance. In summary, Table 3.1 presents a comparative overview of various expression systems, detailing their respective advantages and limitations in terms of protein quality, scalability, and production cost. Their remarkable ability to thrive in high cell-density cultures positions them as living bioreactors, capable of delivering

Table 3.1 Various protein expression systems and their advantages and limitations

Expression system	Advantages[a]	Disadvantages[a]	Cost	Quality	Scalability
E. coli	Reliable and widely used system Grows quickly Produces a large amount of product Can be easily scaled up for industrial use	Inability to undergo post-translational modifications and disulfide bonds Risk of contamination such as endotoxin Formation of inclusion bodies	L	L	H
Yeast	Soluble proteins Grows quickly Produces a large amount of product Available multigene expression Easy to scalable process	Difficulty in assembling multimeric protein complexes Need for molecular chaperonin for proper protein folding Inappropriate glycosylation patterns	L	M	H
Insect cells	Grow quickly Suitable glycosylation patterns Possible coinfection or coexpression A relatively high amount of protein Easy-to-scale process	Requires construction and delivery of recombinant baculovirus for gene expression Demands high-cost infrastructure for insect cell culture Expensive specialized media for optimal cell growth	M	M	M
Silkworms	Grow quickly Suitable glycosylation patterns Supports coinfection or coexpression A relatively high amount of protein Affordable and easy to large-scale production	Variability in protein expression levels among individual larvae	L–M	M	H

(continued)

Table 3.1 (continued)

Expression system	Advantages[a]	Disadvantages[a]	Cost	Quality	Scalability
Plants	Affordable and easy to large-scale production Suitable glycosylation patterns	Primarily supports transient gene expression rather than stable integration Limited flexibility and capacity of available expression vector systems	M	M	M
Mammalian cells	Accurate glycosylation Available hybrid virus-like particles Triggers strong immune responses Supports coinfection or coexpression	Complexity in constructing expression vectors and practical application Low protein expression yield High production cost due to resource-intensive processes	H	H	L

[a]Advantages and disadvantages of expression systems were adapted with permission (Park and Minkner 2025), Copyright 2025 Elsevier, and edited. The symbols L, M, and H denote low, moderate, and high levels of cost, protein quality, and scalability of the expression system, respectively

significant quantities of valuable proteins (see Table 3.1 for a comprehensive comparison).

3.1.2 *Silkworm as a Bioreactor*

In evaluating the silkworm as a bioreactor, how does its efficiency in recombinant protein production compare with that of conventional bioreactor systems? The protein production rate (P_d, $\frac{g\ \text{product}}{L \cdot h}$) is defined as follows:

$$P_d = q_p \times x$$

Here, q_p represents the protein production rate per cell mass per hour, expressed as either ($\frac{g\ \text{product}}{h \cdot g\ \text{cell}}$) or ($\frac{g\ \text{product}}{h \cdot \text{cell number}}$). This rate reflects the amount of product produced per hour per cell mass or per cell and is assumed to remain constant across cells. The variable x denotes the cell concentration, given in units of ($\frac{g}{L}$), or ($\frac{\text{cell number}}{L}$).

For example, if a bioreactor achieves a cell concentration that is twice as high ($x_2 = 2x_1$), the resulting protein production rate (P_d) will also double. Thus, when using the same cell strain, P_d is directly influenced by the cell concentration within the bioreactor. Insect and animal cell cultures typically reach a maximum cell concentration of approximately 10^6 cells/cm^3 in conventional bioreactors:

$$P_{\mathrm{d}}\left(\frac{g}{\mathrm{cm}^3 \cdot h}\right) = q_{\mathrm{p}}\left(\frac{g}{\mathrm{cells}\cdot h}\right)\times 10^6\left(\frac{\mathrm{cells}}{\mathrm{cm}^3}\right)$$

Considering a silkworm larva approximately 1 cm in diameter and 5 cm in length, its total body volume can be estimated at around 4 cm^3 (equivalent to 4×10^{12} μm^3). Assuming each cell has a diameter of about 20 μm, the volume of a single cell is roughly 4×10^3 μm^3. If we further assume that only half of the larva's body volume is occupied by these cells, the total number of packed cells can be estimated as follows:

$$\frac{4\times 10^{12}\,\mu\mathrm{m}^3 \times 0.5\,/\,\mathrm{larva}}{4\times 10^3\,\mu\mathrm{m}^3\,/\,\mathrm{cell}} = 5\times 10^8\ \mathrm{cells}\,/\,\mathrm{larva}.$$

The silkworm larva volume is measured to be approximately 4 cm^3, and the cell concentration based on the silkworm larva volume is expected to be significant, $\frac{5\times 10^8\ \mathrm{cells}\,/\,\mathrm{larva}}{4\,\mathrm{cm}^3\,/\,\mathrm{larva}} \cong 1\times 10^8\ \frac{\mathrm{cells}}{\mathrm{cm}^3}$. The P_{d} value of the silkworm is

$$P_{\mathrm{d\,silkworm}}\left(\frac{g}{\mathrm{cm}^3 \cdot h}\right) = q_{\mathrm{p}}\left(\frac{g}{\mathrm{cells}\cdot h}\right)\times 10^8\left(\frac{\mathrm{cells}}{\mathrm{cm}^3}\right).$$

When comparing the P_{d} value of silkworms as bioreactors to that of conventional bioreactors, the calculation shows a ratio of $\frac{q_{\mathrm{p}}\times 10^8\ \mathrm{cells}}{q_{\mathrm{p}}\times 10^6\ \mathrm{cells}} \cong 1\times 10^2$, indicating that silkworms can produce at very least 100 times more efficiently than conventional bioreactors, as illustrated in Fig. 3.1. Thanks to their high productivity, silkworms are capable of efficiently producing recombinant proteins at a large scale.

While conventional bioreactors require high-cost equipment to control pH, temperature, and oxygen and use expensive media with additives such as serum or antibiotics, silkworms can be reared in low-cost setups that require only temperature and humidity control. This approach can make recombinant protein production significantly more cost-effective than in bioreactors. The quality of recombinant proteins is biochemically important, and silkworms are capable of performing post-translational modifications that are similar to those done by mammalian cells—something microorganisms cannot do. Silkworms have been extensively studied for their physiological and biological properties, and the techniques for rearing them are highly developed in many Asian countries. For these reasons, silkworms are an excellent choice for a biofactory for recombinant protein production. By utilizing silkworms as a production platform, we can unlock the potential for innovative biopharmaceutical solutions that address the critical needs of today's healthcare landscape.

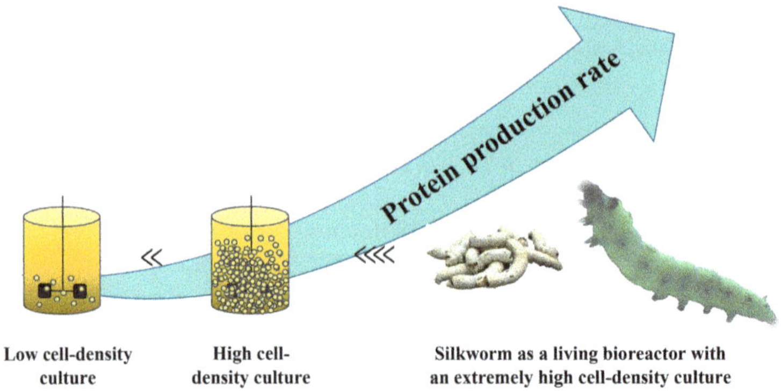

Fig. 3.1 Comparative analysis of protein production rates between conventional bioreactors and silkworms functioning as living bioreactors with high cell-density culture. The small round spheres in the reactor represent cells. (Adapted with permission (Park 2025). Copyright 2025 Elsevier)

3.2 Baculovirus Expression System

Baculoviruses are a diverse group of viruses that naturally infect insects. They are characterized by rod-shaped capsids, virus particles measuring 40–50 nm in diameter and 200–400 nm in length (Harrap 1972). These capsids contain a large double-stranded DNA genome, typically ranging from 80 to 200 kbp (Burgess 1977). Two specific baculoviruses, *Autographa californica* multiple nucleopolyhedrovirus (AcMNPV) and *Bombyx mori* nucleopolyhedrovirus (BmNPV), have been developed for targeting gene expression. Sequence analysis reveals that BmNPV shares approximately 90% nucleotide identity with AcMNPV. Despite this high degree of similarity, the two viruses exhibit distinct host specificities. AcMNPV commonly infects insect species such as *Trichoplusia ni*, *Heliothis virescens*, and *Spodoptera frugiperda*, whereas BmNPV specifically infects *B. mori*, targeting cell lines including Bm5, BmN4, and BmN, as well as silkworm larvae and pupae. The nucleopolyhedrosis viruses (NPVs) of AcMNPV and BmNPV have a unique ability to produce large amounts of polyhedra, which shield viral particles in the environment from ultraviolet light and other factors that could inactivate them. Interestingly, these polyhedra are not directly involved in the formation of the virus itself, meaning that the virus can still survive even if the polyhedra are replaced with foreign genes. This characteristic serves as a starting point for the development of insect biotechnology. One notable example is AcMNPV, which was among the first viruses to have its genome sequenced in 1994 (Ayres et al. 1994). It was also the first virus to undergo genetic engineering, in which its polyhedrin gene was replaced with foreign genes, and the resulting protein was expressed. Following this, genome sequencing of the BmNPV also progressed in a similar manner and was completed in 1999 (Gomi et al. 1999). Genetic manipulation successfully replaced the polyhedrin gene with foreign genes, laying the groundwork for protein production at both the cellular level and individual silkworm larva.

3.2.1 *Insertion of a Heterologous Gene into the Viral Genome*

One of the main challenges in developing expression vectors for insect biotechnology was the precise integration of foreign genes into specific sites within the large viral genome, as well as ensuring their stable replication in insect cells. Initially, when AcMNPV was being developed as an expression vector, researchers encountered difficulties due to the lack of restriction endonuclease recognition sites in its genome. To overcome this issue, allelic replacement (Patel et al. 1987) was adopted for inserting heterologous genes. This method involves two separate homologous recombination events, each occurring in the regions of viral DNA flanking the foreign gene on the transfer plasmid (Fig. 3.2). These recombination events are independent of one another. Following co-transfection, the most frequently observed recombinant viruses are single-crossover variants, characterized by the integration of the entire plasmid DNA into the viral genome. However, these recombinants are generally unstable due to the presence of duplicated viral genome segments, which originate from the flanking regions in the transfer plasmid (Fig. 3.2). Single-crossover recombination does not disrupt the function of the viral gene at the insertion site, as long as the plasmid includes sufficient flanking sequence to maintain an intact copy of the viral gene at one end. Depending on where the second homologous recombination occurs, a single-crossover recombinant can either lead to true allelic replacement (resulting in a double-crossover recombinant) or revert to the parental virus (Fig. 3.2).

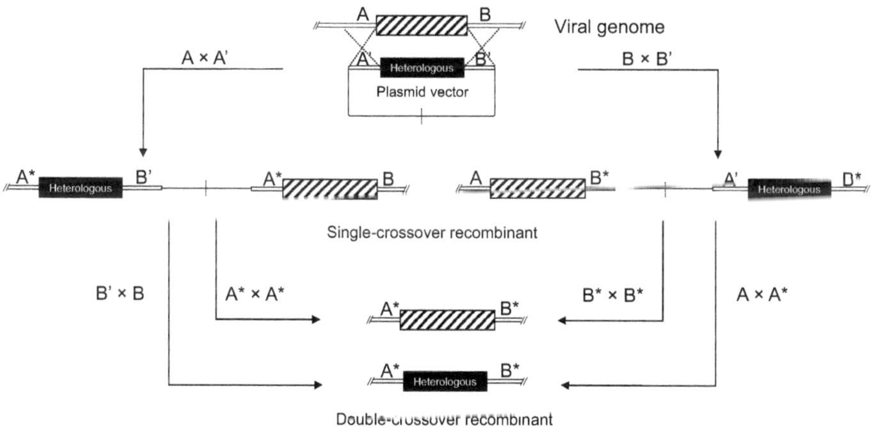

Fig. 3.2 Graphical depiction of allelic replacement within genomic DNA. Diagonal-lined boxes and black boxes indicate nonessential genes of the viral genome and heterogeneous DNA of the plasmid vector, respectively. A and B denote the flanking sequences in the viral genome; A′ and B′ denote the flanking sequences in the same sequence in the plasmid vector. The top, center, and bottom of the figure illustrate the structure of the viral genome and the plasmid vector, as well as single-crossover and double-crossover recombinant reactions, respectively. A* and B* are sequences derived from the recombination between A and A′ or B and B′, respectively. (Adapted with permission (O'reilly et al. 1994). Copyright 1994 Oxford University Press)

The transfer plasmid can be designed to replace a nonessential gene with a heterologous gene through allelic replacement. In this scenario, selection marker genes, such as the plasmid-carried *lacZ* gene, are combined with a screening method to identify the loss of a virus-borne gene function. Using *lacZ* gene and screening exclusively for a blue plaque phenotype will produce a single-crossover, blue plaque recombinant. However, this virus stock must undergo rescreening for the desired gene replacement, double-crossover recombinants using another technique, such as DNA hybridization. Therefore, it is recommended to use a screening method that enables direct identification of double-crossover recombinants. In allelic replacement reactions that involve 2–4 kb of viral DNA flanking each side of the heterologous DNA in the transfer plasmid, obtaining the double-recombinant viruses is likely only 0.2–1% of the total virus stock (O'reilly et al. 1994). Consequently, efficient screening for these rare viruses within a mixture is essential.

3.2.2 Promoters for the Baculovirus Expression System

Strong promoters are crucial for the efficient production of proteins during gene expression. In the baculovirus expression system, a promoter responsible for the synthesis of polyhedrin, a nonessential occlusion body protein, is preferable. There are two types of promoters in baculoviruses: very strong and very late promoters, polyhedrin (*polh*) and *p10* promoters.

The polh promoter is 50 kb, making it easy to use in vector construction in either its original or reverse orientation. To take advantage of this system, the first series of nonfusion transfer plasmids for AcMNPV *polh* promoter-based expression was created: a construct called pAc373 (Smith et al. 1983), which contained an oligonucleotide with a BamHI site inserted at −7 with respect to the *polh* translational initiation codon ATG. Plasmids such as pAcYM1 (Matsuura et al. 1987) and pEVmXIV (Wang et al. 1991) from AcMNPV, and pBM030 (Horiuchi et al. 1987) from BmNPV, contain the *polh* promoter, including the original *polh* ATG, and have been created. Using these promoters, their expression levels can reach the same levels of *polh* expression. Some recombinant proteins are reported to account for 20–50% of the total cellular protein. A different transfer vector, pVL941, was challenged; the translation initiation codon of *polh* was changed from ATG to ATT (Luckow and Summers 1989). Foreign genes can be inserted at a restriction site within the N-terminal polyhedrin coding sequence. To facilitate foreign gene insertion, pVL1393 and pBlueBac2 have been created, adding a multicloning site to pVL941.

The *p10* Promoter A large amount of the *p10* protein is produced very late in infection and forms large fibrillar structures in infected cells. It was suggested that when the p10 gene was deleted, the lysis of the cells infected by this virus was delayed (Williams et al. 1989). The mutation of the *p10* gene does not affect the formation of the budded virus or the formation of occlusion bodies. Thus, the vector replacing the *p10* gene with a heterologous gene does not lead to a noticeable phenotype in homologous recombination, similar to the plasmid used for *polh* replacement. The *p10* promoter is regulated in a similar manner to the *polh* promoter and contains an essential TAAG sequence at its transcriptional start site (Weyer et al. 1990). The relative strengths of the two promoters are quite similar. The *p10*-based plasmids, including pAcAS2, pAcUW1, and pEP252, are used with a parent virus that carries *lacZ* at the *p10* locus. Transfer plasmid vectors are categorized into three types: vectors for the expression of a single foreign gene, coexpression vectors that express one foreign gene and contain a screenable genetic marker, such as the *E. coli lac*Z gene, and dual-expression vectors engineered to express two distinct foreign genes simultaneously. A summary of these categories is presented in Table 3.2.

The *polh* and *p10* promoters are very strong for the high-level expression of heterologous genes; however, the levels of post-translational modifications, such as phosphorylation and N-glycosylation, decline at very late times post-infection. Generally, *polh*- and *p10*-based transfer plasmid vectors are used to evaluate the expression capability. However, for post-translational modification and protein secretion, a transfer plasmid vector that can be expressed in the early stage is preferable. For example, the promoter of the major capsid protein vp39 is used in some plasmids. This *vp39* promoter maximally expressed reporter genes between 12 and 24 h post-infection (Hill-Perkins and Possee 1990), while the *polh* promoter did so between 24 and 60 h post-infection. From 12 to 24 h post-infection, the *vp39* promoter expressed a higher level than the *polh* promoter, but by 48 h post-infection, the *polh* promoter expressed a level exceeding that of the *vp39* promoter (Thiem and Miller 1990). So, a promoter that can be expressed actively at both late and very late times was designed, referred to as a hybrid *vp39/polh* fusion promoter (Thiem and Miller 1990). Although the expression level of the *vp39* promoter is lower than that of the *polh* promoter, it is possible to improve the postmodification process, and the total amount of properly postmodified protein may increase. This is because the protein movement rate through the secretory pathway declines at very late times (Jarvis et al. 1990). Some proteins move slowly through the endoplasmic reticulum (ER) membrane even under normal conditions, so it takes an additional 12 h to move through the ER to the Golgi. On the other hand, the early expressible promoters are not strong, but they are useful for driving marker genes, such as the *lacZ* gene, in recombinant viruses. Recently, various promoters have been developed and can be chosen depending on target protein production, as shown in Table 3.2.

3.3 Transient Expression System in Silkworm

3.3.1 Baculovirus Expression System

The AcMNPV expression vector system has been developed and thoroughly validated (Jarvis 2009; O'reilly et al. 1994). Similarly, an effective expression system for silkworms has also been established using a conventional vector approach, as

Table 3.2 Baculovirus transfer vectors

Plasmid		Reference (Source)
Expression of a single gene		
AcMNPV polh*-based, no ATG supplied*		
	pAc373	Smith et al. (1985)
	pAcYM1	Invitrogen
	pBacPAK1	Kitts and Possee (1993)
	pEV55	David et al. (1986)
	pVL941	Luckow and Summers (1989)
	pVL1393	O'reilly et al. (1994)
AcMNPV polh*-based, ATG fused to polyhedrin coding sequence*		
	pAC360	Smith et al. (1983)
	pAC700	Luckow and Summers (1988)
AcMNPV polh*-based, ATG fused to metal-binding domain*		
	pBlueBacHIS	Invitrogen
AcMNPV polh*-based, ATG fused to honeybee mellitin signal peptide*		
	pVT-Bac	Tessier et al. (1991)
AcMNPV polh*-based, ATG supplied*		
	pAcCL29-8	Livingstone and Jones (1989)
AcMNPV p10*-based, no ATG supplied*		
	pAcAs2	Vlak et al. (1990)
	pAcUW1	Weyer et al. (1990)
BmNPV polh*-based, no ATG supplied*		
	pBM030	Horiuchi et al. (1987)
Coexpression of a single gene		
AcMNPV polh*-based,* lacZ *marker gene, no ATG supplied*		
	pAcDZ1	Zuidema et al. (1990)
	pBlueBac, pBlueBac2, pBlueBac3	Richardson (2008)
AcMNPV p10*-based, no ATG supplied*		
	pAcUW2B	Weyer et al. (1990)
AcMNPV *polh- and p10-based* dual expression of two heterologous genes		
AcMNPV polh*- and* p10*-based*		
	p2Bac;2	Invitrogen
	pAcUW3	Weyer and Possee (1991)
	pAcUW31	cleantech

This table is compiled from data excerpted from the previously published research (Luckow 1994; O'reilly et al. 1994) and reorganized. Adapted with permission (Luckow 1994; O'reilly et al. 1994). Copyright 1994 Wiley-Liss and Oxford University Press

illustrated in Fig. 3.3 (Maeda et al. 1985). At first, a transfer vector to insert the target protein gene is created. The size is approximately 10 kb, and the backbone is a commonly used vector, such as pUC9, which contains an ampicillin resistance gene and a bacterial DNA replication origin. HindIII fragments, each approximately 3 kb in length and corresponding to the 5′ and 3′ regions of the full-length *polh* gene

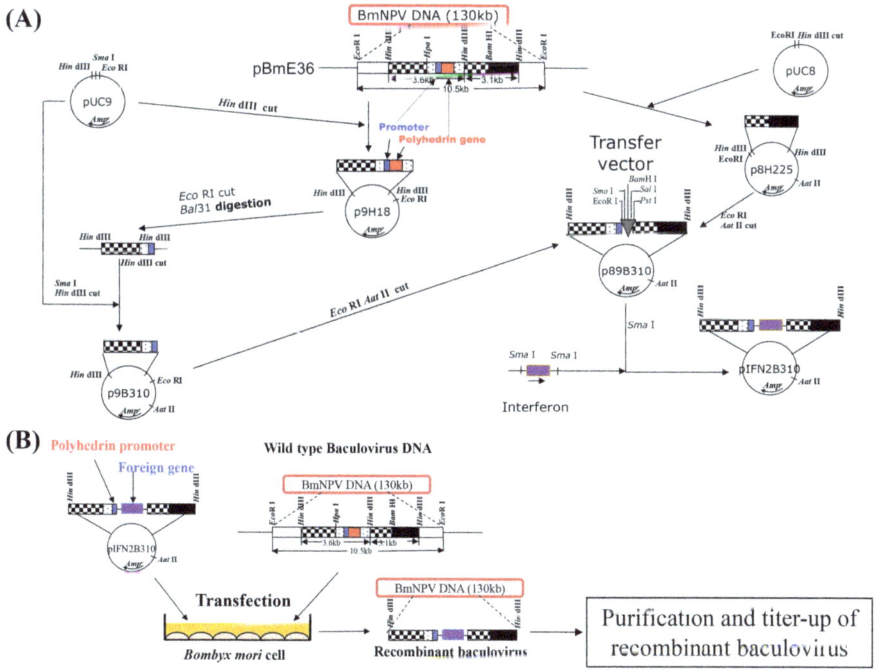

Fig. 3.3 (**a**) Construction vectors: A 3.9 kb HindIII fragment containing the *polh* gene from plasmid pBmE36 was first cloned into pUC9, generating plasmid p9H18. A 2.7 kb blunt-ended fragment, obtained through EcoRI digestion, followed by Bal31 treatment, was then inserted into the HindIII-SmaI site of pUC9 to produce p9B310. Subsequently, an EcoRI-AatII fragment from p8H225 was ligated to the corresponding 4.7 kb fragment from p9B310. Finally, the IFN-α-J gene was inserted into the SmaI site of p89B310, resulting in the construction of the recombinant plasmid pIFN2B310. (**b**) Preparation of recombinant baculovirus: The pIFN2B310 plasmid was co-transfected with wild-type baculoviral genomic DNA into *B. mori* cells. This led to the generation of a recombinant BmNPV expressing the IFN-α-J gene. (Modified from permission (Maeda et al. 1985). Copyright 1985 Springer Nature)

from the plasmid pBmE36, were cloned into the pUC9 plasmid. These regions are necessary for replacing the target gene with the *polh* gene during transfection. These 5′ and 3′ regions were artificially created by removing the original *polh* gene and inserting a nucleotide sequence recognized by the linker in its place. The generating plasmid p9H18 was first modified by excising the *polh* gene using EcoRI digestion, followed by Bal31 exonuclease treatment, resulting in the derivative plasmid p9B310. The downstream region of the *polh* gene, isolated as a HindIII fragment from pBmE36, was initially cloned into the pUC8 vector, resulting in the construction of plasmid p8H225. This plasmid was subsequently digested with EcoRI and AatII, and the excised fragment, designated p89B310, incorporates both polylinker sequences and the *polh* promoter gene (Fig. 3.3a).

The DNA fragment to be inserted should be cDNA or a gene lacking introns. The presence of introns impairs the transcription and translation efficiency, reducing expression levels to less than one hundredth (Maeda 1993). Furthermore, the untranslated region upstream of the translation start site of the inserted gene should be relatively short, ideally 50 bp or less. If the target gene has the same restriction enzyme sites at both ends as the linker, these can be utilized for insertion.

The human α-interferon (IFN-α) gene was inserted into the p89B310 vector, yielding the recombinant plasmid pIFN2B310. This plasmid was then co-transfected with baculoviral genomic DNA into *B. mori* cells to generate a recombinant baculovirus (BmNPV) (Fig. 3.3b). The supernatant from transfected cultured cells primarily contains wild-type baculoviruses, which can form polyhedra. In contrast, recombinant baculoviruses do not form polyhedra because they contain a foreign gene inserted into the polyhedron translation region. As a result, recombinant baculoviruses are present in smaller quantities within the cell broth. They can be isolated and purified using a plaque assay and subsequently amplified to generate a high-titer stock for infecting silkworms. This process typically takes several weeks. Usually, approximately 10^8 infectious recombinant BmNPV particles per mL of cell culture broth are collected and stored at either 4 °C or −80 °C. The fifth instar silkworm larvae are commonly used for recombinant BmNPV infections. A diluted solution of recombinant BmNPV—approximately ten-fold—is administered subcutaneously at a volume of 10–20 μL to the silkworms using a 1 mL tuberculin syringe. The silkworms are then reared for 4–5 days. When symptoms of a viral disease appear, such as translucent skin, it is important to collect the infected individuals promptly. For secretory proteins, gather the hemolymph (body fluid) while taking care to prevent melanization, which can darken the fluid. Melanization is an insect defense response, and once it occurs, purifying the target protein becomes extremely difficult. To suppress melanization, it is advisable to add a small amount of a chemical, such as phenylthiourea, to the hemolymph prior to hemolymph collection. After obtaining the hemolymph samples, lightly centrifuge them to remove blood cells and other debris, and then store them at −80 °C. Proteins that accumulate in the cytoplasm or nucleus and are not secreted into hemolymph accumulate in the fat bodies, which are the primary sites of viral replication. Therefore, it is necessary to dissect the silkworms in order to recover these fat bodies. Using recombinant baculovirus encoding IFN-α gene, approximately 20 μg of IFN-α was successfully produced from a single silkworm larva (Maeda et al. 1985).

3.3.2 Bacmid Expression System

The baculovirus expression system (BES) is extensively used for the high-level production of recombinant proteins in insect cell cultures. This is mainly due to the powerful polyhedrin promoter that drives gene expression. The BES has significant advantages for expressing eukaryotic proteins, as insect cells can perform post-translational modifications similar to those found in higher organisms. Additionally, insect cells are often easier to manipulate compared to bacterial systems (Luckow et al. 1993; Luckow and Summers 1989; O'reilly et al. 1994).

The AcMNPV bacmid system, functioning as a baculovirus shuttle vector, has a significantly simplified protein expression in insect cells. This plasmid can replicate as a plasmid in *E. coli*, enabling the generation of recombinant BmNPV DNA via site-specific transposition. This process renders the bacmid infectious to insect cells. Importantly, the use of the bacmid shortens the time and reduces the complexity of selecting and purifying recombinant viruses to approximately 10 days. In contrast, conventional methods for producing recombinant baculoviruses encoding foreign genes typically require at least 40 days due to multiple rounds of purification and amplification. However, the AcMNPV bacmid does not directly infect silkworms (*B. mori*) or *B. mori* cell lines due to differences in host specificity. Instead, a hybrid baculovirus combining BmNPV and AcMNPV can infect both *B. mori* cell lines and silkworms (Zhu et al. 1998). Unfortunately, this hybrid has not been widely adopted because of its low infection efficiency in *B. mori cell* lines. As a result, there has been a longstanding need for a next-generation expression system capable of effectively utilizing silkworms as biofactories for protein expression.

3.3.2.1 Preparation of BmNPV Transfer Vector

The first practical BmNPV bacmid system was developed by creating a BmNPV bacmid and transforming it into *E. coli* DH10B (Motohashi et al. 2005). To achieve this, essential elements such as the miniF replicon and Tn7 insertion sites, necessary for replication and transposition in *E. coli,* respectively, were combined to form the "miniF-Kan" construct. This was done by ligating these elements to a digested version of the pBKblue DNA, which lacked both the *polh* promoter and the *lacZα* gene. A fragment of 8.4 kb containing the miniF replicon, a kanamycin resistance gene, the *lacZα* gene, and Tn7 insertion sites was generated through PCR. The primers used for PCR were 5′-TTTTGCTTTGCCACGGAA CGGTC TGC-3′ and 5′-TTGCCGGGTCCCAGGAAAGGAT-3′, sourced from the AcNPV bacmid within DH10Bac competent cells. This "miniF-Kan" construct was subsequently introduced into the baculovirus transfer vector, pBKblue (Fig. 3.4).

3.3.2.2 Construction of BmNPV Bacmid

In *B. mori* cells, the plasmid, pBKblue/miniF-Kan, underwent homologous recombination with wild-type BmNPV DNA to generate the BmNPV bacmid. Following this, a modified strain of *E. coli* DH10Bac, referred to as *E. coli* Bm DH10Bac (Fig. 3.5), was developed. These engineered cells harbor the BmNPV bacmid along with a helper plasmid, pMON7124, which facilitates the insertion of foreign genes into the BmNPV bacmid via a transfer vector system.

To enable homologous recombination between the pBKblue/miniF-Kan transfer vector and wild-type BmNPV viral DNA, 1 μg of wild-type viral BmNPV DNA and

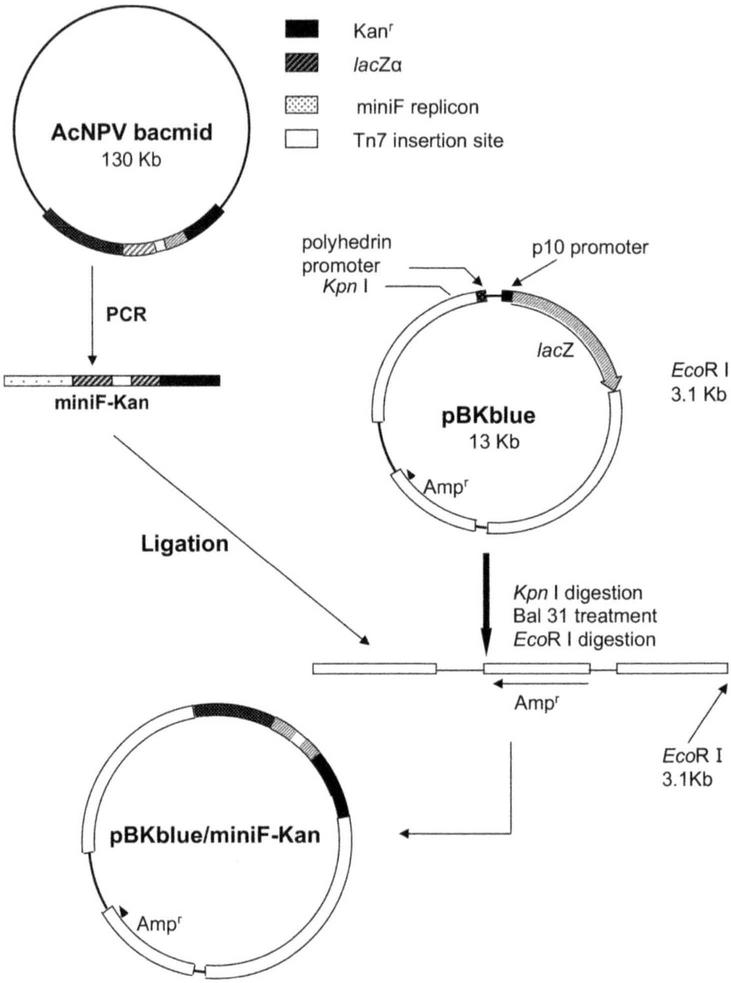

Fig. 3.4 Schematic outline for the generation of BmNPV transfer vector. It involves introducing a miniF replicon, a kanamycin resistance gene, and a *lacZα* gene. A DNA fragment containing the miniF replicon, the kanamycin resistance gene, and the *lacZα* gene was obtained from the AcNPV bacmid and inserted into a modified pBKblue vector lacking the *lacZα*. The resulting construct was designated as pBKblue/miniF-Kan. (Adapted with permission (Motohashi et al. 2005). Copyright 2005 Elsevier)

5 μg of pBKblue/miniF-Kan vector were suspended in 100 μL of Sf900 II medium. This mixture was then combined with 120 μL of Sf900 II medium containing 20 μL of Lipofectin reagent (GIBCO BRI), and the resulting solution was added to 1 mL of a Bm5 cell culture at a concentration of 7×10^5 cells/mL. The culture was incubated at 27 °C for 8 h. Following incubation, the cells were harvested and centrifuged at 3000 rpm and 4 °C for 5 min. The supernatant, containing both recombinant and wild-type viruses, was centrifuged at 25,000 g for 1 h. To extract the viral DNA,

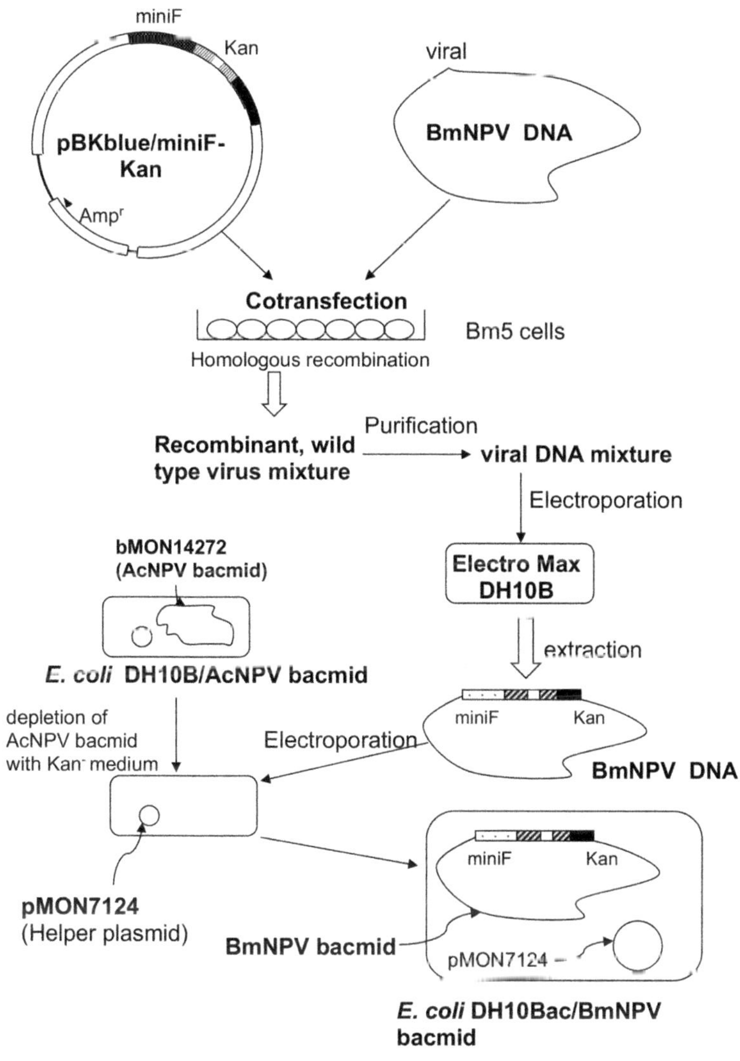

Fig. 3.5 Construction of BmNPV bacmid. Homologous recombination between pBKblue/miniF-Kan and wild-type BmNPV DNA was carried out in insect cells. The purified recombinant viral DNA was subsequently introduced into ElectroMax DH10B cells by electroporation. Kanamycin-resistant colonies that were ampicillin-sensitive and positive for the *lacZα* gene were selected, indicating successful recombination. The resulting BmNPV bacmid was introduced into a derivative of *E. coli* DH10Bac that had lost the AcNPV bacmid. Colonies resistant to both kanamycin and tetracycline were subsequently isolated and designated as Bm DH10Bac. (Adapted with permission (Motohashi et al. 2005). Copyright 2005 Elsevier)

10 μL of proteinase K (20 mg/ml) was added to the obtained supernatant and incubated at 50 °C for 1 h. The resulting viral DNA, comprising both recombinant and wild-type genomes, was introduced into Electro Max DH10B cells via electroporation. Colonies exhibiting kanamycin resistance, *lacZα* activity, and sensitive to ampicillin were selected. PCR analysis confirmed successful acquisition of the

BmNPV bacmid. Separately, when DH10Bac cells harboring the AcNPV bacmid (bMON14272) and the helper plasmid (pMON7124) were cultured in a medium lacking kanamycin, a subset of cells lost the AcNPV bacmid while retaining only the helper plasmid. The BmNPV bacmid was subsequently introduced into these cells via electroporation. Colonies were selected on a medium containing both kanamycin and tetracycline and designated as *E. coli* Bm DH10Bac (Motohashi et al. 2005).

Homologous recombination between the pBKblue/miniF-Kan and wild-type viral BmNPV DNA was performed in insect cells. The purified virus DNA was subsequently introduced into ElectroMax DH10B cells via electroporation. Colonies exhibiting kanamycin resistance, ampicillin sensitivity, and positive *lacZ*α activity were selected, indicating successful recombination. The resulting BmNPV bacmid was then transferred into a derivative of *E. coli* DH10Bac, and colonies resistant to both kanamycin and tetracycline were subsequently isolated and designated Bm DH10Bac.

3.3.2.3 Construction of BmNPV Bacmid Expressing GFP_{uv}

The BmNPV bacmid system was validated by expressing green fluorescent protein (GFP_{uv}). To achieve this, a donor plasmid GFP_{uv}/pDEST8 was transposed into Bm DH10Bac cells, resulting in the generation of a recombinant BmNPV bacmid encoding GFP_{uv} (BmNPV bacmid/GFP_{uv}). The GFP_{uv} gene was obtained through PCR using the following primers: 5′-C ACC ATG AGT AAA GGA GAA GAA CTT TTC-3′ and 5′-TTATTT GTA GAG CTC ATC CAT GCC ATG-3′ from pBlueBacHis2-GFP_{uv}/β-3GnT (Kato et al. 2004). The amplified PCR product was cloned into the Gateway entry vector pENTER/D-TOPO according to the protocol provided with the Topo cloning system. The resulting plasmid, named pENTER/D--TOPO/GFP_{uv}, was subsequently recombined with the destination vector pDEST8, yielding the final construct pDEST8/GFP_{uv} (Fig. 3.6).

Transposition was performed by introducing the donor plasmid pDEST8/GFP_{uv} into Bm DH10Bac cells, which harbor a transposition helper plasmid, as illustrated in Figs. 3.5 and 3.6. White colonies resistant to kanamycin and gentamicin were selected, and the BmNPV bacmid was subsequently isolated and designated BmNPV bacmid/GFP_{uv}. Approximately 100 ng of BmNPV bacmid/GFP_{uv} DNA was transfected into Bm5 cells using the Lipofectin method, as previously described. Following incubation, the Bm5 cell culture was harvested and centrifuged at 3000 rpm and 5 min at 4 °C. The supernatant was collected, filter-sterilized, and stored at 4 °C as the recombinant virus, BmNPV/GFP_{uv} (vBmNPV/GFP_{uv}).

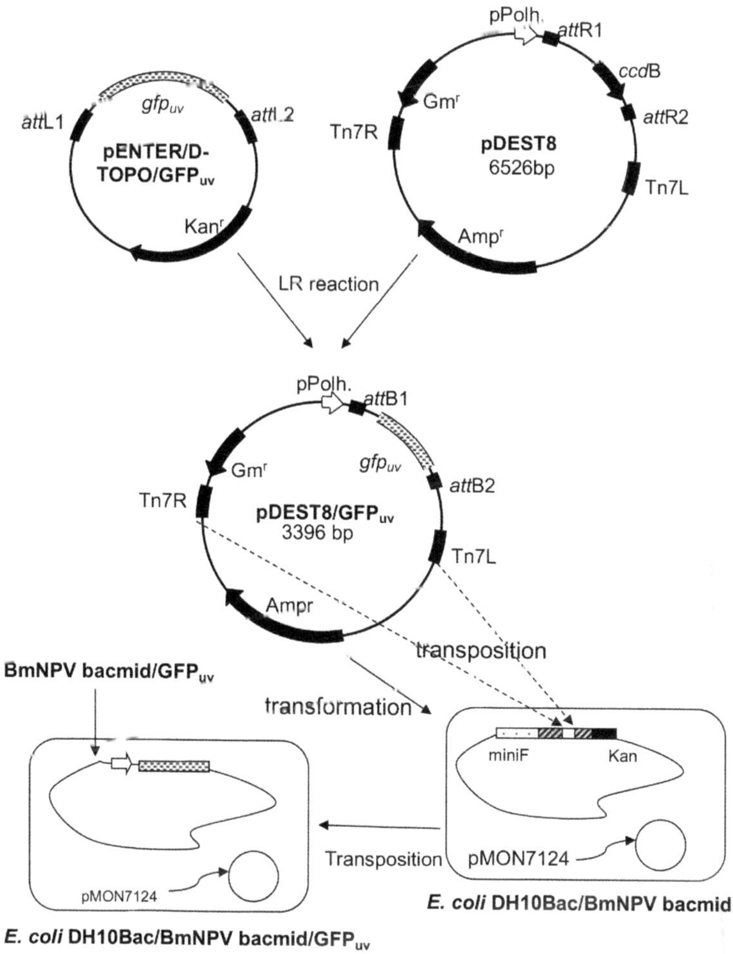

Fig. 3.6 Construction of BmNPV transfer vector containing GFP$_{uv}$ gene. The GFP$_{uv}$ gene from pENTER/D-TOPO/GFP$_{uv}$ was introduced into pDEST8 by LR reaction. (Adapted with permission (Motohashi et al. 2005). Copyright 2005 Elsevier)

3.3.2.4 Expression of GFP$_{uv}$ in *B. mori* Cells, Larvae, and Pupae Using BmNPV Bacmid/GFP$_{uv}$ DNA

To evaluate GFPuv expression in *B. mori* cells, *B. mori* Bm5 cells were infected with vBmNPV/GFPuv (Fig. 3.7a). The infected cells were harvested at 24, 48, and 72 h post-infection and examined under a fluorescence microscope. GFP$_{uv}$ fluorescence was first detected at 24 h, and by 48 h, the majority of cells exhibited strong green fluorescence (Fig. 3.7b). These findings demonstrate the effective functionality of the developed BmNPV bacmid expression system.

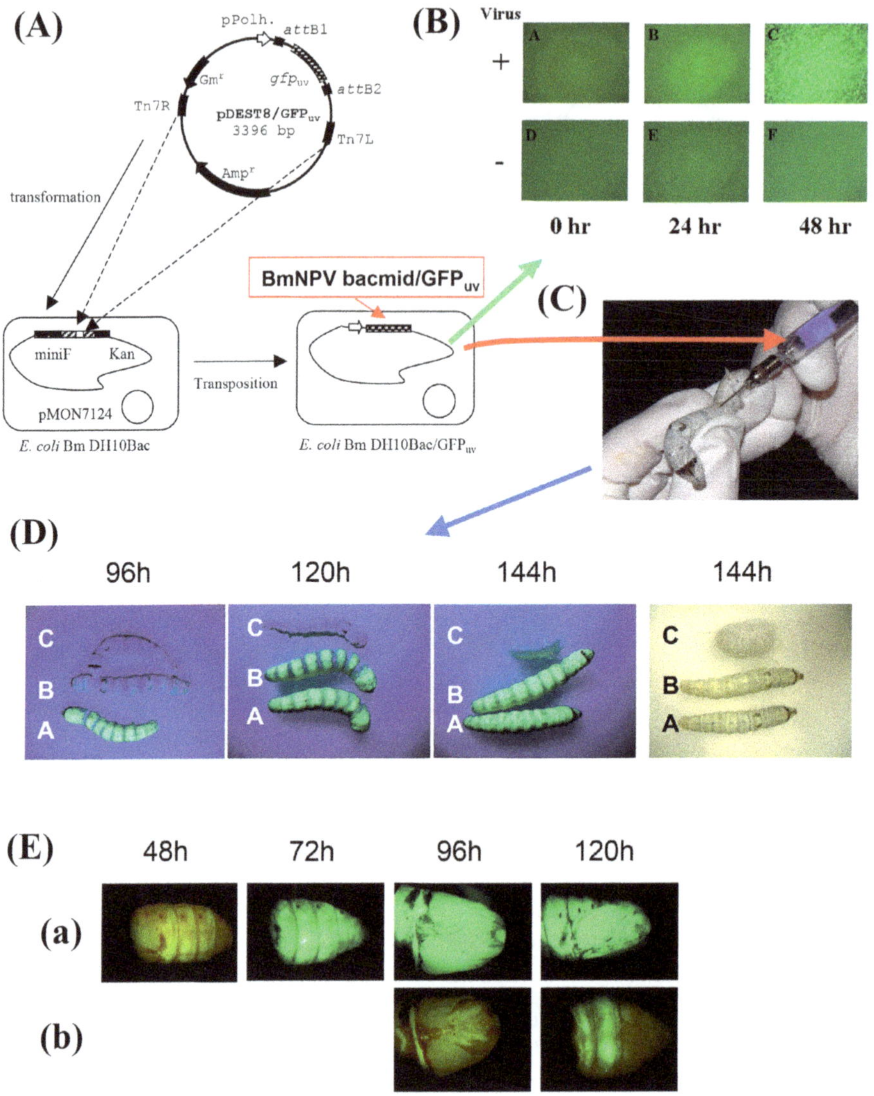

Fig. 3.7 (**a**) Construction and expression of GFP_{uv} using the BmNPV bacmid system. The donor plasmid pDEST8/GFP_{uv}, containing the GFP_{uv} gene, was introduced into *E. coli* Bm DH10Bac competent cells to facilitate transposition. The resulting recombinant DNA was designated as BmNPV bacmid/GFP_{uv}. (**b**) GFP_{uv} expression in *B. mori* cells. Cells were transfected with BmNPV bacmid/GFP_{uv} DNA (panels A–C) or mock-transfected (panels D–F). Fluorescence microscopy images were captured at 0, 24, and 48 h post-transfection under complete darkness. (**c**) BmNPV bacmid/GFP_{uv} DNA or its recombinant virus was injected into silkworm larvae or pupae using a syringe (larvae) or pipette (pupae), as shown. (**d**) GFP_{uv} expression in *B. mori* larvae. Larvae were infected via direct syringe injection with either the recombinant virus vBmNPV/GFP_{uv} (A), BmNPV bacmid/GFP_{uv} DNA, (B) or mock solution (C). Fluorescence images were taken at 96, 120, and 144 h post-infection under UV illumination in complete darkness. An image of larvae

The silkworm larvae (Ehime Sansyu Co., Ltd., Ehime, Japan) were on the first day of the fifth instar stage, and the pupae (Ehime Sansyu Co., Ltd., Ehime, Japan) were on the fourth day after cocoon formation, having developed from the fifth instar larvae. GFP_{uv} expression in silkworms was performed using two methods: (1) direct injection of BmNPV bacmid/GFP_{uv} DNA and (2) infection with recombinant virus vBmNPV/GFP_{uv}. For the direct injection method, 0.2–0.3 μg of BmNPV bacmid/GFP_{uv} DNA was suspended in 2 μL of 1,2-dimyristyloxypropyl-3-dimethyl-hydroxyethyl ammonium bromide (DMRI-C; Invitrogen) reagent and incubated at room temperature for 45 min. The DNA mixture was then diluted with 16 μL of Sf-900II serum-free medium to achieve a final volume of 20 μL. For administration, the DNA solution was injected into the dorsal side of larvae using a syringe fitted with a 26-gauge beveled needle and into pupae using a pipette (Fig. 3.7c). For infection with the recombinant virus vBmNPV/GFP_{uv}, 10 μL of virus solution (2.7×10^7 pfu/mL) was injected into the larvae using a syringe and into the pupae using a pipette. In pupal injection, the needle tip was first dipped into the virus solution and then used to puncture the dorsal side of the pupae. Following infection, both larvae and pupae were dissected, and their tissues were examined under a fluorescence microscope (MZFL III, LEICA Microsystems Inc., Wetzlar, Germany).

Upon infection with vBmNPV/GFP_{uv}, *B. mori* larvae exhibited visible green fluorescence by 96 h post-infection (Fig. 3.7d-A). In comparison, larvae injected with BmNPV bacmid/GFP_{uv} DNA showed fluorescence approximately 24 h later, at around 120 h post-injection (Fig. 3.7d-B). Although slightly delayed, the direct injection method offers a lower biohazard risk. Mock-infected larvae displayed no fluorescence and proceeded to spin cocoons by 144 h post-infection (Fig. 3.7d-C). Strong GFP_{uv} expression was observed in both the fat body and hemolymph. Quantitative analysis indicated that approximately 2–3 mg of GFP_{uv} protein could be obtained from a single larva (Motohashi et al. 2005). *B. mori* pupae have been demonstrated to be an effective alternative host system for large-scale production of recombinant proteins.

A key advantage of using pupae lies in their low endogenous protein synthesis during the early pupal stage, which allows for a highly efficient synthesis of viral proteins (Mikhailov et al. 1992). Therefore, GFP_{uv} expression in *B. mori* pupae was available through two methods: (1) injection of vBmNPV/GFP_{uv} virus using a needle tip and (2) direct injection of bacmid DNA. Notably, even a very small amount of virus introduced via needlepoint resulted in strong GFP_{uv} expression within 48–72 h post-infection (Fig. 3.7E-a). In addition, direct injection of BmNPV bacmid/GFP_{uv} DNA into *B. mori* pupae was carried out. Fluorescence was not observed until 96 h post-injection; however, by 120 h, the pupae exhibited strong GFP_{uv}

Fig. 3.7 (continued) infected with the recombinant virus vBmNPV/GFP_{uv}, injected BmNPV bacmid/GFP_{uv} DNA, or mock solution was taken under visible light at 144 h post-infection. (**E**) GFP_{uv} expression in *B. mori* pupae. The pupae were infected either by pricking with a needle tip immersed in vBmNPV/GFP_{uv} virus solution (panel a) or by direct injection of 20 μL BmNPV bacmid/GFP_{uv} DNA using a pipette (panel b). Fluorescence microscopy images were captured at 48, 72, 96, and 120 h post-infection. (Adapted with permission (Motohashi et al. 2005). Copyright 2005 Elsevier)

expression (Fig. 3.7E-b). While direct injection of bacmid DNA into *B. mori* pupae resulted in detectable GFP_{uv} expression, the efficiency was notably lower compared to that observed in larvae. This suggests that further optimization of experimental conditions for pupal expression is still required.

At 96 h post-infection, the silkworm larvae were dissected to isolate tissues such as the fat body and hemolymph for the analysis of GFP_{uv} expression. Although GFP_{uv} is not secreted as a soluble protein, strong green fluorescence was observed in the hemolymph (Fig. 3.8a-1). This phenomenon is likely due to the insect's open

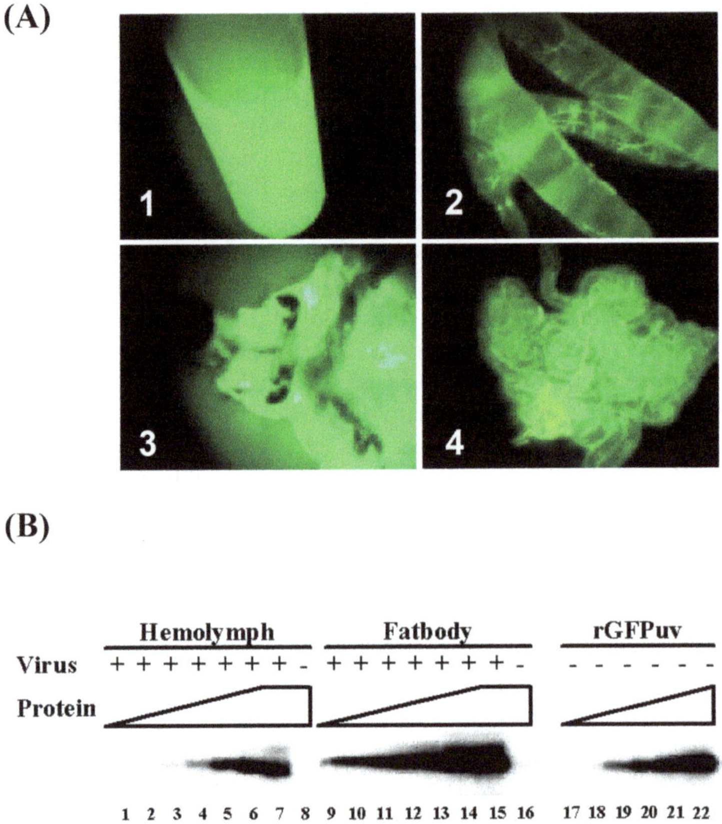

Fig. 3.8 (**a**) GFP_{uv} expression in tissues of *B. mori* larvae. The silkworm larvae were examined at 96 h post-infection. Dissection revealed fluorescence signals in the hemolymph (1), fat body (2), head (3), and silk gland (4), as observed under a fluorescence microscope (original magnification × 20). (**b**) Western blot analysis was performed to detect GFP_{uv} expression in the fat body and hemolymph of *B. mori* larvae injected with BmNPV bacmid/GFP_{uv} DNA. GFP_{uv} protein was identified using an anti-GFP antibody. Total protein samples from infected larvae were loaded in lanes 1–7 (hemolymph: 0.001, 0.003, 0.01, 0.03, 0.1, 0.3, and 1 mg) and lanes 9–15 (fat body: same concentrations). As negative controls, 1 mg of total protein from non-infected larvae was loaded in lanes 8 (hemolymph) and 16 (fat body). Recombinant GFP_{uv} protein was used as a positive control and loaded in lanes 17–22 (0.001, 0.003, 0.01, 0.03, 0.1, and 0.3 mg). (Adapted with permission (Motohashi et al. 2005). Copyright 2005 Elsevier)

circulatory system, which allows proteins from the fat body to diffuse into the hemolymph. Additionally, during the late stages of viral infection, lysis of the fat body may further contribute to the release of proteins into the hemolymph. The fat body itself exhibited intense green fluorescence (Fig. 3.8a-2). To quantify GFP_{uv} expression, Western blot analysis was conducted using protein expressed from the hemolymph and fat body of larvae injected with BmNPV bacmid/GFP_{uv} DNA. High levels of GFP_{uv} expression were detected in both tissues—lanes 1–7 for hemolymph and lanes 9–15 for fat body (Fig. 3.8b). Based on comparison with the recombinant GFP_{uv} standards (lanes 17–22 in Fig. 3.8b), the estimated yield was approximately 2–3 mg of GFP_{uv} protein per larva, following injection with the recombinant virus (1.4×10^6 pfu; 50 μL of 2.7×10^7 pfu/mL). Comparable expression levels were also achieved via direct injection of a small amount of bacmid DNA (0.2–0.3 μg). Furthermore, similar yields (2–3 mg/larva) of glutathione S-transferase (GST) were obtained using the same BmNPV bacmid system (data not shown).

The BmNPV bacmid expression system holds considerable promise for the large-scale expression of foreign genes in silkworms. Unlike conventional methods, it eliminates the need for lengthy procedures to amplify virus titers in *B. mori* cells and offers a high level of biosafety. Furthermore, its cost-effectiveness—approximately 10% of the expense associated with expression systems based on SF9 or Bm cells—makes it an economically attractive alternative. Overall, the BmNPV bacmid system represents a powerful and scalable platform for the expression of eukaryotic proteins in silkworms.

3.4 Transgenic Silkworm

3.4.1 piggyBac Vector

Although a transient expression system has been established, it only allows for the production of recombinant protein for a single generation, and viral infection is necessary each time expression is required. To address this issue, foreign genes need to be introduced into the silkworm genome so that they can function stably. For this purpose, the piggyBac vector is employed. piggyBac was originally discovered as a causative agent of few polyhedra mutations in the cultured cells of the moth *Trichoplusia ni*, commonly known as the cabbage looper (Fraser et al. 1983). This genetic element, which is dispersed and repeated throughout the genome of this cell line, was isolated from this few-polyhedra mutant strain (Cary et al. 1989) and showed autonomous function in a transposon assay. When used for germline transformation, the piggyBac system consists primarily of two elements. These are typically introduced into cells as separate plasmids. One is a transposon vector, which contains the target gene to be introduced. Special DNA sequences called Inverted Terminal Repeats (ITRs) are placed at both ends of this gene. The other is the piggyBac transposase, an enzyme that recognizes the ITRs, accurately excises the target gene flanked by these ITRs from the DNA, and inserts it into a specific location in the host genome. The most significant feature distinguishing piggyBac from other transposons is the target sequence for genome insertion. PiggyBac specifically recognizes the target four-base "TTAA" sequence (Beames and Summers 1990; Fraser et al. 1995) and inserts the gene only at this precise location. This "TTAA" sequence is abundant in the human genome, providing numerous potential insertion sites.

Compared with other transposons or viral vectors, piggyBac can carry very large DNA fragments ranging from 10 to 14 kb or more (Li et al. 2011), with some reports suggesting it can theoretically exceed 100 kb. This capability allows for the simultaneous introduction of multiple genes or complex gene circuits. When piggyBac excises from the inserted genome, it fully repairs the insertion site's "TTAA" sequence back to its original state, leaving no genetic footprint (Elick et al. 1996). This characteristic is commonly used to efficiently remove reprogramming factors introduced during the generation of induced pluripotent stem (iPS) cells after the establishment of iPS cells (Zhu et al. 1998). Furthermore, the piggyBac system operates across a wide range of species, including yeast (Zhu et al. 2018), insects or silkworms (Jiang et al. 2014; Uchino et al. 2008; Wang et al. 2015), plants (Nishizawa-Yokoi and Toki 2023), and mammals (Liu et al. 2013).

3.4.2 Protein Expression in Transgenic Silkworm

In silkworms, the piggyBac vector has been used to stably produce recombinant proteins across multiple generations, eliminating the need for repeated genetic modification for each expression cycle. This was achieved by inserting the target gene into the piggyBac vector's ITRs and injecting it, along with a helper vector encoding the piggyBac transposase, into silkworm eggs at the preblastodermal stage. As a result, transgenic silkworms were successfully generated. A key advancement was achieved through germline transformation using a vector derived from piggyBac transposon (Tamura et al. 2000). The vector, named pPIGA3GFP (Fig. 3.9a), is derived from the wild-type piggyBac transposon found in *T. ni* (Cary et al. 1989). In this construct, the central region was replaced with a transgene composed of the BmA3 promoter, GFP, and SV40 regulatory elements (Mange et al. 1997). The helper plasmid accompanying pPIGA3GFP encodes the transposase enzyme necessary for transposition. To initiate embryonic development, silkworm eggs were collected within 1–2 h after oviposition and microinjected with both the transfer vector and the helper plasmid. This method successfully produced G0 adult silkworms carrying transformed gametes, with a reported transformation efficiency of 1.8% (Tamura et al. 2000). The inserted piggyBac DNA was stably inherited across successive generations without any detectable alterations.

This genetic engineering method is especially effective for producing recombinant proteins within the silk glands of transgenic silkworms. It allows these proteins to be incorporated either into the insoluble fibroin core or onto the hydrophilic sericin outer layer of the silk fiber. The silk gland itself consists of three distinct regions:

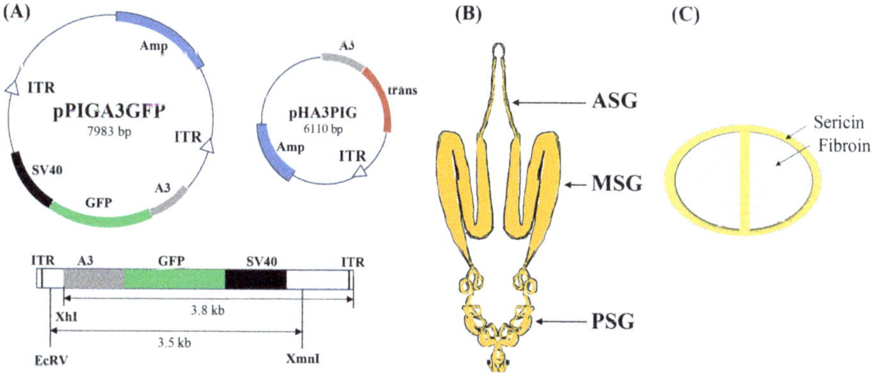

Fig. 3.9 (**a**) Design of piggyBac vectors for germline transformation. The pPIGA3GFP vector includes the *B. mori* A3 promoter, GFP reporter gene, and SV40 3′ untranslated region. The helper plasmid, pHA3PIG, provides the piggyBac transposase. Key elements of this system include inverted terminal repeats of piggyBac (ITR) and an ampicillin resistance gene for bacterial selection. (Adapted with permission (Tamura et al. 2000). Copyright 2000 Springer-Nature). (**b**) Illustration of silk gland structure. The silk gland is anatomically divided into three regions: PSG, MSG, and ASG. (**c**) Cross-sectional view of a silk fiber. The diagram illustrates the internal fibroin core surrounded by a sericin layer, which acts as a hydrophilic adhesive. ((**b**) and (**c**) were reproduced with permission (Tomita 2011). Copyright 2011 Springer-Nature)

the posterior silk gland (PSG), the middle silk gland (MSG), and the anterior silk gland (ASG) (Fig. 3.9b) (Grzelak 1995). Fibroin proteins are synthesized in the PSG and then transported into the lumen of the MSG, where they form the inner core of the silk fiber. This core is subsequently coated with sericin, a hydrophilic adhesive protein that binds the fibroin layers and solidifies upon exposure to air. When foreign genes are expressed in the PSG, the resulting recombinant proteins are incorporated into the inner fibroin core. In contrast, gene expression in the MSG results in protein localization in the outer sericin layer (Fig. 3.9c). This system offers several advantages. It enables the expression of recombinant proteins in their native, nonfused form, which can be easily extracted from cocoons using a gentle aqueous solution. For instance, enhanced green fluorescent protein was produced at a yield of 7–8 μg per mg of cocoon (Tomita 2011; Tomita et al. 2007; Tomita et al. 2003). A wide range of recombinant proteins has been successfully expressed using this approach (Tomita 2011).

References

Ayres MD, Howard SC, Kuzio J, Lopez-Ferber M, Possee RD (1994) The complete DNA sequence of *Autographa californica* nuclear polyhedrosis virus. Virology 202:586–605. https://doi.org/10.1006/viro.1994.1380

Beames B, Summers MD (1990) Sequence comparison of cellular and viral copies of host cell DNA insertions found in *Autographa californica* nuclear polyhedrosis virus. Virology 174:354–363. https://doi.org/10.1016/0042-6822(90)90089-a

Burgess S (1977) Molecular weights of lepidopteran baculovirus DNAs: derivation by electron microscopy. J Gen Virol 37:501–510. https://doi.org/10.1099/0022-1317-37-3-501

Cary LC, Goebel M, Corsaro BG, Wang HG, Rosen E, Fraser MJ (1989) Transposon mutagenesis of baculoviruses: analysis of *Trichoplusia ni* transposon IFP2 insertions within the FP-locus of nuclear polyhedrosis viruses. Virology 172:156–169. https://doi.org/10.1016/0042-6822(89)90117-7

Cederbaum SD, Fareed GC, Lovett MA, Shapiro LJ (1984) Recombinant DNA in medicine. West J Med 141:210–222

Cox MM (2012) Recombinant protein vaccines produced in insect cells. Vaccine 30:1759–1766. https://doi.org/10.1016/j.vaccine.2012.01.016

David WM, Pennina S, Lois KM (1986) An insect baculovirus host-vector system for high-level expression of foreign genes. Genet Eng 8:277–298. https://doi.org/10.1007/978-1-4615-9456-7_14

Elick TA, Bauser CA, Fraser MJ (1996) Excision of the piggyBac transposable element in vitro is a precise event that is enhanced by the expression of its encoded transposase. Genetica 98:33–41

Fraser MJ, Smith GE, Summers MD (1983) Acquisition of host cell DNA sequences by baculoviruses: relationship between host DNA insertions and FP mutants of *Autographa californica* and *galleria mellonella* nuclear polyhedrosis viruses. J Virol 47:287–300. https://doi.org/10.1128/JVI.47.2.287-300.1983

Fraser MJ, Cary L, Boonvisudhi K, Wang HG (1995) Assay for movement of lepidopteran transposon IFP2 in insect cells using a baculovirus genome as a target DNA. Virology 211:397–407. https://doi.org/10.1006/viro.1995.1422

Gomi S, Majima K, Maeda S (1999) Sequence analysis of the genome of *Bombyx mori* nucleopolyhedrovirus. J Gen Virol 80(Pt 5):1323–1337. https://doi.org/10.1099/0022-1317-80-5-1323

Grzelak K (1995) Control of expression of silk protein genes. Comp Biochem Physiol B Biochem Mol Biol 110:671–681. https://doi.org/10.1016/0305-0491(94)00215-g

Harrap KA (1972) The structure of nuclear polyhedrosis viruses. II The virus particle. Virology 50:124–132. https://doi.org/10.1016/0042-6822(72)90352-2

Hill-Perkins MS, Possee RD (1990) A baculovirus expression vector derived from the basic protein promoter of *Autographa californica* nuclear polyhedrosis virus. J Gen Virol 71(Pt 4):971–976. https://doi.org/10.1099/0022-1317-71-4-971

Horiuchi T, Marumoto Y, Saeki Y, Sato Y, Furusawa M, Kondo A, Maeda S (1987) High-level expression of the human-a-interferon gene through the use of an improved baculovirus vector in the silkworm, *Bombyx mori*. Agric Biol Chem 51:1573–1580. https://doi.org/10.1080/00021369.1987.10868243

Jarvis DL (2009) Chapter 14 Baculovirus–insect cell expression systems. In: Burgess RR, Deutscher MP (eds) Methods in enzymology. Academic, pp 191–222

Jarvis DL, Oker-Blom C, Summers MD (1990) Role of glycosylation in the transport of recombinant glycoproteins through the secretory pathway of lepidopteran insect cells. J Cell Biochem 42:181–191. https://doi.org/10.1002/jcb.240420402

Jiang L, Sun Q, Liu W, Guo H, Peng Z, Dang Y, Huang C, Zhao P, Xia Q (2014) Postintegration stability of the silkworm piggyBac transposon. Insect Biochem Mol Biol 50:18–23. https://doi.org/10.1016/j.ibmb.2014.03.006

Kato T, Murata T, Usui T, Park EY (2004) Comparative analysis of GFP(UV)-beta1,3-N-acetylglucosaminyltransferase 2 production in two insect-cell-based expression systems. Protein Expr Purif 35:54–61. https://doi.org/10.1016/j.pep.2003.12.018

Kato T, Kajikawa M, Maenaka K, Park EY (2010) Silkworm expression system as a platform technology in life science. Appl Microbiol Biotechnol 85:459–470. https://doi.org/10.1007/s00253-009-2267-2

Kitts PA, Possee RD (1993) A method for producing recombinant baculovirus expression vectors at high frequency. BioTechniques 14:810–817

Krammer F, Schinko T, Palmberger D, Tauer C, Messner P, Grabherr R (2010) *Trichoplusia ni* cells (High Five) are highly efficient for the production of influenza A virus-like particles: a comparison of two insect cell lines as production platforms for influenza vaccines. Mol Biotechnol 45:226–234. https://doi.org/10.1007/s12033-010-9268-3

Lalonde ME, Durocher Y (2017) Therapeutic glycoprotein production in mammalian cells. J Biotechnol 251:128–140. https://doi.org/10.1016/j.jbiotec.2017.04.028

Li MA, Turner DJ, Ning Z, Yusa K, Liang Q, Eckert S, Rad L, Fitzgerald TW, Craig NL, Bradley A (2011) Mobilization of giant piggyBac transposons in the mouse genome. Nucleic Acids Res 39:e148. https://doi.org/10.1093/nar/gkr764

Liu X, Li N, Hu X, Zhang R, Li Q, Cao D, Liu T, Zhang Y, Liu X (2013) Efficient production of transgenic chickens based on piggyBac. Transgenic Res 22:417–423. https://doi.org/10.1007/s11248-012-9642-y

Livingstone C, Jones I (1989) Baculovirus expression vectors with single strand capability. Nucleic Acids Res 17:2366. https://doi.org/10.1093/nar/17.6.2366

Luckow VA (1994) Baculovirus expression systems and biopesticides. In: Shuler ML, Wood HA, Granados RR, Hammer DA (eds) . Wiley-Liss, Inc., New York, pp 51–90

Luckow VA, Summers MD (1988) Trends in the development of baculovirus expression vectors. Bio/Technology 6:47–55. https://doi.org/10.1038/nbt0188-47

Luckow VA, Summers MD (1989) High level expression of nonfused foreign genes with *Autographa californica* nuclear polyhedrosis virus expression vectors. Virology 170:31–39. https://doi.org/10.1016/0042-6822(89)90348-6

Luckow VA, Lee SC, Barry GF, Olins PO (1993) Efficient generation of infectious recombinant baculoviruses by site-specific transposon-mediated insertion of foreign genes into a baculovirus genome propagated in *Escherichia coli*. J Virol 67:4566–4579. https://doi.org/10.1128/JVI.67.8.4566-4579.1993

Maeda S (1993) Konntyuuuirusutobalotekunoroji-(insect virus and biotechnology). Saiennsuhausu (science house), Tokyo

Maeda S, Kawai T, Obinata M, Fujiwara H, Horiuchi T, Saeki Y, Sato Y, Furusawa M (1985) Production of human alpha-interferon in silkworm using a baculovirus vector. Nature 315:592–594. https://doi.org/10.1038/315592a0

Mange A, Julien E, Prudhomme JC, Couble P (1997) A strong inhibitory element down-regulates SRE-stimulated transcription of the A3 cytoplasmic actin gene of Bombyx mori. J Mol Biol 265:266–274. https://doi.org/10.1006/jmbi.1996.0734

Marsian J, Lomonossoff GP (2016) Molecular pharming - VLPs made in plants. Curr Opin Biotechnol 37:201–206. https://doi.org/10.1016/j.copbio.2015.12.007

Matsuura Y, Possee RD, Overton HA, Bishop DH (1987) Baculovirus expression vectors: the requirements for high level expression of proteins, including glycoproteins. J Gen Virol 68(Pt 5):1233–1250. https://doi.org/10.1099/0022-1317-68-5-1233

Mikhailov VS, Zemskov EA, Abramova EB (1992) Protein synthesis in pupae of the silkworm Bombyx mori after infection with nuclear polyhedrosis virus: resistance to viral infection acquired during pupal period. J Gen Virol 73(Pt 12):3195–3202. https://doi.org/10.1099/0022-1317-73-12-3195

Mitsudome T, Xu J, Nagata Y, Masuda A, Iiyama K, Morokuma D, Li Z, Mon H, Lee JM, Kusakabe T (2014) Expression, purification, and characterization of endo-beta-*N*-acetylglucosaminidase H using baculovirus-mediated silkworm protein expression system. Appl Biochem Biotechnol 172:3978–3988. https://doi.org/10.1007/s12010-014-0814-5

Morikawa Y, Goto T, Yasuoka D, Momose F, Matano T (2007) Defect of human immunodeficiency virus type 2 Gag assembly in *Saccharomyces cerevisiae*. J Virol 81:9911–9921. https://doi.org/10.1128/JVI.00027-07

Motohashi T, Shimojima T, Fukagawa T, Maenaka K, Park EY (2005) Efficient large-scale protein production of larvae and pupae of silkworm by *Bombyx mori* nuclear polyhedrosis virus bacmid system. Biochem Biophys Res Commun 326:564–569. https://doi.org/10.1016/j.bbrc.2004.11.060

Nishizawa-Yokoi A, Toki S (2023) Precise genetic engineering with piggyBac transposon in plants. Plant Biotechnol (Tokyo) 40:255–262. https://doi.org/10.5511/plantbiotechnology.23.0525a

O'reilly DR, Miller LK, Luckow VA (1994) Baculovirus expression vectors: a laboratory manual. Oxford University Press, New York
Park EY (2025) From silkroad to bioroad-silkworm biotechnology. Protein Expr Purif 234:106762. https://doi.org/10.1016/j.pep.2025.106762
Park EY, Minkner R (2025) A systematic approach for scalable purification of virus-like particles. Protein Expr Purif 228:106664. https://doi.org/10.1016/j.pep.2025.106664
Patel AH, Nowlan P, Weavers ED, Foster T (1987) Virulence of protein A-deficient and alpha-toxin-deficient mutants of *Staphylococcus aureus* isolated by allele replacement. Infect Immun 55:3103–3110
Pietrzyk AJ, Bujacz A, Lochynska M, Jaskolski M, Bujacz G (2011) Isolation, purification, crystallization and preliminary X-ray studies of two 30 kDa proteins from silkworm haemolymph. Acta Crystallogr Sect F Struct Biol Cryst Commun 67:372–376. https://doi.org/10.1107/S1744309110054564
Richardson CD (2008) Baculovirus expression protocols. Humana Press
Smith GE, Summers MD, Fraser MJ (1983) Production of human beta interferon in insect cells infected with a baculovirus expression vector. Mol Cell Biol 3:2156–2165. https://doi.org/10.1128/mcb.3.12.2156-2165.1983
Smith GE, Ju G, Ericson BL, Moschera J, Lahm HW, Chizzonite R, Summers MD (1985) Modification and secretion of human interleukin 2 produced in insect cells by a baculovirus expression vector. Proc Natl Acad Sci USA 82:8404–8408. https://doi.org/10.1073/pnas.82.24.8404
Tamura T, Thibert C, Royer C, Kanda T, Abraham E, Kamba M, Komoto N, Thomas JL, Mauchamp B, Chavancy G, Shirk P, Fraser M, Prudhomme JC, Couble P (2000) Germline transformation of the silkworm *Bombyx mori* L. using a piggyBac transposon-derived vector. Nat Biotechnol 18:81–84. https://doi.org/10.1038/71978
Tessier DC, Thomas DY, Khouri HE, Laliberte F, Vernet T (1991) Enhanced secretion from insect cells of a foreign protein fused to the honeybee melittin signal peptide. Gene 98:177–183. https://doi.org/10.1016/0378-1119(91)90171-7
Thiem SM, Miller LK (1990) Differential gene expression mediated by late, very late and hybrid baculovirus promoters. Gene 91:87–94. https://doi.org/10.1016/0378-1119(90)90166-o
Tomita M (2011) Transgenic silkworms that weave recombinant proteins into silk cocoons. Biotechnol Lett 33:645–654. https://doi.org/10.1007/s10529-010-0498-z
Tomita M, Munetsuna H, Sato T, Adachi T, Hino R, Hayashi M, Shimizu K, Nakamura N, Tamura T, Yoshizato K (2003) Transgenic silkworms produce recombinant human type III procollagen in cocoons. Nat Biotechnol 21:52–56. https://doi.org/10.1038/nbt771
Tomita M, Hino R, Ogawa S, Iizuka M, Adachi T, Shimizu K, Sotoshiro H, Yoshizato K (2007) A germline transgenic silkworm that secretes recombinant proteins in the sericin layer of cocoon. Transgenic Res 16:449–465. https://doi.org/10.1007/s11248-007-9087-x
Uchino K, Sezutsu H, Imamura M, Kobayashi I, Tatematsu K-I, Iizuka T, Yonemura N, Mita K, Tamura T (2008) Construction of a piggyBac-based enhancer trap system for the analysis of gene function in silkworm Bombyx mori. Insect Biochem Mol Biol 38:1165–1173. https://doi.org/10.1016/j.ibmb.2008.09.009
Vlak JM, Schouten A, Usmany M, Belsham GJ, Klinge-Roode EC, Maule AJ, Van Lent JW, Zuidema D (1990) Expression of cauliflower mosaic virus gene I using a baculovirus vector based upon the p10 gene and a novel selection method. Virology 179:312–320. https://doi.org/10.1016/0042-6822(90)90299-7
Wang XZ, Ooi BG, Miller LK (1991) Baculovirus vectors for multiple gene expression and for occluded virus production. Gene 100:131–137. https://doi.org/10.1016/0378-1119(91)90358-i
Wang F, Wang R, Wang Y, Xu H, Yuan L, Ding H, Ma S, Zhou Y, Zhao P, Xia Q (2015) Remobilizing deleted piggyBac vector post-integration for transgene stability in silkworm. Mol Gen Genomics 290:1181–1189. https://doi.org/10.1007/s00438-014-0982-6
Weyer U, Possee RD (1991) A baculovirus dual expression vector derived from the *Autographa californica* nuclear polyhedrosis virus polyhedrin and p10 promoters: co-expression of

two influenza virus genes in insect cells. J Gen Virol 72(Pt 12):2967–2974. https://doi.org/10.1099/0022-1317-72-12-2967

Weyer U, Knight S, Possee RD (1990) Analysis of very late gene expression by *Autographa californica* nuclear polyhedrosis virus and the further development of multiple expression vectors. J Gen Virol 71(Pt 7):1525–1534. https://doi.org/10.1099/0022-1317-71-7-1525

Williams GV, Rohel DZ, Kuzio J, Faulkner P (1989) A cytopathological investigation of Autographa californica nuclear polyhedrosis virus p10 gene function using insertion/deletion mutants. J Gen Virol 70(Pt 1):187–202. https://doi.org/10.1099/0022-1317-70-1-187

Zhu Y, Qi Y, Liu D, Joshua MN, Wang Y (1998) A novel expression system based on host-range expansion of baculovirus. J Virol Methods 76:101–108. https://doi.org/10.1016/S0166-0934(98)00128-1

Zhu J, Zhu Q, Gong R, Xu Q, Cai M, Jiang T, Zhou X, Zhou M, Zhang Y (2018) PiggyBac transposon-mediated mutagenesis and application in yeast Komagataella phaffii. Biotechnol Lett 40:1365–1376. https://doi.org/10.1007/s10529-018-2592-6

Zuidema D, Schouten A, Usmany M, Maule AJ, Belsham GJ, Roosien J, Klinge-Roode EC, Van Lent JW, Vlak JM (1990) Expression of cauliflower mosaic virus gene I in insect cells using a novel polyhedrin-based baculovirus expression vector. J Gen Virol 71(Pt 10):2201–2209. https://doi.org/10.1099/0022-1317-71-10-2201

Chapter 4
Improvement of BmNPV Bacmid

Abstract The baculovirus expression system utilizing *Bombyx mori* nucleopolyhedrovirus (BmNPV) in silkworm larvae is a powerful tool for recombinant protein production, but it is often limited by viral-induced host liquefaction and proteolytic degradation. To address this, a double-deletion BmNPV bacmid lacking both cysteine protease and chitinase genes (BmNPV-*CP*⁻-*Chi*⁻) was constructed. This modification successfully prevented larval liquefaction, abolished the degradation of the target protein, and significantly improved the stability and yield of secreted protein. Furthermore, transcriptional efficiency was enhanced by engineering the very late polyhedrin (*polh*) promoter with multiple burst sequences (BSs). While two BSs were optimal for expression in cultured insect cells, a modified promoter harboring nine BSs maximized β1,3-*N*-acetylglucosaminyltransferase 2 (β3GnT2) expression in silkworm larvae, achieving a 6.8-fold increase in enzymatic activity. Additionally, the coexpression of very late transcription factor 1 (VLF-1) synergistically augmented target protein production by further stimulating the engineered *polh* promoter. In conclusion, the integration of protease/chitinase gene deletions with targeted *polh* promoter modifications establishes a highly robust, high-yielding platform for large-scale eukaryotic recombinant protein production in silkworm biofactories.

Keywords BmNPV bacmid • Proteolytic degradation • Polyhedrin promoter • Cysteine protease • Chitinase • Burst sequence • Very late transcription factor 1

4.1 Introduction

The baculovirus expression system is commonly used to produce recombinant proteins in cultured insect cells, utilizing the strong polyhedrin promoter. This system offers significant advantages for expressing eukaryotic proteins, as insect cells can perform proper folding and post-translational modifications, including signal peptide processing (Davis et al. 1992), phosphorylation (HÉricourt et al. 2000), and glycosylation (James et al. 1995). Traditionally, *Autographa californica* multiple

E. Y. Park, *Silkworm Biofactory*, https://doi.org/10.1007/978-981-92-0056-6_4

nucleopolyhedrovirus (AcMNPV) or *Bombyx mori* nucleopolyhedrovirus (BmNPV) has been employed for protein expression in insect cells. A practical BmNPV bacmid system was developed for direct application in silkworm larvae (Motohashi et al. 2005). This system enables faster protein expression than previous methods because the bacmid can replicate as a large plasmid in *Escherichia coli*, generate recombinant viral DNA via site-specific transposition, and remain infectious to insect cells. The system enables rapid protein production in silkworms for up to 10 days without posing biohazard risks, as it does not require a baculovirus infection step. The bacmid technology eliminates the need for viral amplification, making it a powerful tool for future large-scale production of recombinant eukaryotic proteins.

However, the silkworm larvae typically undergo liquefaction around 5 d post-infection (dpi). This process is triggered by a cysteine protease with cathepsin L-like properties, known as v-cath (Slack et al. 1995), which plays a key role in host-tissue degradation during virus infection. v-cath is initially synthesized as 35.5 and 32 kDa precursors, which are processed into a 27.5 kDa mature form. In insect cells infected with baculoviruses, v-cath accumulates as an inactive propeptide and is activated upon cell death (Hom et al. 2002). The BmNPV genome also encodes a v-cath-like cysteine protease (Ohkawa et al. 1994) that shares 96.3% sequence identity with the putative viral cysteine protease of AcMNPV (Whitford et al. 1989). This protease significantly contributes to the degradation of recombinant proteins, thereby limiting the efficiency of baculovirus expression systems (Pham et al. 1999; Pyle et al. 1995; Yamada et al. 1990). For example, a human glycoprotein, β-1,3-*N*-acetylglucosaminyltransferase-2 (β3GnT2), fused with GFP_{uv} and expressed in Sf9 and Tn-5B1-4 cells, exhibited a marked decrease in enzymatic activity by 3 dpi due to v-cath-mediated degradation (Kato et al. 2004). These findings indicate that v-cath activity represents a significant limitation for glycoprotein expression in baculovirus systems. To address this issue, a BmNPV mutant lacking the cysteine protease gene was developed, which enabled a highly efficient production of firefly luciferase and human growth factor in silkworms by markedly reducing protein degradation (Suzuki et al. 1997). This improvement underscores the importance of eliminating v-cath activity to enhance recombinant protein yield in baculovirus-based expression systems.

In baculovirus-based expression systems using insect cells or silkworm larvae, high-level expression of foreign genes is typically driven by the very late polyhedrin (*polh*) or p10 promoter. During the very late phase of baculovirus infection, strong expression of these genes depends on the presence of an A/T-rich sequence downstream of the very late promoter (Rankin et al. 1988; Weyer and Possee 1988). Additionally, the 5′ noncoding leader sequence of the p10 promoter, approximately 50 bp in length, is essential for maximal expression and is referred to as the burst sequence (BS), as it mediates the expression surge during this phase. Mutations within the BS reduce gene expression by 10- to 20-fold, significantly lowering both *polh* RNA levels and the rate of transcription initiation from the *polh* promoter (Ooi et al. 1989). Deletion analysis of the p10 promoter further confirmed that the BS is indispensable for high-level expression during the very late phase (Weyer and

Possee 1988, 1989). In contrast, mutations in sequences upstream of the TAAG motif do not significantly affect very late gene expression (Ooi et al. 1989; Rankin et al. 1988).

Another key factor influencing strong expression during the very late phase is very late transcription factor 1 (VLF-1), which is essential for *polh* expression during the final stage of infection when occlusion bodies are formed (Mclachlin and Miller 1994). VLF-1 specifically transactivates the *polh* promoter, indicating that it serves as a primary regulator of very late gene expression (Todd et al. 1996). Studies using recombinant viruses with altered *vlf-1* expression demonstrated that *polh* expression is controlled by both the timing and concentration of VLF-1 in Sf-21 cells (Yang and Miller 1998b). VLF-1 is therefore a limiting factor for *polh* expression, and elevated VLF-1 levels significantly enhance *polh* transcription under the *polh* promoter (Yang and Miller 1998a). Furthermore, VLF-1 interacts with the burst sequence (BS) within the regulatory regions of p10 and *polh,* implicating its role in selective upregulation of very late genes (Yang and Miller 1999). To confirm its function as a transcriptional activator, VLF-1 was overexpressed, purified, and added to transcription assays containing baculovirus RNA polymerase, which resulted in the stimulation of *polh* promoter transcription (Mistretta and Guarino 2005). In addition to its transcriptional role, VLF-1 is required for normal capsid assembly and plays an essential role in DNA packaging during the final stages of infection (Li et al. 2005; Vanarsdall et al. 2004, 2006).

A hyper-expression system in the filamentous fungi of the genus *Aspergillus* using an improved promoter was developed (Kanamasa et al. 2007; Kanamasa et al. 2003). This promoter was engineered by introducing 12 copies of the cis-acting element Region III from the amylase gene into an *Aspergillus niger* promoter (Minetoki et al. 1998). Using this system, recombinant β-mannosidase was successfully produced in *Aspergillus aculeatus* at a concentration of 2.7 g/L, which is among the highest yields reported for a eukaryotic expression system. Given that the burst sequence (BS) in the *polh* promoter functions similar to Region III in the amylase promoter, the *polh* promoter can be further optimized by incorporating multiple BS copies. To date, no modified *polh* promoter containing multiple BS elements has been reported for insect cell or silkworm expression systems.

A BmNPV bacmid lacking both the cysteine protease and chitinase genes (BmNPV-CP^{-}-Chi^{-}) was constructed using the lambda recombination system. This modification significantly improved recombinant protein production in silkworm larvae, indicating that the BmNPV-CP^{-}-Chi^{-} bacmid is suitable to produce recombinant eukaryotic proteins. In a separate approach, multiple BSs were artificially inserted into the *polh* promoter region of BmNPV to evaluate their effect on recombinant protein expression. Furthermore, coexpression of multiple BSs with VLF-1 in silkworm larvae was performed to enhance the transcriptional activity of target genes under the control of the *polh* promoter. This represents a novel strategy for improving recombinant protein yield in BmNPV-based expression systems.

4.2 Cysteine Protease-Deleted BmNPV (BmNPV-*CP*⁻) Bacmid

4.2.1 Construction of BmNPV-CP⁻ Bacmid

The construction of the BmNPV-*CP*⁻ bacmid is illustrated with a flowchart in Fig. 4.1. A cysteine protease-deleted chloramphenicol acetyltransferase (CPDcat) fragment was obtained through PCR amplification of the CAT gene from the pKD3 plasmid. This was done using the primers Bm98758Fcat and Bm99687Rcat (Table 4.1), which include 50 bp sequences corresponding to the start and stop codons of the v-cath gene, respectively. The pKD3 template, mixed with the PCR product, was then digested with the DpnI restriction enzyme.

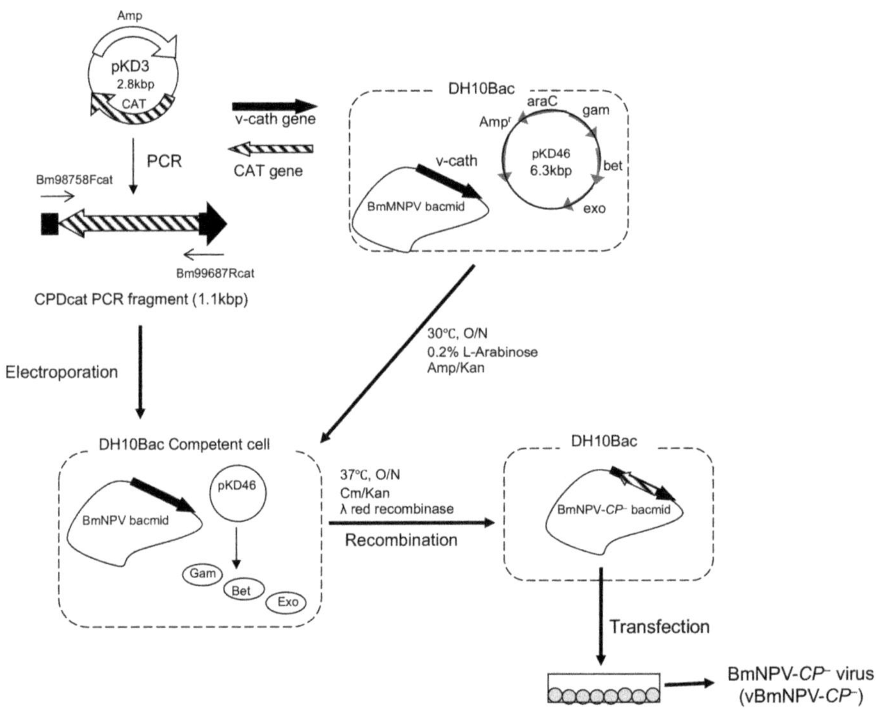

Fig. 4.1 Schematic outline for the construction of the BmNPV-*CP*⁻ bacmid. DH10Bac cells resistant to ampicillin and kanamycin were grown on LB agar with 0.2% L-arabinose at 30 °C to induce Gam, Bet, and Exo expression from pKD46. The CPDcat PCR fragment, amplified from the pKD3 plasmid, was then introduced into DH10Bac cells carrying BmNPV bacmid DNA and pKD46 by electroporation. Homologous recombination occurred between the CPDcat fragment and BmNPV bacmid DNA. After overnight incubation at 37 °C, chloramphenicol- and kanamycin-resistant colonies were selected and designated as the BmNPV-*CP*⁻ bacmid. (Adapted with permission (Hiyoshi et al. 2007). Copyright 2007 Elsevier)

Table 4.1 Primers used to prepare the CPDcat fragment and signal sequence (Hiyoshi et al. 2007)

Primer name	Sequence (5′ → 3′)
Bm98758Fcat	GAACAAAATTTTGTTTTATTTGTTTGTGTACGCCGTTGTAA v-cath AGAGCGCGG TTGTGTAGGCTGGAGCTGCT chloramphenicol acetyltransferase
Bm99687Rcat	TAATAAATGACTGCAGTAGACGCAAGTTCGTTTCTCATAC v-cath CACAAGCGT TCCATATGAATATCCTCCTTA chloramphenicol acetyltransferase
bx-GFP_{uv}- β3GnT2	CACCATGAAGATACTCCTTGCTATTGCATTAATGTTGTCA Signal sequence ACAGTAATGTGGGTGTCAACACAACCGCGGGGTTCTCATCATC GFP_{uv}-β3GnT2
Reverse primer	CGGAATTCTGAAGGGTTTAGAGGCCCTCAAATGGG

Next, the red recombinase plasmid pKD46 was transformed into DH10Bac competent cells containing the BmNPV bacmid DNA (Motohashi et al. 2005). After overnight incubation at 30 °C on LB agar with kanamycin and ampicillin, colonies were transferred to SOB (20 g tryptone, 5 g yeast extract, 0.5 g NaCl, 5 g $MgSO_4 \cdot 7H_2O$ per 1 L) agar supplemented with 0.2% L-arabinose and incubated again at 30 °C. A purified CPDcat fragment (500 ng) was then electroporated into these cells. Colonies resistant to kanamycin and chloramphenicol were selected at 37 °C, and PCR confirmed replacement of the v-cath gene with CAT, generating the BmNPV-*CP*⁻ bacmid. Approximately 100 ng of this bacmid DNA was transfected into Bm5 cells by lipofection. After 72 h at 27 °C, recombinant BmNPV-*CP*⁻ was harvested, diluted ten-fold, and used to reinfect Bm5 cells for amplification. The final amplified virus was stored for future silkworm larval infections.

4.2.2 Protease Activity in BmNPV-CP⁻-Infected Bm5 Cells and Silkworm Larvae

When BmNPV and BmNPV-*CP*⁻ were used to infect Bm5 cells, intracellular protease activity was assessed at 72 h post-injection. Protease activity in BmNPV-infected Bm5 cells was 0.34 U/mL, compared to 0.02 U/mL in BmNPV-*CP*⁻-infected cells (Fig. 4.2a). The low activity in BmNPV-*CP*⁻ cells was similar to that observed with the cysteine protease inhibitor E64, indicating that the reduction resulted from the deletion of the v-cath gene.

In BmNPV-*CP*⁻-infected silkworm larvae, hemolymph protease activity was 0.32 U/mL (Fig. 4.2b), representing only 15% of the activity observed in BmNPV-infected larvae (2.03 U/mL). Compared with BmNPV infection, BmNPV-*CP*⁻ reduced protease activity by 94% in Bm5 cells and by 85% in silkworm larvae, due to the deletion of the cysteine protease gene in the BmNPV bacmid. Larvae infected

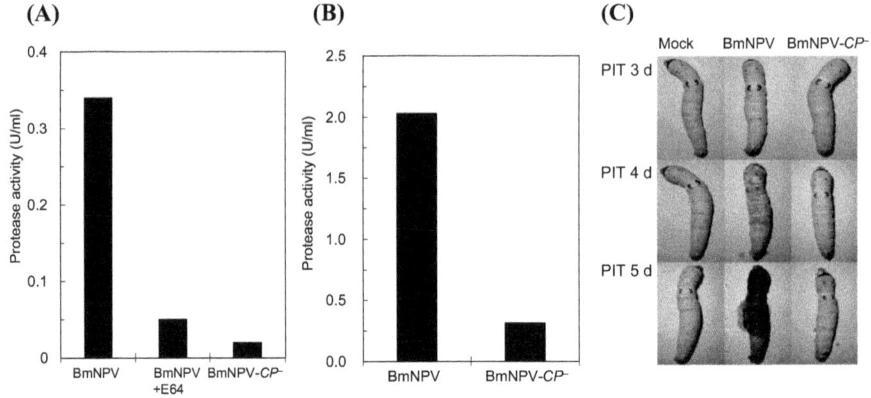

Fig. 4.2 Intracellular protease activity of BmNPV- and BmNPV-*CP*$^-$-infected Bm5 cells (**a**), protease activity in the hemolymph of BmNPV- and BmNPV-*CP*$^-$-injected silkworm larvae (**b**), and larval liquefaction (**c**). E64 in (**a**) denotes trans-epoxysuccinyl-l-leucylamido-4-guanidino butane used as a cysteine protease inhibitor, and the left column in (**c**) indicates post-injection time (dpi). (Adapted with permission (Hiyoshi et al. 2007). Copyright 2007 Elsevier)

with BmNPV underwent liquefaction at 5 dpi. In contrast, those infected with BmNPV-CP$^-$ retained their normal white appearance and showed neither melanization nor liquefaction at 5 dpi (Fig. 4.2c).

4.2.3 Improved β3GnT2 in Silkworm Larvae

A GFP$_{uv}$-fused β1,3-*N*-acetylglucosaminyltransferase 2 (β3GnT2) (abbreviated GGT2) fusion protein expression was compared using BmNPV and BmNPV-*CP*$^-$ bacmids. At 7 dpi, silkworm larvae infected with BmNPV exhibited complete liquefaction. In contrast, those injected with the BmNPV-*CP*$^-$/bx (bombyxin signal sequence)-GGT2 bacmid remained intact and showed no signs of melting (right of Fig. 4.3a). Protease activity in larval hemolymph was 0.74 U/mL for BmNPV/bx-GGT2 and 0.09 U/mL for BmNPV-*CP*$^-$/bx-GGT2 (Fig. 4.3b). About 88% of cysteine protease activity was lost in BmNPV-*CP*$^-$/bx-GGT2 infections, consistent with Fig. 4.2b. Activity in mock-infected larvae was similar to that in BmNPV-*CP*$^-$/bx-GGT2 infections. The difference between BmNPV and BmNPV-*CP*$^-$ may reflect infectivity variations.

The maximum β3GnT activity in the hemolymph of the silkworm larvae injected with the BmNPV-*CP*$^-$/bx-GGT2 bacmid reached 39.3 mU/mL at 4 dpi. In contrast, larvae injected with the BmNPV/bx-GGT2 bacmid exhibited 29.7 mU/mL (Fig. 4.3c). The lower β3GnT activity in the BmNPV/bx-GGT2 bacmid-injected larvae may result from the proteolytic degradation of β3GnT2. Consistent with this, SDS-PAGE analysis revealed the disappearance of the full-length fusion protein and the appearance of low-molecular-weight bands at 4 dpi (Fig. 4.3d). In contrast, the

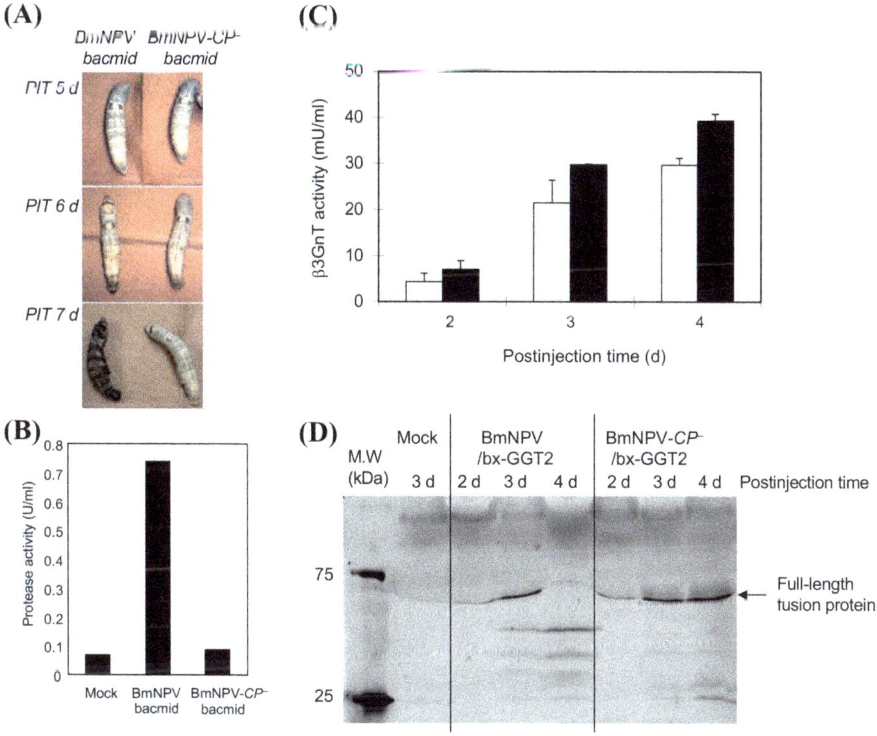

Fig. 4.3 Silkworm larval liquefaction (**a**) and protease activity (**b**) of BmNPV and BmNPV-CP^- bacmid-injected silkworm. PIT in (**a**) denotes post-injection time. Protease activity in the larval hemolymph of mock, BmNPV, and BmNPV-CP^- bacmid-injected silkworm was measured at 6 d post-injection. β3GnT activity (**c**) in the larval hemolymph of BmNPV (white bars) and BmNPV-CP^- bacmid (black bars)-injected silkworm. Bars in (**c**) denote standard deviations ($n = 3$). Protein bands in (**d**) were detected by GFP_{uv} fluorescence. (Adapted with permission (Hiyoshi et al. 2007). Copyright 2007 Elsevier)

fusion protein band in the hemolymph from BmNPV-CP^-/bx-GGT2-injected larvae was markedly thicker and showed no signs of degradation.

Many studies (Kaba et al. 2004; Lee et al. 2006; Wang et al. 2005) have shown that foreign gene products in silkworm larvae are frequently degraded by cysteine proteases encoded by the BmNPV gene. This presents a significant challenge for the large-scale production of target proteins in silkworm larvae. The BmNPV-CP^- bacmid, in which the v-cath gene was replaced by the chloramphenicol acetyltransferase gene, reduced protease activity by 94% in Bm5 cells and by 85% in silkworm larvae. Larvae infected with BmNPV-CP^- retained their normal white appearance and did not undergo liquefaction even at 5 dpi. In contrast, liquefaction in this study was delayed by 2 d compared with that induced by conventional viral infection. The fusion protein was expressed at a higher level than with the standard BmNPV bacmid. These results indicate that the BmNPV-CP^- bacmid is an effective platform for expressing eukaryotic genes in the silkworm hemolymph. For the successful production of secretory proteins, hemolymph should be collected before product degradation begins. Using the BmNPV-*CP*⁻ bacmid as an expression vector may facilitate the recovery of intact proteins and prevent their proteolytic degradation during purification from the hemolymph.

4.3 Cysteine Protease and Chitinase-Deleted BmNPV (BmNPV-*CP*⁻-*Chi*⁻) Bacmid

*4.3.1 Construction of BmNPV-*CP⁻-Chi⁻ *Bacmid*

Both protease and chitinase genes were extracted from wild-type BmNPV DNA by PCR using the oligonucleotide primers BmChi-F (5′-GTTCATCTTTAGGTTTAAACTGTGCGTTTA-3′) and BmCP-R (5′-CGATATGATTATGTTCCAAATAGCGAGT-3′), as shown Fig. 4.4. A PCR product containing partial cysteine protease and chitinase genes (Bm-*CP*-*Chi*)

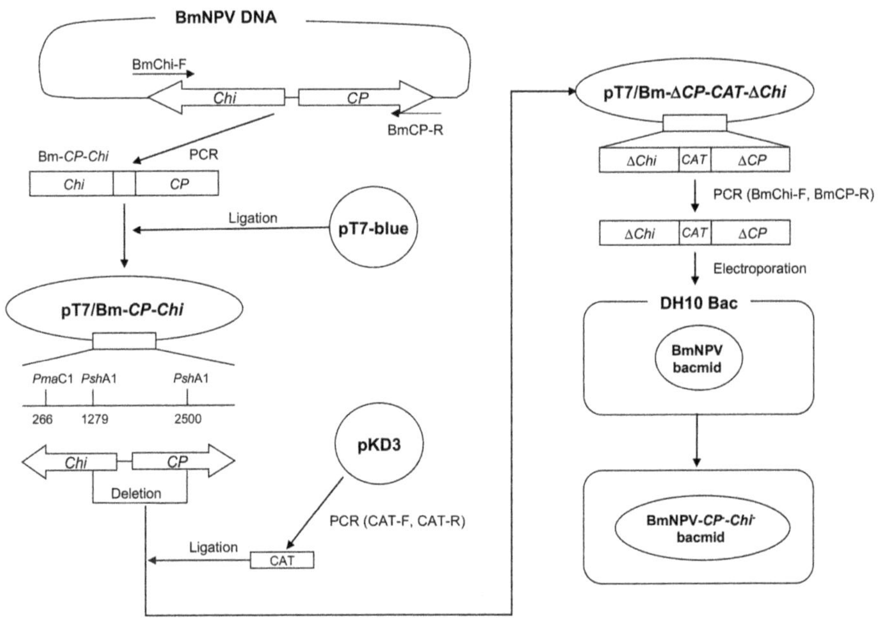

Fig. 4.4 Cysteine protease- and chitinase-deficient BmNPV preparation. Cysteine protease and chitinase genes were PCR-amplified from BmNPV DNA and cloned into pT7-blue. The CAT gene from pKD3 was inserted into the PshAI-digested fragment, creating a Δ*Chi*-*CAT*-Δ*CP* construct. This fragment was electroporated into DH10Bac cells, and chloramphenicol-resistant colonies were selected as the BmNPV-*CP*⁻-*Chi*⁻ bacmid. The fragment was electroporated into DH10Bac cells, and chloramphenicol-resistant colonies were designated as the BmNPV-*CP*⁻-*Chi*⁻ bacmid. (Adapted with permission (Park et al. 2008). Copyright 2008 Portland Press)

was amplified and cloned into the pT7-blue vector, generating pT7/Bm-*CP-Chi*. *E. coli* DH5α cells were transformed with pT7/Bm-*CP-Chi*, and transformants were screened on LB agar plates containing ampicillin, IPTG, and X-Gal at 37 °C. The fragment containing partial chitinase and cysteine protease genes was excised from pT7/Bm-*CP-Chi* by digestion with PshAI. The chloramphenicol acetyltransferase (CAT) gene, amplified from pKD3 using primers CAT-F (5′-GCATTACACGTCTTGAGCGATTGTGTA-3′) and CAT-R (5′-GACATGGGAATTAGCCATGGTCCATAT-3′), was inserted into the PshAI-digested vector via blunt-end ligation, yielding pT7/Bm-Δ*CP-CAT*-Δ*Chi*. To modify the chitinase and cysteine protease genes in the BmNPV bacmid, a 2.4 kb DNA fragment containing both genes, partially replaced by CAT, was amplified from pT7/Bm-Δ*CP-CAT*-Δ*Chi* using primers BmChi-F and BmCP-R. This fragment was introduced into DH10Bac cells harboring the BmNPV bacmid and the recombinase plasmid. Chloramphenicol-resistant colonies were selected and designated as BmNPV-CP^--Chi^- bacmid.

4.3.2 *Protease and Chitinase Activities in the Hemolymph of BmNPV-*CP⁻-Chi⁻*-Injected Larvae and Improved β3GnT2 Expression in Silkworm Larvae*

Fifth-instar silkworm larvae were injected with either the BmNPV bacmid or the BmNPV-CP^--Chi^- bacmid with 0.3 μg DNA per larva. By 7 dpi, more than 90% of the larvae injected with the BmNPV bacmid had liquefied (Fig. 4.5A-b), whereas those injected with the BmNPV-CP^--Chi^- bacmid remained intact and exhibited no symptoms of liquefaction (Fig. 4.5A-c).

Protease activity dropped from 0.73 U/mL in BmNPV bacmid-injected larvae to just 0.03 U/mL in BmNPV-CP^--Chi^- bacmid-injected larvae (Fig. 4.5b), indicating that only 4% of the cysteine protease activity remained in BmNPV-CP^--Chi^- infected larvae. Chitinase activity in the larvae infected with BmNPV- and BmNPV-CP^--Chi^- bacmids was 0.28 U/mL and 0.12 U/mL, respectively. In mock-injected larvae, chitinase activity was 0.09 U/mL, which was comparable to that observed in BmNPV-CP^--Chi^--infected larvae. These findings indicate that baculovirus-derived chitinase activity was significantly reduced in the hemolymph of larvae injected with the BmNPV-CP^--Chi^- bacmid (Fig. 4.5b).

Silkworm larvae were injected with either the BmNPV/me-GGT2 bacmid or the BmNPV-CP^--Chi^-/me-GGT2 bacmid to evaluate the expression of the GGT2 fusion protein. At 7 dpi, expression of the fusion protein in larval hemolymph was analyzed by SDS-PAGE. In the hemolymph from larvae injected with the BmNPV-CP^--Chi^-/me-GGT2 bacmid, a distinct fluorescent band of approximately 70 kDa was observed in lanes 1 and 3, with no detectable low-molecular-weight degradation products (Fig. 4.5c). In contrast, hemolymph from larvae injected with the BmNPV/me-GGT2 bacmid showed additional small-molecular-weight bands in

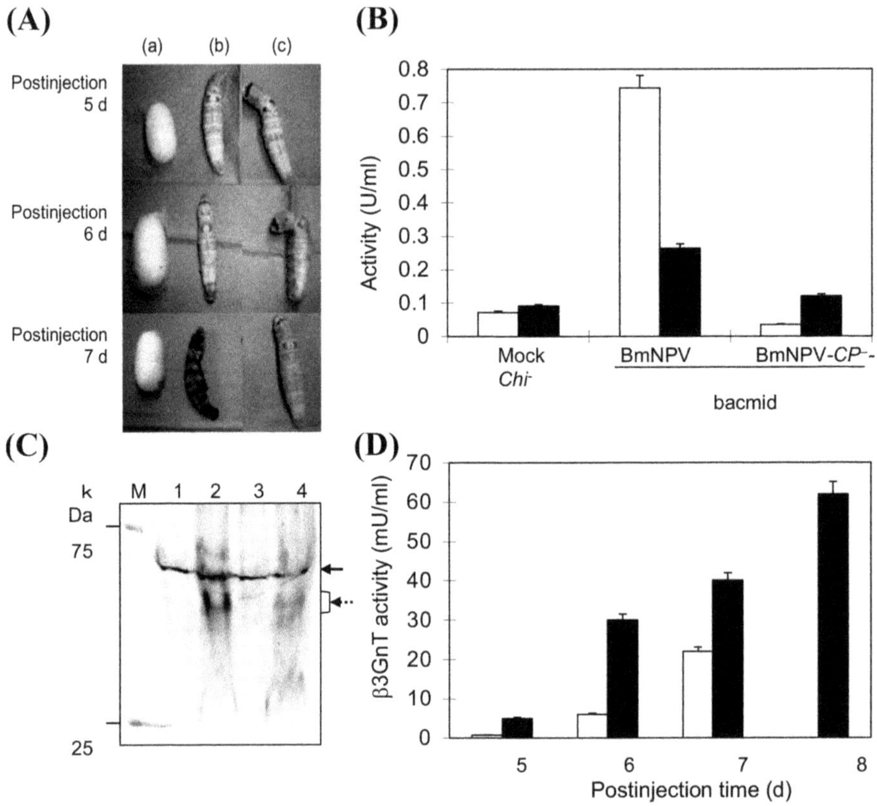

Fig. 4.5 (**A**) Comparison of larval liquefaction. a, b, and c denote mock, BmNPV, and BmNPV-CP^{-}-Chi^{-} bacmid-injected larvae. (**b**) Comparison of the protease (open bars) and chitinase activities (black bars) in the hemolymph of larvae injected with the BmNPV bacmid or the BmNPV-CP^{-}-Chi^{-} bacmid. Error bars indicate standard deviations ($n = 5$). (**c**) GFP_{uv} fluorescence analysis of GGT2 fusion protein in the hemolymph of larvae injected with BmNPV/me-GGT2 and BmNPV-CP^{-}-Chi^{-} bacmid /me-GGT2 bacmid. Lanes 1,3: BmNPV-CP^{-}-Chi^{-}/me-GGT2-injected larvae; lanes 2,4: BmNPV/me-GGT2-injected larvae. Samples collected at 7 dpi show a ~70 kDa fluorescent band (solid arrows) and degraded fragments (dotted arrows). (**d**) β3GnT2 activity in the hemolymph from larvae injected with BmNPV/me-GGT2 (white bars) or BmNPV-CP^{-}-Chi^{-}/me-GGT2 (black bars). Error bars represent standard deviations ($n = 5$). (Adapted with permission (Park et al. 2008). Copyright 2008 Portland Press)

lanes 2 and 4, indicating degradation of the fusion protein by baculovirus-derived proteases (Ohkawa et al. 1994).

When silkworm larvae were injected with the BmNPV/me-GGT2 bacmid, β3GnT2 activity reached 22 mU/mL at 7 dpi. Still, it was undetectable by 8 dpi (Fig. 4.5d), likely due to proteolytic degradation of the fusion protein. In contrast, larvae injected with the BmNPV-CP^{-}-Chi^{-}/me-GGT2 bacmid showed a continuous increase in β3GnT2 activity, reaching 62 mU/mL by 8 dpi (Fig. 4.5d).

4.3.3 *Summary*

Silkworm larvae represent an attractive and cost-effective platform for producing eukaryotic proteins, as they can express functional proteins at higher levels than insect cell cultures or mammalian systems. However, a major limitation of using silkworms for recombinant protein expression has been the proteolytic degradation of target proteins by viral-encoded enzymes, particularly cysteine proteases and chitinases from BmNPV (Ohkawa et al. 1994). These enzymes are thought to facilitate host liquefaction and viral transmission during the late stages of infection, but they significantly compromise protein stability and yield (Suzuki et al. 1997).

Previous studies have demonstrated that deletion of the cysteine protease gene in BmNPV markedly reduces protein degradation, thereby enabling efficient production of proteins such as human growth hormone. Similarly, removal of chitinase from AcMNPV or BmNPV prevented larval liquefaction and improved protein stability. The combined deletion of both genes further enhanced the expression of secreted proteins, as shown by increased cellulase production and protection of *Theileria parva* p67 from degradation in insect cells (Kaba et al. 2004).

In this chapter, these findings were applied to a practical BmNPV bacmid system for silkworm larvae. The double-deletion mutant (BmNPV-*CP*$^-$-*Chi*$^-$) exhibited a dramatic reduction in protease and chitinase activity, resulting in sustained secretion and stability of the GGT2 fusion protein for up to 8 days post-injection. Compared with the conventional bacmid, the modified system achieved a 2.8-fold increase in enzymatic activity and prevented larval liquefaction, thereby maintaining normal morphology throughout the expression period (Park et al. 2008). These results confirm that the simultaneous deletion of cysteine protease and chitinase significantly improves recombinant protein yield and stability in silkworm larvae.

The BmNPV-*CP*$^-$-*Chi*$^-$ bacmid offers a robust and safe expression platform for large-scale production of eukaryotic proteins, particularly those susceptible to proteolytic degradation. For secretory proteins, the timely collection of hemolymph remains critical to maximize recovery before degradation begins. Overall, this system represents a promising tool for industrial-scale production of recombinant proteins using silkworm larvae as a biofactory.

4.4 Promoter Modification

4.4.1 *Design and Construction of Recombinant AcMNPV and BmNPV Bacmids Incorporating Modified* polh *Promoters*

The burst sequence (BS) was amplified from the *polh* promoter of AcMNPV by PCR using primers polh F and polh-R (Table 4.2), synthesized (Sigma-Genosys Japan, K.K. Hokkaido, Japan). The amplified fragment was flanked with compatible

Table 4.2 Primers used (Manohar et al. 2010)

Primer name	Sequence (5′ → 3′)
Opie2-F	TGTCTCGAGCATGATGATAAACAATGTATGGTGC
Opie2-R	CACACTGGACTAGTGGATCCGAG
DEST-F	CTATAAAGCTTCCGGATTATTCATACCGTCCCACCATC
DEST-R	CTTCTCGAGAAAACCGAGGATGCGAACCACTTCATC
polh-F	TGACTCGAGCATCGTTTGTTCGCCCAGGAC
polh-R	TGTCAAGCTTATGGTCGAAGTGTCTAGATATTTATAG GTTTTTTTATTACAAAACTGTTACGAAAACAGACTAGTG GTGGGACGGTATGAATAATCC
DEST-160-R	CTCGTTCAGCTTTTTTGTACAAACTTGTG
VLF1-F	CACCATGAACGGTTTTAATGTTCGCAACG
VLF1-R	TGCTATTCGTTGCGATAGTACAACAACGATTC
β3GNT2-F	CTCCACATTTTGCCAGAAGG
β3GNT2-R	AGCTTCTTATCCCGATGAGG
Actin 1-F	CAACAGGGAGAAGATGACC
Actin 1-R	CAACGGAACCTCTCGTTACC

Underline indicates the part of the *polh* promoter that includes the BS

restriction sites and contained one BS. To construct pDB2, the pDEST vector was PCR-amplified with DEST-F/DEST-R to remove part of the *polh* promoter and ligated with a fragment containing the promoter plus one burst sequence, creating a promoter with two BSs (Fig. 4.6a). Using pDB2 as a template, PCR with polh-F and polh-R generated a fragment that was digested and inserted to create pDB3 with three BSs. By repeating this strategy, pDB5, pDB9, and pDB17 were constructed, containing five, nine, and seventeen BSs, respectively. Thus, DB2, DB3, DB5, DB9, and DB17 have 2, 3, 5, 9, and 17 BSs in the *polh* promoter (Fig. 4.6b). Each modified promoter was confirmed by PCR and sequencing using polh-F and DEST-160-R primers (Table 4.2) (Fig. 4.6c).

pENTR-β-glucuronidase (GUS) and pENTR/D/GFP$_{uv}$-β3GnT2 (Kato et al. 2003) were used to construct recombinant AcMNPV and BmNPV-*CP*$^{-}$ bacmids with a modified *polh* promoter. A 1264 bp DNA fragment comprising a truncated segment and the 3′-untranslated region of β3GnT2 was amplified by PCR from Quick-Clone human fetal brain cDNA using oligonucleotide primers containing BamHI and EcoRI restriction sites. The amplified fragment of β3GnT2 was inserted between the *Bam*H I and *Eco*R I sites of pBlueBacHis2-GFP$_{uv}$ (Cha et al. 1999) to yield pBlueBacHis2-GFP$_{uv}$/β3GnT2. To enable the introduction of a new signal peptide coding region, the GFP$_{uv}$-β3GnT2 (GGT2) fusion fragment was fused to a honeybee prepromelittin signal peptide (Vlasak et al. 1983). The GGT2 sequence, including its signal peptide, was inserted into the entry vector pENTR/D-TOPO, and the resulting construct was designated pENTR/D/GGT2.

The *gus* and GGT2 genes were placed under the control of *polh* promoter. Each AcMNPV bacmid (AcMNPV-DB1/GUS, AcMNPV-DB2/GUS, AcMNPV-DB3/GUS, AcMNPV-DB5/GUS, AcMNPV-DB9/GUS, AcMNPV-DB17/GUS) was constructed using the destination vector and *E. coli* DH10bac and transfected into High

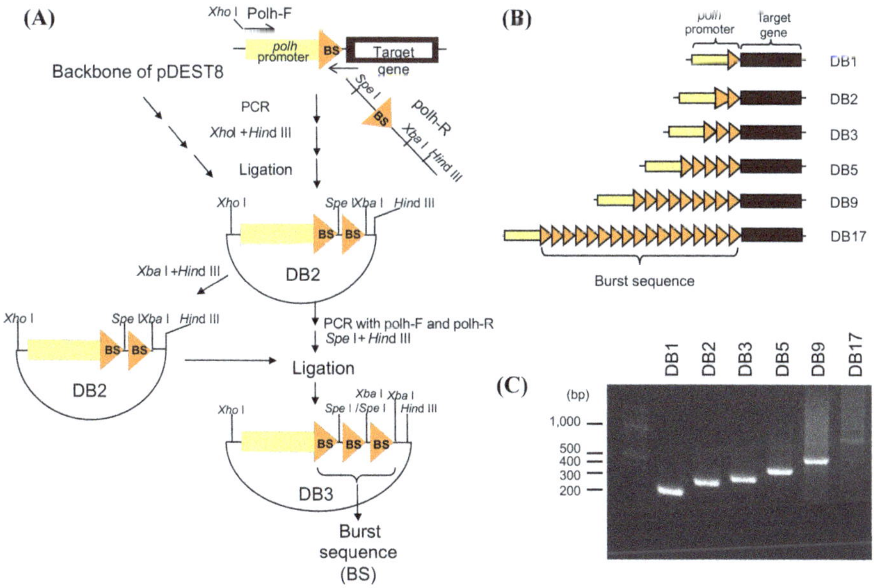

Fig. 4.6 (**a**) Construction of pDB2 and pDB3. The BS was amplified from the AcMNPV *polh* promoter using *polh* primers (Table 4.2) and flanked with restriction sites (Spe I, Xba I, Hind III). The *polh* promoter in pDEST was replaced with the PCR fragment containing one BS, generating pDB2 with two BSs. Using pDB2 as a template, an additional BS was amplified and ligated into pDB2 digested with Xba I and Hind III, resulting in pDB3 with three BSs. Each triangle represents one BS. Constructs pDB5, pDB9, and pDB17 were generated using the same strategy. (**b**) Schematic representation of modified *polh* promoters. Each triangle indicates one BS. (**c**) Verification of BS-containing *polh* promoters in pDB1, pDB2, pDB3, pDB5, pDB9, and pDB17 by PCR using polh-F and DEST-160-R primers. PCR products were analyzed by agarose gel electrophoresis. (Adapted with permission (Manohar et al. 2010). Copyright 2010 Wiley Periodicals, Inc.)

Five cells and amplified twice. Baculovirus titer was measured according to laboratory protocol (O'reilly et al. 1994). Each BmNPV-*CP*⁻ (BmNPV-*CP*⁻-DB1/GGT2, BmNPV-*CP*⁻-DB2/GGT2, BmNPV-*CP*⁻-DB3/GGT2, BmNPV-*CP*⁻-DB5/GGT2, BmNPV-*CP*⁻-DB9/GGT2, BmNPV-*CP*⁻-DB17/GGT2) bacmid was constructed using *E. coli* BmDH10bac (Hiyoshi et al. 2007).

4.4.2 Construction of VLF-1 Expression Recombinant Bacmid

The *polh* promoter was replaced with the ie-2 early promoter from *Orgyia pseudotsugata* multiple nucleopolyhedrovirus (Theilmann and Stewart 1992). The ie-2 promoter and a promoter-deleted pDEST fragment were PCR-amplified, digested, and ligated to create pDie. After transformation into *ccdB*-survival cells, the *vlf-1* gene from AcMNPV was inserted into pDie using Gateway cloning (Hiyoshi et al. 2007).

4.4.3 *Effect of the Modified* polh *Promoter in Insect Cells and on β3GnT Expression in Silkworm Larvae*

High Five cells were infected with six different recombinant AcMNPVs at a multiplicity of infection (MOI) of 1. A recombinant AcMNPV lacking the *gus* gene served as the negative control. GUS activity in cell lysates was qualitatively evaluated using X-Gluc staining (Fig. 4.7a). Blue coloration was observed in all samples except the control, confirming the absence of GUS activity in cells infected with the control AcMNPV. Among the six constructs, AcMNPV-DB2/GUS produced the most intense blue staining, indicating the highest level of GUS expression. As the number of BS elements increased beyond two, both color intensity and GUS

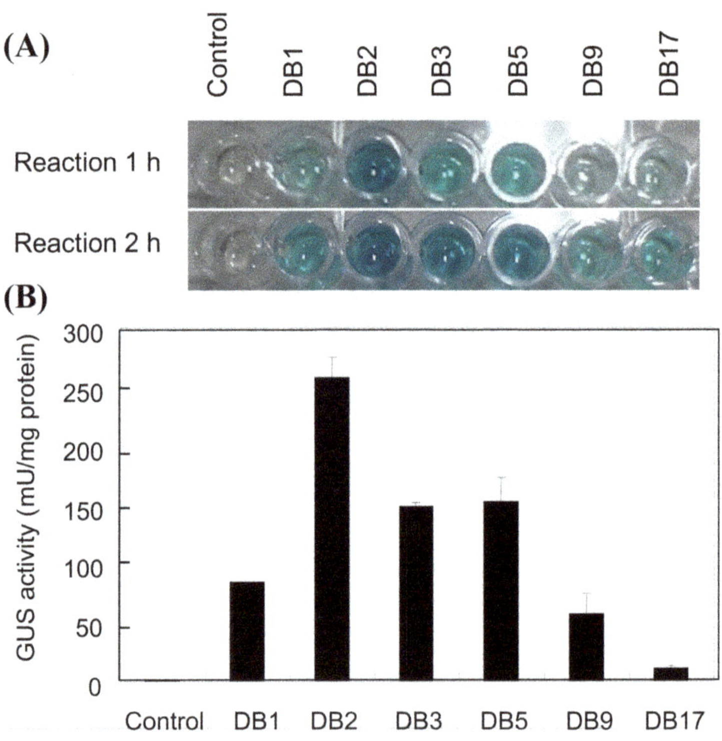

Fig. 4.7 Qualitative and quantitative analyses of GUS activity in High Five cells infected with recombinant baculoviruses using X-Gluc (**a**) and pNP-Gluc (**b**). High Five cells were infected with recombinant baculoviruses at an MOI of 1 and cultured for 4 d. GUS activity was measured using X-Gluc or pNP-Gluc. DB1–DB17 indicate AcMNPV-(DB1–DB17)/GUS-infected cells; the control was AcMNPV without *gus* gene. Error bars show standard deviation ($n = 3$). (Adapted with permission (Manohar et al. 2010). Copyright 2010 Wiley Periodicals, Inc.)

activity declined. GUS activity in AcMNPV-DB2/GUS infected cells was about three times higher than that in AcMNPV-DB1/GUS (Fig. 4.7b). DB3 and DB5 also showed higher activity than DB1, whereas DB9 and DB17 showed the lowest. Similar results were observed in Sf-9 cells, indicating that two BS elements in the *polh* promoter provide optimal expression, while adding more reduces activity.

In silkworm larvae, the GGT2 fusion protein was used to assess the impact of modified *polh* promoters on expression. Bacmids carrying GGT2 were injected, and hemolymph (Fig. 4.8a1) and fat body (Fig. 4.8a2) samples were analyzed by 12% SDS-PAGE and GFP_{uv} fluorescence. Strong bands appeared in all samples except those injected with BmNPV-CP^--DB17/GGT2. Quantification (Fig. 4.8b) showed that BmNPV-CP^--DB9/GGT2 achieved 6.8-fold higher activity than BmNPV-CP^--DB1/GGT2. Adding BS elements enhanced expression, but β3GnT activity peaked at nine BSs and declined thereafter. Thus, an optimal number of BSs maximizes expression and activity. In silkworm larvae, nine BSs showed the highest expression, in contrast to the AcMNPV results in High Five cells.

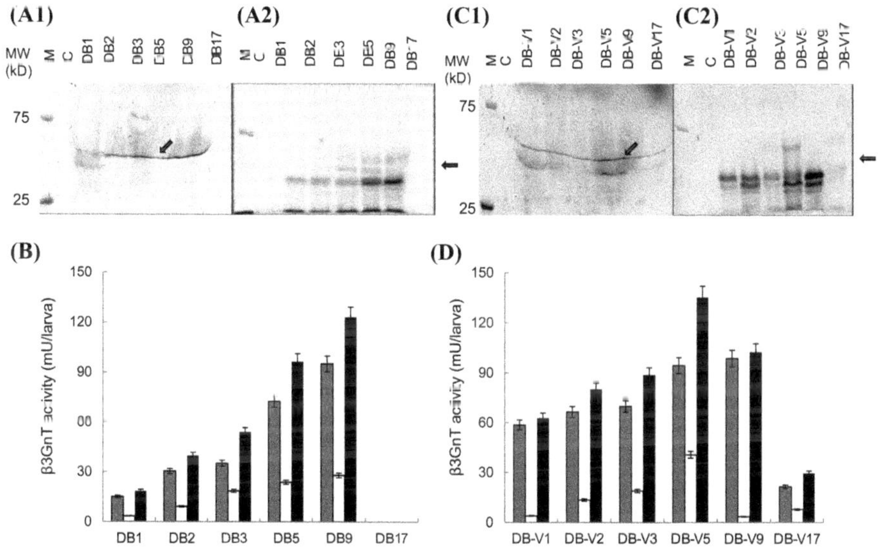

Fig. 4.8 (**a1, a2**) GFP_{uv} fluorescence of SDS-PAGE showing GGT2 fusion protein in the hemolymph (**a1**) and fat body (**a2**) of silkworm larvae injected with BmNPV-CP^--(DB1-DB17)/GGT2 bacmids. (**b**) β3GnT activity in the hemolymph and fat body of larvae injected with BmNPV-CP^--DBs/GGT2 bacmids. (**c1, c2**) GFP_{uv} fluorescence of SDS-PAGE of hemolymph (**c1**) and fat body (**c2**) of larvae co-injected with BmNPV-CP^--VLF-1 bacmid. (**d**) β3GnT activity in the hemolymph and fat body of larvae coexpressing BmNPV-CP^--VLF-1. M, C, DB, and DB-V indicate molecular marker, control, burst sequence, and burst sequence with VLF-1 coexpression, respectively. Numbers (1, 2, 3, 5, 9, 17) represent BS copy numbers. Arrows indicate the GGT2 fusion protein. Gray, open, and black bars in (**b**) and (**d**) show activity in the hemolymph, fat body, and the combined activity (mean ± standard deviation, $n = 3$). (Adapted with permission (Manohar et al. 2010). Copyright 2010 Wiley Periodicals, Inc.)

Hemolymph (Fig. 4.8c1) and fat body (Fig. 4.8c2) samples were collected from larvae co-injected with BmNPV-*CP*⁻-(DB1–DB9)/GGT2 bacmids and BmNPV-*CP*⁻-VLF-1 and analyzed by 12% SDS-PAGE and GFP_{uv} fluorescence to confirm GGT2 expression. Coexpression with VLF-1 increased β3GnT2 activity by 3.5-fold, 2-fold, 1.6-fold, 1.1-fold, and 1-fold compared to samples without VLF-1, respectively (Fig. 4.8d). The reduction in activity suggests that VLF-1 provides significant enhancement only at low BS counts within the promoter.

However, the coexpression of BmNPV-*CP*⁻-DB17/GGT2 with BmNPV-*CP*⁻-VLF-1 yielded 29.6 mU/larva, despite showing no activity without VLF-1. The mechanism by which VLF-1 activates the very late gene promoter is still unknown; however, our findings suggest that a moderate increase in the number of BSs and their coexpression with VLF-1 may facilitate the recruitment of RNA polymerase to the very late promoter or stabilize the transcription complex to enhance transcription initiation (Yang and Miller 1999). Coexpression of BmNPV-*CP*⁻-DB5/GGT2 with BmNPV-*CP*⁻-VLF-1 increased β3GnT activity 2.1-fold compared to DB1/GGT2 coexpression. These results suggest that a *polh* promoter with 5–9 BSs provides optimal expression in silkworm larvae under both single and coexpression conditions.

4.4.4 Effect of VLF-1 Coexpression on Protein Expression

Notably, the coexpression of BmNPV-*CP*⁻-DB5/GGT2 with VLF-1, driven by the ie-2 early promoter, resulted in a significantly higher specific β3GnT activity (Fig. 4.9a). The *polh* promoter is regulated by the intracellular concentration of VLF1 (Yang and Miller 1998a), and increasing its concentration enhances β3GnT activity in the hemolymph, as the coexpression of VLF-1 acted in silkworm larvae. Early expression of VLF-1 increased the expression from the *polh* promoter but caused premature cellular disintegration (Yang and Miller 1998b). Earlier activation of the *polh* promoter via VLF-1 coexpression is advantageous for optimizing the post-translational modification of the expressed foreign protein, thereby improving β3GnT2 folding and glycosylation. Notably, β3GnT2 is known to undergo glycosylation at four of its five *N*-glycosylation sites when expressed in insect cells (Kato et al. 2005). These results indicate that VLF-1 coexpression enhances the efficient production of structurally complex proteins.

RT-PCR was performed using specific primers for β3GnT2 and silkworm actin (Table 4.2), with 15 cycles for BS samples, 20 cycles for VLF-1 coexpressed samples, and 30 cycles for actin. Enhanced β3GnT2 activity in BmNPV-*CP*⁻-DB5/GGT2 and its coexpression with BmNPV-*CP*⁻-VLF-1 correlated with increased transcription (Fig. 4.9b). Baculovirus RNA polymerase promotes high mRNA synthesis and capping, and transcriptional analysis confirmed that DB5/GGT2 and VLF-1 coexpressed samples (Figs. 4.9a, d) showed higher transcription levels, consistent with β3GnT enzymatic activity.

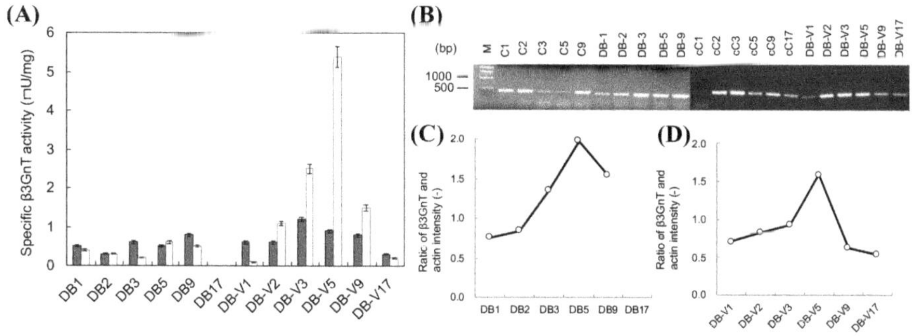

Fig. 4.9 (**a**) Specific β3GnT activity in the hemolymph (gray bars) and fat body (white bars) of silkworm larvae injected with BmNPV-*CP*⁻-(DB1-DB17)/GGT2 and coexpressed with BmNPV-*CP*⁻-VLF-1. Specific activity was calculated by dividing β3GnT activity (Fig. 4.8b, d) by the protein concentration. Data are mean ± standard deviation ($n = 3$). (**b**) Semiquantitative RT-PCR of larvae injected with BmNPV-*CP*⁻-(DB1-DB17)/GGT2 and BmNPV-*CP*⁻-VLF-1 coexpressed bacmids, visualized on 1% agarose gel. C and cC indicate actin amplification without and with VLF-1 co-expression, respectively. (**c**, **d**) Transcriptional ratio of β3GnT2 to actin in fat body samples from larvae injected with BmNPV-*CP*⁻-DBs/GGT2 (**c**) and coexpressed with BmNPV-*CP*⁻ VLF-1 (**d**). Numbers indicate BS copy numbers. (Adapted with permission (Manohar et al. 2010). Copyright 2010 Wiley Periodicals, Inc.)

4.5 Summary

Cytoplasmic actin is widely employed as a reference gene in silkworm studies; however, its expression stability can be influenced by developmental stage and experimental conditions. Although previous research indicated that the a3 (actin) promoter remains largely unaffected by viral factors for up to 3–4 dpi of recombinant AcMNPV (Zhang et al. 2008), the present study analyzed samples collected during the larval-to-pupal transition, approximately 7 days after recombinant BmNPV infection. This period is characterized by significant physiological changes, and aging organisms exhibit reduced homeostatic functions, including diminished mRNA and protein synthesis (Labuhn and Brack 1997; Shikama and Brack 1996). Furthermore, actin gene expression has been reported to fluctuate markedly during silkworm development (Wang et al. 2008). Additional variability may arise from repeated incorporation of BSs, which could affect actin transcription. Overall, reference gene stability is subject to dose- and time-dependent effects induced by experimental agents (Toegel et al. 2007).

For recombinant protein production, the baculovirus *polh* promoter plays a crucial role during the very late phase of infection. It requires a sequence between the TAAG motif and the translation initiation site, known as the BS. Expression from the *polh* promoter is enhanced through the specific interaction of VLF-1 with the BS. To improve protein yield, a modified *polh* promoter system incorporating multiple BSs was developed. In High Five insect cells, the inclusion of two BSs increased GUS activity threefold compared to the native promoter. Similarly, in the silkworm expression system, a promoter containing nine BSs resulted in a 6.8-fold increase in β3GnT2 activity per larva. Coexpression of VLF-1 further augmented β3GnT activity. These findings demonstrate that optimizing the number of BSs and coexpressing VLF-1 significantly enhances gene expression in insect cells and silkworm larvae. The engineered promoter described herein represents the most potent system for overexpressing foreign proteins in eukaryotic hosts, marking a substantial advancement in baculovirus–insect cell and silkworm biotechnology.

References

Cha HJ, Dalal NG, Vakharia VN, Bentley WE (1999) Expression and purification of human interleukin-2 simplified as a fusion with green fluorescent protein in suspended Sf-9 insect cells. J Biotechnol 69:9–17. https://doi.org/10.1016/s0168-1656(98)00190-4

Davis TR, Trotter KM, Granados RR, Wood HA (1992) Baculovirus expression of alkaline phosphatase as a reporter gene for evaluation of production, glycosylation and secretion. Bio/Technology 10:1148–1150. https://doi.org/10.1038/nbt1092-1148

HÉricourt F, Blanc S, Redeker V, Jupin I (2000) Evidence for phosphorylation and ubiquitinylation of the turnip yellow mosaic virus RNA-dependent RNA polymerase domain expressed in a baculovirus–insect cell system. Biochem J 349:417–425. https://doi.org/10.1042/bj3490417

Hiyoshi M, Kageshima A, Kato T, Park EY (2007) Construction of a cysteine protease deficient *Bombyx mori* multiple nucleopolyhedrovirus bacmid and its application to improve expression of a fusion protein. J Virol Methods 144:91–97. https://doi.org/10.1016/j.jviromet.2007.04.005

Hom LG, Ohkawa T, Trudeau D, Volkman LE (2002) *Autographa californica* M Nucleopolyhedrovirus ProV-CATH is activated during infected cell death. Virology 296:212–218. https://doi.org/10.1006/viro.2002.1378

James DC, Freedman RB, Hoare M, Ogonah OW, Rooney BC, Larionov OA, Dobrovolsky VN, Lagutin OV, Jenkins N (1995) N-glycosylation of recombinant human interferon-γ produced in different animal expression systems. Bio/Technology 13:592–596. https://doi.org/10.1038/nbt0695-592

Kaba SA, Salcedo AM, Wafula PO, Vlak JM, Van Oers MM (2004) Development of a chitinase and v-cathepsin negative bacmid for improved integrity of secreted recombinant proteins. J Virol Methods 122:113–118. https://doi.org/10.1016/j.jviromet.2004.07.006

Kanamasa S, Mochizuki M, Takada G, Kawaguchi T, Sumitani J-I, Araii M (2003) Overexpression of *Aspergillus aculeatus* cellobiohydrolase I in *Aspergillus oryzae*. J Biosci Bioeng 95:627–629. https://doi.org/10.1016/S1389-1723(03)80174-3

Kanamasa S, Kawaguchi T, Takada G, Kajiwara S, Sumitani J, Arai M (2007) Development of an efficient production method for β-mannosidase by the creation of an overexpression system in *Aspergillus aculeatus*. Lett Appl Microbiol 45:142–147. https://doi.org/10.1111/j.1472-765X.2007.02160.x

Kato T, Murata T, Usui T, Park EY (2003) Improvement of GFPuv-beta 3GnT2 fusion protein production by suppressing protease in baculovirus expression system. Biosci Biotechnol Biochem 67:2388–2395. https://doi.org/10.1271/bbb.67.2388

Kato T, Murata T, Usui T, Park EY (2004) Comparative analysis of GFP(UV)-beta1,3-N-acetylglucosaminyltransferase 2 production in two insect-cell-based expression systems. Protein Expr Purif 35:54–61. https://doi.org/10.1016/j.pep.2003.12.018

Kato T, Suzuki M, Murata T, Park EY (2005) The effects of N-glycosylation sites and the N-terminal region on the biological function of beta1,3-N-acetylglucosaminyltransferase 2 and its secretion. Biochem Biophys Res Commun 329:699–705. https://doi.org/10.1016/j.bbrc.2005.02.033

Labuhn M, Brack C (1997) Age-related changes in the mRNA expression of actin isoforms in *Drosophila melanogaster*. Gerontology 43:261–267. https://doi.org/10.1159/000213861

Lee KS, Je YH, Woo SD, Sohn HD, Jin BR (2006) Production of a cellulase in silkworm larvae using a recombinant *Bombyx mori* nucleopolyhedrovirus lacking the virus-encoded chitinase and cathepsin genes. Biotechnol Lett 28:645–650. https://doi.org/10.1007/s10529-006-0030-7

Li Y, Wang J, Deng R, Zhang Q, Yang K, Wang X (2005) vlf-1 deletion brought AcMNPV to defect in nucleocapsid formation. Virus Genes 31:275–284. https://doi.org/10.1007/s11262-005-3242-3

Manohar SL, Kanamasa S, Nishina T, Kato T, Park EY (2010) Enhanced gene expression in insect cells and silkworm larva by modified polyhedrin promoter using repeated Burst sequence and very late transcriptional factor-1. Biotechnol Bioeng 107:909–916. https://doi.org/10.1002/bit.22896

Mclachlin JR, Miller LK (1994) Identification and characterization of vlf-1, a baculovirus gene involved in very late gene expression. J Virol 68:7746–7756. https://doi.org/10.1128/jvi.68.12.7746-7756.1994

Minetoki T, Kumagai C, Gomi K, Kitamoto K, Takahashi K (1998) Improvement of promoter activity by the introduction of multiple copies of the conserved region III sequence, involved in the efficient expression of *Aspergillus oryzae* amylase-encoding genes. Appl Microbiol Biotechnol 50:459–467. https://doi.org/10.1007/s002530051321

Mistretta TA, Guarino LA (2005) Transcriptional activity of baculovirus very late factor 1. J Virol 79:1958–1960. https://doi.org/10.1128/JVI.79.3.1958-1960.2005

Motohashi T, Shimojima T, Fukagawa T, Maenaka K, Park EY (2005) Efficient large-scale protein production of larvae and pupae of silkworm by *Bombyx mori* nuclear polyhedrosis virus bacmid system. Biochem Biophys Res Commun 326:564–569. https://doi.org/10.1016/j.bbrc.2004.11.060

O'reilly DR, Miller LK, Luckow VA (1994) Baculovirus expression vectors: a laboratory manual. Oxford University Press, New York

Ohkawa T, Majima K, Maeda S (1994) A cysteine protease encoded by the baculovirus *Bombyx mori* nuclear polyhedrosis virus. J Virol 68:6619–6625. https://doi.org/10.1128/jvi.68.10.6619-6625.1994

Ooi BG, Rankin C, Miller LK (1989) Downstream sequences augment transcription from the essential initiation site of a baculovirus polyhedrin gene. J Mol Biol 210:721–736. https://doi.org/10.1016/0022-2836(89)90105-8

Park EY, Abe T, Kato T (2008) Improved expression of fusion protein using a cysteine- protease- and chitinase-deficient *Bombyx mori* (silkworm) multiple nucleopolyhedrovirus bacmid in silkworm larvae. Biotechnol Appl Biochem 49:135–140. https://doi.org/10.1042/BA20070098

Pham M-Q, Naggie S, Wier M, Cha HJ, Bentley WE (1999) Human interleukin-2 production in insect (*Trichoplusia ni*) larvae: effects and partial control of proteolysis. Biotechnol Bioeng 62:175–182. https://doi.org/10.1002/(SICI)1097-0290(19990120)62:2<175::AID-BIT7>3.0.CO;2-J

Pyle LE, Barton P, Fujiwara Y, Mitchell A, Fidge N (1995) Secretion of biologically active human proapolipoprotein A-I in a baculovirus-insect cell system: protection from degradation by protease inhibitors. J Lipid Res 36:2355–2361. https://doi.org/10.1016/S0022-2275(20)39716-9

Rankin C, Ooi BG, Miller LK (1988) Eight base pairs encompassing the transcriptional start point are the major determinant for baculovirus polyhedrin gene expression. Gene 70:39–49. https://doi.org/10.1016/0378-1119(88)90102-3

Shikama N, Brack C (1996) Changes in the expression of genes involved in protein synthesis during *Drosophila* aging. Gerontology 42:123–136. https://doi.org/10.1159/000213783

Slack JM, Kuzio J, Faulkner P (1995) Characterization of v-cath, a cathepsin L-like proteinase expressed by the baculovirus *Autographa californica* multiple nuclear polyhedrosis virus. J Gen Virol 76(Pt 5):1091–1098. https://doi.org/10.1099/0022-1317-76-5-1091

Suzuki T, Kanaya T, Okazaki H, Ogawa K, Usami A, Watanabe H, Kadono-Okuda K, Yamakawa M, Sato H, Mori H, Takahashi S, Oda K (1997) Efficient protein production using a *Bombyx mori* nuclear polyhedrosis virus lacking the cysteine proteinase gene. J Gen Virol 78(Pt 12):3073–3080. https://doi.org/10.1099/0022-1317-78-12-3073

Theilmann DA, Stewart S (1992) Molecular analysis of the trans-activating IE-2 gene of *Orgyia pseudotsugata* multicapsid nuclear polyhedrosis virus. Virology 187:84–96. https://doi.org/10.1016/0042-6822(92)90297-3

Todd JW, Passarelli AL, Lu A, Miller LK (1996) Factors regulating baculovirus late and very late gene expression in transient-expression assays. J Virol 70:2307–2317

Toegel S, Huang W, Piana C, Unger FM, Wirth M, Goldring MB, Gabor F, Viernstein H (2007) Selection of reliable reference genes for qPCR studies on chondroprotective action. BMC Mol Biol 8:13. https://doi.org/10.1186/1471-2199-8-13

Vanarsdall AL, Okano K, Rohrmann GF (2004) Characterization of a baculovirus with a deletion of vlf-1. Virology 326:191–201. https://doi.org/10.1016/j.virol.2004.06.003

Vanarsdall AL, Okano K, Rohrmann GF (2006) Characterization of the role of very late expression factor 1 in baculovirus capsid structure and DNA processing. J Virol 80:1724–1733. https://doi.org/10.1128/JVI.80.4.1724-1733.2006

Vlasak R, Unger-Ullmann C, Kreil G, Frischauf AM (1983) Nucleotide sequence of cloned cDNA coding for honeybee prepromelittin. Eur J Biochem 135:123–126. https://doi.org/10.1111/j.1432-1033.1983.tb07626.x

Wang F, Zhang CX, Shyam Kumar V, Wu XF (2005) Influences of chitinase gene deletion from BmNPV on the cell lysis and host liquefaction. Arch Virol 150:981–990. https://doi.org/10.1007/s00705-004-0458-9

Wang G-H, Xia Q-Y, Cheng D-J, Duan J, Zhao P, Chen J, Zhu L (2008) Reference genes identified in the silkworm *Bombyx mori* during metamorphism based on oligonucleotide microarray and confirmed by qRT-PCR. Insect Sci 15:405–413. https://doi.org/10.1111/j.1744-7917.2008.00227.x

Weyer U, Possee RD (1988) Functional analysis of the plO gene 5′ leader sequence of the *Autographaz californica* nuclear polyhedrosis virus. Nucleic Acids Res 16:3635–3653

Weyer U, Possee RD (1989) Analysis of the promoter of the *Autographa californica* nuclear polyhedrosis virus pl0 gene. J Gen Virol 70:203–208

Whitford M, Stewart S, Kuzio J, Faulkner P (1989) Identification and sequence analysis of a gene encoding gp67, an abundant envelope glycoprotein of the baculovirus *Autographa californica* nuclear polyhedrosis virus. J Virol 63:1393–1399. https://doi.org/10.1128/jvi.63.3.1393-1399.1989

Yamada K, Nakajima Y, Natori S (1990) Production of recombinant sarcotoxin IA in *Bombyx mori* cells. Biochem J 272:633–636. https://doi.org/10.1042/bj2720633

Yang S, Miller LK (1998a) Expression and mutational analysis of the baculovirus very late factor 1 (vlf-1) gene. Virology 245:99–109

Yang S, Miller SK (1998b) Control of baculovirus polyhedrin gene expression by very late factor 1. Virology 248:131–138. https://doi.org/10.1006/viro.1998.9272

Yang S, Miller LK (1999) Activation of baculovirus very late promoters by interaction with very late factor 1. J Virol 73:3404–3409. https://doi.org/10.1128/jvi.73.4.3404-3409.1999

Zhang Y, Zhang X, Xia H, Xue Y, Wang J, Tian B, Wei Z, Lu C (2008) Quantitative analysis of cytoplasmic actin gene promoter and nuclear polyhedrosis virus immediate-early promoter activities in various tissues of silkworm *Bombyx mori* using recombinant *Autographa californica* nuclear polyhedrosis virus as vector. Acta Biochim Biophys Sin (Shanghai) 40:533–538. https://doi.org/10.1111/j.1745-7270.2008.00425.x

Chapter 5
Post-Translational Modifications in Silkworms

Abstract This chapter highlights the critical role of the silkworm (*Bombyx mori*) expression system in producing complex eukaryotic proteins with proper post-translational modifications, a major advantage over prokaryotic hosts. The system's capacity for accurate disulfide bond formation was demonstrated using porcine insulin-like peptide 3 (pINSL3), which correctly folded and exhibited full biological activity. Furthermore, the system supports complex regulatory modifications, evidenced by the successful expression of human acetyl-CoA carboxylase 2 (hACC2) and malonyl-CoA decarboxylase (hMCD). These lipid metabolism enzymes underwent appropriate biotinylation and phosphorylation, ensuring the functional integrity and precise allosteric regulation of their catalytic activities. Finally, the chapter addresses the inherent limitations of insect *N*-glycosylation pathways, which typically yield paucimannosidic structures. By coexpressing human immunoglobulin G (hIgG) with human glycosyltransferases (hGnT II and hGalT I) under the *polh* promoter, the silkworm system was successfully engineered to produce biantennary, terminally galactosylated complex *N*-glycans. Collectively, these findings validate the silkworm as a highly effective, scalable biofactory capable of executing advanced post-translational modifications—including disulfide bridge formation, phosphorylation, and mammalian-like glycosylation—essential for the structural and functional viability of biopharmaceuticals.

Keywords Post-translational modification • Porcine insulin-like peptide 3 • Human acetyl-CoA carboxylase 2 • Malonyl-CoA decarboxylase • Phosphorylation • *N*-Glycosylation

5.1 Introduction

The efficient production of recombinant proteins remains a central objective in biotechnology, biopharmaceutical development, and structural biology. Among available hosts, the *Escherichia coli* expression system is widely adopted owing to its genetic tractability, rapid growth, well-characterized molecular biology, and its

E. Y. Park, *Silkworm Biofactory*, https://doi.org/10.1007/978-981-92-0056-6_5

capacity to generate high protein yields at low cost. Although expressing eukaryotic proteins in *E. coli* has advantages, it often leads to the formation of insoluble inclusion bodies, necessitating denaturation and refolding protocols that are often protein-specific and yield variable recovery of biological activity (Baneyx and Mujacic 2004). This limitation is fundamentally linked to the paucity of efficient post-translational modification machinery in *E. coli*, including constraints on disulfide bond formation in the cytosol, the absence of eukaryotic glycosylation pathways, and limited capacity for regulatory modifications such as phosphorylation. Post-translational modifications are indispensable determinants of protein structure and function. Disulfide bonds stabilize tertiary and quaternary architecture, guide proper folding, and can influence secretion and receptor binding; their mispairing or absence frequently produces misfolded, aggregation-prone species with compromised activity (Fig. 5.1a). Phosphorylation is a key reversible regulatory mechanism that modulates enzymatic activity, protein–protein interactions, signal transduction cascades, and subcellular localization (Fig. 5.1b). Glycosylation affects a wide array of biophysical and biological properties, including conformational stability, solubility, protease resistance, trafficking, and interactions with lectins and immune receptors; critically, glycan composition and branching can modulate pharmacokinetics, serum half-life, and immunogenicity of therapeutic proteins

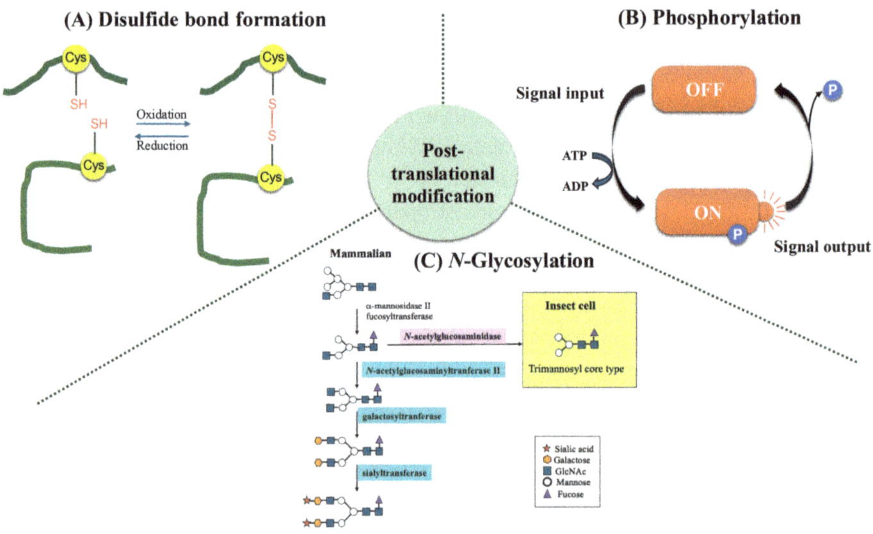

Fig. 5.1 Post-translational modifications in eucaryotic cells. (**a**) Disulfide bond formation: Within the endoplasmic reticulum, -SH groups on cysteine residues in the folded polypeptide undergo oxidation to form disulfide bonds, contributing to structural stability. (**b**) Protein phosphorylation: A protein kinase catalyzes the transfer of the terminal phosphate group from ATP to specific residues on the target protein. This modification is later reversed by a protein phosphatase. (**c**) N-glycosylation: Typical primary structures of N-linked glycans found in mammalian and insect cells, illustrating differences in glycan composition and branching patterns. (Adopted with permission (Park and Minkner 2025). Copyright 2025 Elsevier)

(Fig. 5.1c). The absence or improper installation of these modifications can yield misfolded or nonfunctional proteins, altered binding affinities, or unfavorable pharmacological profiles or outcomes, which are particularly consequential for drug candidates, enzymes, and signaling proteins intended for translational use.

To address these challenges, eukaryotic expression platforms—particularly silkworms and insect cell-based systems—offer environments conducive to the post-translational modification and protein maturation. Silkworm (*Bombyx mori*) and other insect hosts can form disulfide bonds, execute phosphorylation, and perform glycosylation pathways that more closely approximate those of mammalian cells (Kato et al. 2010). In practice, these systems have proven advantageous for producing complex, disulfide-rich proteins (e.g., antibodies, cytokines, and secreted enzymes), kinases and phosphoproteins where regulatory phosphorylation is functionally relevant, and glycoproteins whose glycans contribute to bioactivity and pharmacokinetics. While insect glycosylation patterns may differ from fully humanized profiles, the ability of these hosts to provide core glycan structures and an oxidative folding environment can substantially enhance yield and functional integrity relative to *E. coli*, reducing reliance on post-expression refolding and minimizing loss of activity. Consequently, silkworm and insect cell systems constitute compelling platforms for research and preclinical production of biologically active eukaryotic proteins, as elaborated in the sections below.

5.2 Disulfide Bond Formation

Insulin-like peptide 3 (INSL3) was successfully expressed in silkworm larvae as a small-sized model protein. INSL3, first identified in a boar testis cDNA library (Adham et al. 1993), is a major product of fetal and adult Leydig cells. It is constitutively expressed and secreted either as an A–B heterodimer (Bullesbach and Schwabe 2002) or as a single-chain A–B–C form (Minagawa et al. 2012; Siqin et al. 2013) (Fig. 5.2). INSL3 initiates intracellular signalling by binding with high affinity to its specific receptor, relaxin family peptide receptor 2 (RXFP2; formerly leucine-rich G protein-coupled receptor 8 [LGR8]) (Kumagai et al. 2002). NSL3 is essential for fetal testicular descent, as demonstrated by knockout models targeting *Insl3* (Nef and Parada 1999) or *Rxfp2* (Bogatcheva et al. 2003). Evidence also suggests a role in spermatogenesis, acting as a germ cell survival or anti-apoptotic factor in rats (Del Borgo et al. 2006) and boars (Minagawa et al. 2014), and as a germ cell proliferation factor in zebrafish (Assis et al. 2016). Furthermore, RXFP2 expression in the spermatozoa of rodents, humans (Granberg et al. 2012), boar (Feugang et al. 2015), and other livestock species (Pitia et al. 2017) indicates a potential role in sperm function.

Previous production methods include solid-phase synthesis of separate A- and B-domains followed by chain recombination (Wade et al. 2009), or oxidative refolding and enzymatic processing of recombinant peptides expressed in *E. coli* (Luo et al. 2009). Biological activity depends on three disulfide bonds linking the A- and

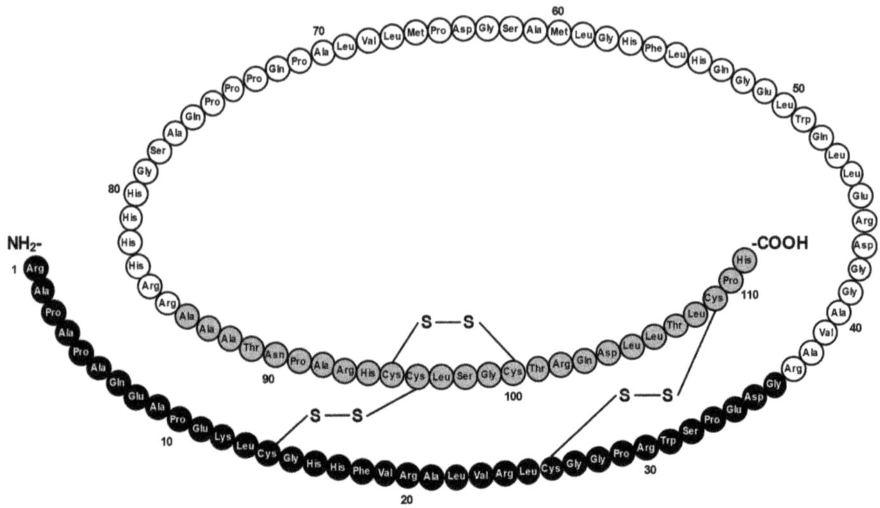

Fig. 5.2 Primary structure of native pINSL3. The B-domain, C-domain, and A-domain residues are indicated by *black filled circles*, *unfilled circles*, and *gray filled circles*, respectively. (Adopted with permission (Miyazaki et al. 2017). Copyright 2017 Springer Nature)

B-domains; substitution with isopeptide bonds reduces receptor affinity (Bullesbach and Schwabe 2012), and analogs lacking the intra-chain disulfide bonds are inactive (Zhang et al. 2010). The difficulty of forming correct disulfide bonds in prokaryotic systems has hindered the production of active INSL3 without post-expression chemical modification.

Here, porcine INSL3 (pINSL3) was expressed in silkworm larvae, purified from hemolymph, and confirmed to possess correct disulfide linkages. The recombinant pINSL3 exhibited RXFP2-stimulating activity comparable to human INSL3 (hINSL3). These findings demonstrate that silkworms can efficiently form disulfide bonds in recombinant proteins, enabling heterologous production of biologically active INSL3 without additional chemical processing.

5.2.1 Expression and Purification of Recombinant pINSL3 in Silkworms and Secondary Structure

FLAG-tagged pINSL3 was produced in silkworms using the BmNPV bacmid system, and its native signal peptide enabled efficient secretion into the larval hemolymph (Fig. 5.3a). Because it was challenging to purify the recombinant protein homogeneously with standard one-step affinity chromatography, the hemolymph was first treated with hydrochloric acid to aggregate silkworm-derived proteins. Following acidification and neutralization, most native silkworm proteins were removed (Fig. 5.3b, lanes 3 and 4). The recombinant pINSL3 was then purified by

FLAG-tag affinity chromatography, yielding a single 14 kDa band on SDS-PAGE and Western blot analysis (Fig. 5.3b, lanes 5 and 6), consistent with a theoretical molecular weight of 13.2 kDa calculated from its amino acid sequence. The final yield of purified recombinant pINSL3 was 8.3 μg from 10 mL of larval hemolymph.

Peptides from the relaxin/insulin family have three α-helices that are important for their function (Bullesbach and Schwabe 2012; Zhang et al. 2010). The circular dichroism (CD) results showed a strong positive peak at 191 nm and a negative peak at 205 nm (Fig. 5.3c), indicating that α-helices are present in the recombinant pINSL3 protein, with 33% of the protein composed of α-helices. According to the prediction, the recombinant protein contains five α-helices (Fig. 5.3d). One α-helix (Gly15-Cys26) is likely in the B-domain, while two medium-length helices (Ala92-Ser98 and Arg102-Thr107) are in the A-domain. Additionally, one α-helix

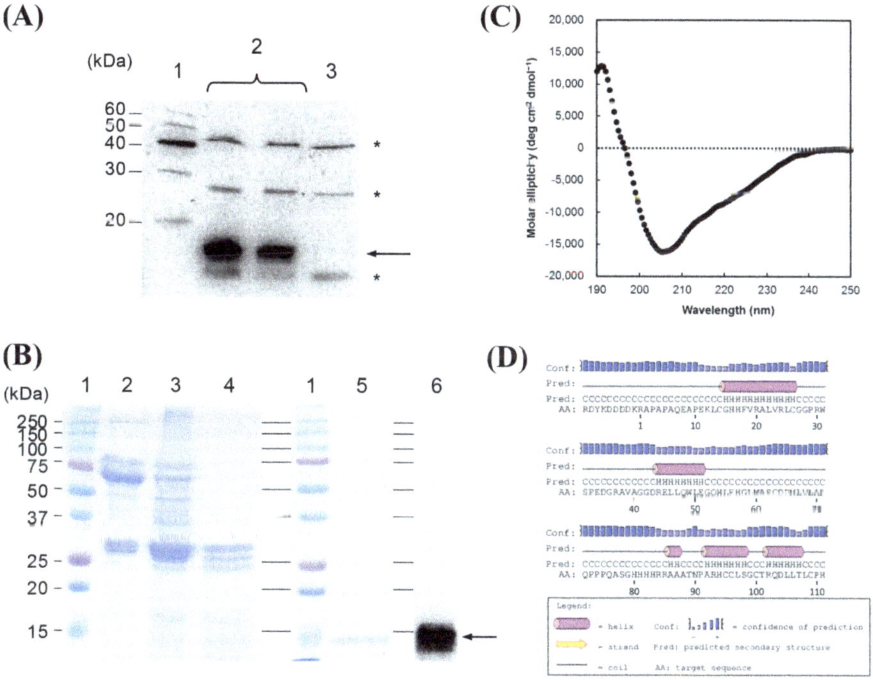

Fig. 5.3 Expression and purification of recombinant pINSL3 and its secondary structure analysis (Miyazaki et al. 2017). (**a**) Western blot of recombinant pINSL3 in silkworm larval hemolymph using an anti-DDDDK-tag antibody: lane 1 is a marker, lane 2 is hemolymph from infected silkworms, and lane 3 is from mock silkworms. The arrow marks recombinant pINSL3; asterisks indicate nonspecific proteins. (**b**) SDS-PAGE showing purification steps for recombinant pINSL3: lane 1, marker; lane 2, hemolymph; lane 3, post-HCl treatment supernatant; lane 4, post-neutralization supernatant; lane 5, after affinity chromatography; lane 6, purified recombinant pINSL3 detected by anti-FLAG antibody. (**c**) CD spectrum analysis of recombinant pINSL3 for secondary structure. (**d**) Secondary structure prediction of recombinant pINSL3 using the PSIPRED server v3.3

(Arg44-Glu51) is predicted in the C-domain, and a short α-helix (Ala86-Ala87) is at the start of the A-domain.

5.2.2 Structural Analysis by LC-MS/MS

The structure of recombinant pINSL3 was determined by LC-MS/MS analysis using both reduced and nonreduced peptide samples. In the reduced sample, two main peaks appeared with deconvoluted masses of 10,254.2 Da and 10,455.3 Da. The 10,254.2 Da peak matches the expected mass of the FLAG-tagged peptide containing the B- and C-domains (Arg-Asp-Tyr-Lys-Asp-Asp-Asp-Asp-Lys-Arg1-His83, 10,254.4 Da), where cysteines were carbamidomethylated by iodoacetamide for mass fingerprinting (Fig. 5.4a, b). The 10,455.3 Da peak likely represents the same peptide but with an extra Leu-Ser at the N-terminus (theoretical mass: 10,454.4 Da), probably due to incorrect cleavage by an endogenous signal peptidase. The LC-MS/MS analysis of the reduced sample also showed a peak at m/z 745.86 ($[M + 4H]^{4+}$), which matches the theoretical mass of the carbamidomethylated A-domain (Ala86-His111, 2979.37 Da), confirmed by MS/MS analysis (Fig. 5.4c, d). These findings suggest that recombinant pINSL3 was cleaved at two peptide bonds: between His83 and Arg84, and between Arg85 and Ala86.

In the nonreduced recombinant pINSL3 sample, two major peaks were observed with deconvoluted masses of 12,887.3 Da and 13,087.5 Da. Both are less than the theoretical mass of mature FLAG-tagged pINSL3 (Arg-Asp-Tyr-Lys-Asp-Asp-Asp-Asp-Lys-Arg1-His111, 13,181.9 Da) (Fig. 5.4e). The 200 Da difference between the two peaks corresponds to the additional N-terminal Leu-Ser residues mentioned above. The 12,887.3 Da mass matches the theoretical mass (12,887.5 Da) of a heterodimer composed of the FLAG-tagged B- and C-domains (Arg-Asp-Tyr-Lys-Asp-Asp-Asp-Asp-Lys-Arg1-His83) and the A-domain (Ala86-His111), linked by three disulfide bonds between six cysteine residues (Fig. 5.4f). These results indicate that recombinant pINSL3 contains the B-, C-, and A-domains and forms three disulfide bonds, similar to native pINSL3. The only difference between the wild-type and recombinant proteins is the internal cleavage between the C- and A-domains.

5.2.3 HCD-MS/MS and ETD-MS/MS Analysis of Disulfide Bond Properties

LC-MS/MS analysis was used to predict the structure of recombinant pINSL3 by examining reduced and nonreduced samples. In the reduced sample, two main peaks were detected: one at 10,254.2 Da, which matches the expected mass of the FLAG-tagged peptide containing the B- and C-domains

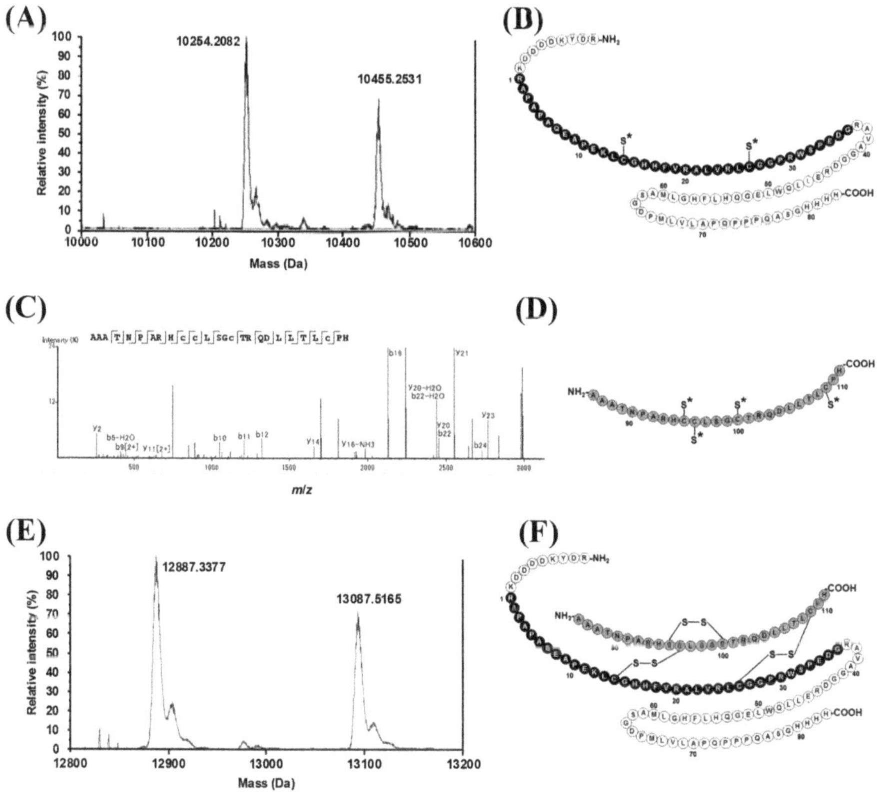

Fig. 5.4 LC-ESI-TOF-MS spectra and predicted structure of recombinant pINSL3 (Miyazaki et al. 2017). Reduced and nonreduced recombinant pINSL3 samples were analyzed using LC-MS/MS to determine their structures. (**a**) The reduced B-C chain showed two mass peaks: 10254.2 Da for the FLAG-tagged B-C chain, and 10455.3 Da for the same chain with an N-terminal Leu-Ser addition. (**b**) The structure of the reduced B-C chain was predicted by its molecular mass. (**c**) The A-chain from the reduced sample had a peak at m/z 745.86, which matched the A-chain amino acid sequence. (**d**) The A-chain structure was predicted from MS/MS product ions. (**e**) The nonreduced sample had two peaks: 12887.3 Da for the heterodimer of FLAG-tagged B-C domains and the A domain, and 13087.5 Da for the same heterodimer with an N-terminal Leu-Ser addition. (**f**) The predicted overall structure of nonreduced recombinant pINSL3 based on these LC-MS/MS results. In the schematic, FLAG-tag, B-domain, C-domain, and A-domain residues are shown as dotted, black-filled, unfilled, and gray-filled circles, respectively. Carbamidomethylated cysteine sulfhydryl groups are marked with an asterisk

(Arg-Asp-Tyr-Lys-Asp-Asp-Asp-Asp-Lys-Arg1-His83, 10254.4 Da) with carbamidomethylated cysteines, and another at 10,455.3 Da (Fig. 5.4a, b), likely representing the same peptide with an extra Leu Ser added to the N-terminus due to miscleavage. Additionally, a peak at m/z 745.86 ($[M + 4H]^{4+}$) was observed, corresponding to the mass of the carbamidomethylated A-domain (Ala86-His111, 2979.37 Da), confirmed by MS/MS analysis (Fig. 5.4c, d). These findings indicate that recombinant pINSL3 was hydrolyzed at two specific peptide bonds: between His83 and Arg84, and between Arg85 and Ala86.

The nonreduced sample of recombinant pINSL3 showed two major peaks at 12,887.3 Da and 13,087.5 Da, both slightly lower than the theoretical mass of mature FLAG-tagged pINSL3 (Arg-Asp-Tyr-Lys-Asp-Asp-Asp-Asp-Lys-Arg1-His111, 13181.9 Da) (Fig. 5.4e). The 200 Da difference between the peaks corresponds to the extra N-terminal Leu-Ser. The 12,887.3 Da peak matches the mass of a heterodimer consisting of the FLAG-tagged B-C-domains (Arg-Asp-Tyr-Lys-Asp-Asp-Asp-Asp-Lys-Arg1-His83) and the A-domain (Ala86-His111), with six cysteines forming disulfide bridges (Fig. 5.4f). Overall, these results suggest that recombinant pINSL3 consists of the B-, C-, and A-domains and contains three disulfide bonds, similar to native pINSL3. The only difference between the wild-type and recombinant proteins is the internal cleavage between the C- and A-domains.

To analyze the disulfide bonds in recombinant pINSL3, the protein was enzymatically digested and subjected to mass spectrometric analysis using higher-energy collisional dissociation (HCD) MS/MS and electron transfer dissociation (ETD) MS/MS. The total ion chromatograms of the resulting peptides are presented in Fig. 5.5a. For the nonreduced sample, two prominent peaks (I and II) appeared at retention times of 18.7 and 32.3 min, respectively. In the reduced sample, two major peaks (III and IV) were detected at 17.4 and 41.4 min. The predominant precursor ions corresponding to these peaks are depicted in Fig. 5.5b. HCD-MS/MS analysis of the reduced digest identified four cysteine-containing peptides: P1 (residues 13–20, LCGHHFVR), P2 (residues 94–102, HCCLSGCTR), P3 (residues 25–30, LCGGPR), and P4 (residues 103–111, QDLLTLCPH). Peaks III and IV correspond to P1 in the B-domain and P4 in the A-domain. Based on accurate mass measurements from HCD-MS/MS, peak I was assigned to a combination of P1 and P2, linked by two disulfide bonds (both inter- and intramolecular). In contrast, peak II was assigned to P3 and P4, connected by one intermolecular disulfide bond.

ETD fragmentation of the triply charged ion with an *m/z* value of 648.62 (derived from peak I) yielded two fragments, P1 and P2, indicating that Cys14 in P1 is linked via a disulfide bond to either Cys95, Cys96, or Cys100 in P2 (Fig. 5.6a). However, this method does not allow for the precise determination of which cysteine in P2 forms the intramolecular disulfide bond. Similarly, ETD analysis of the triply charged ion with an *m/z* value of 546.9 (from peak II) generated two fragments, P3 and P4, demonstrating that these peptides are joined by an intermolecular disulfide bond between Cys26 and Cys109 (Fig. 5.6b). Additionally, unexpected disulfide pairings were detected at *m/z* 502.3 $[M + 3H]^{3+}$ and 672.3 $[M + 3H]^{3+}$, each representing 6% of the total cysteine-containing peptides. Overall, these findings demonstrate that the majority of recombinant pINSL3 peptides contained the correct disulfide bond arrangements.

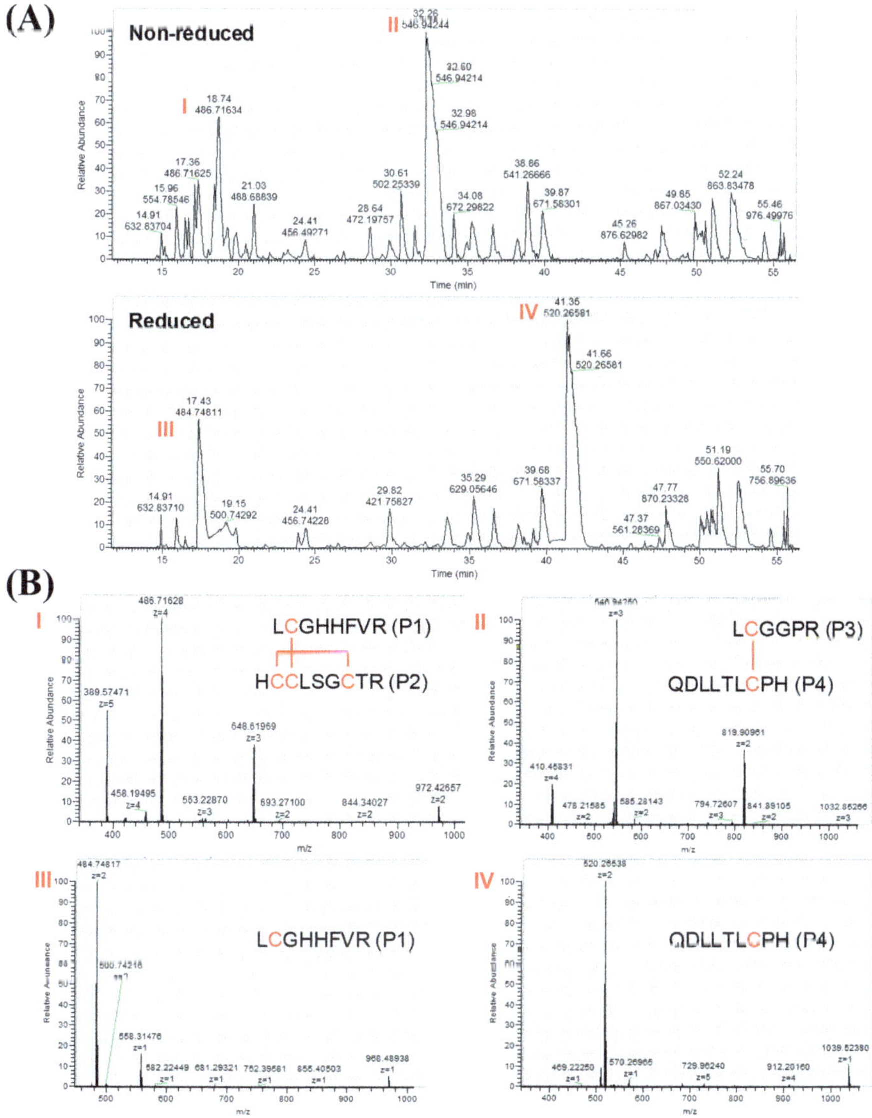

Fig. 5.5 MS analysis of tryptic recombinant pINSL3 (Miyazaki et al. 2017). In panel (**a**), the total ion chromatogram displays major peaks for the nonreduced form at 18.7 and 32.3 min (labeled I and II), and for the reduced form at 17.4 and 41.4 min (labeled III and IV). Panel (**b**) presents MS1 spectra for each peak, highlighting cysteine-containing peptide ions. Peak I shows that peptides P1 (L13CGHHFVR20) and P2 (H94CCLSGCTR102) are linked by two disulfide bonds. Peak II shows peptides P3 (L^{25}CGGPR^{30}) and P4 (Q^{103}DLLTLCPH^{111}) linked by one disulfide bond. Peaks III and IV represent the reduced forms of peptides P1 and P4, respectively

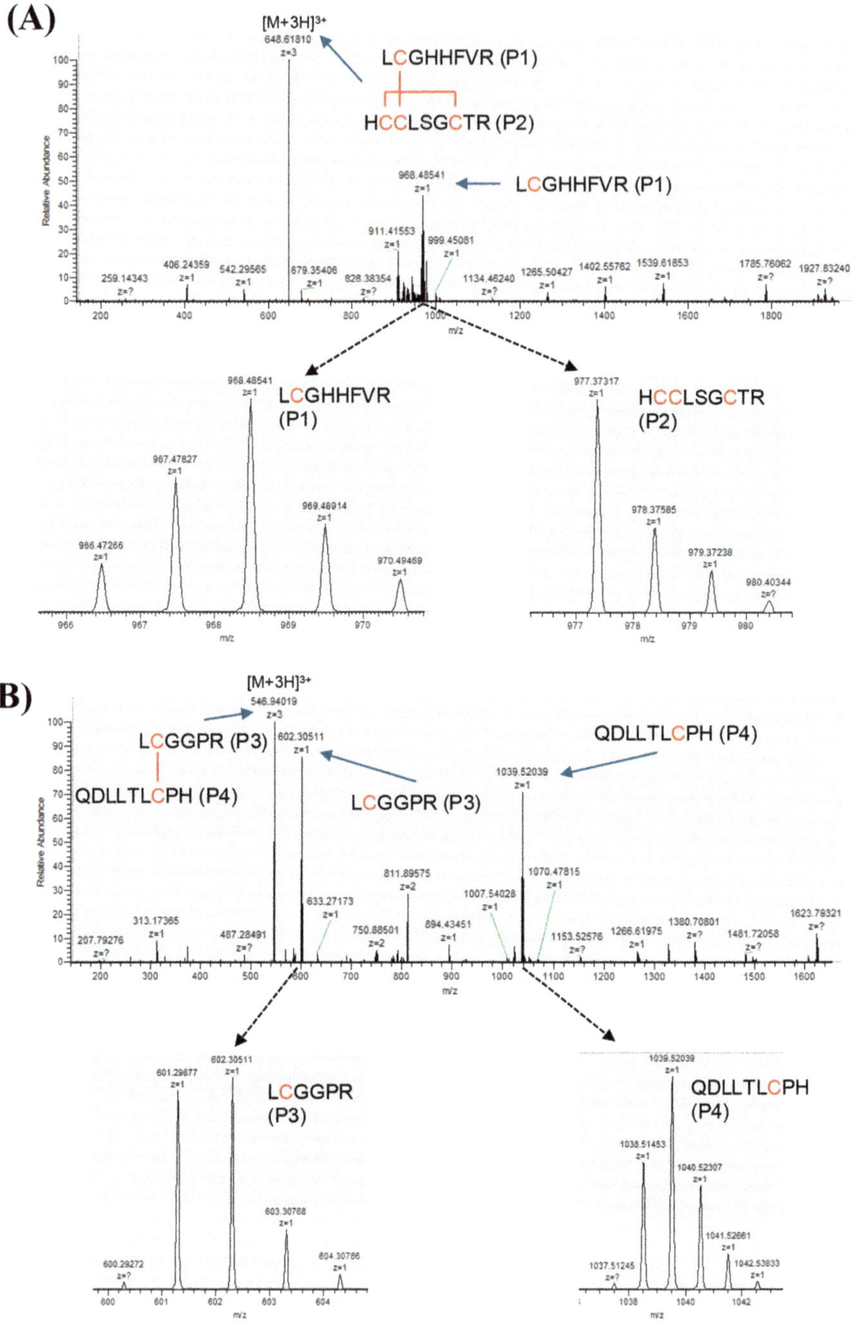

Fig. 5.6 ETD-MS/MS analysis of disulfide-linked peptides (Miyazaki et al. 2017). (**a**) The MS/MS spectrum of the nonreduced ion at *m/z* 648.6 (3+) from peak I reveals peptide fragments P1 (L[13]CGHHFVR[20]) and P2 (H[94]CCLSGCTR[102]), showing that Cys14 in P1 is disulfide-linked to a cysteine in P2. (**b**) The MS/MS spectrum of the nonreduced ion at *m/z* 546.9 (3+) from peak II displays peptide fragments P3 (L[25]CGGPR[30]) and P4 (Q[103]DLLTLCPH[111]), demonstrating that Cys26 in P3 forms an intermolecular disulfide bond with Cys109 in P4

5.2.4 Biological Activity of pINSL3

The biological activity of the purified recombinant pINSL3 peptide was evaluated using a receptor activation assay in HEK293 cells that express RXFP2. Increasing the concentration of recombinant pINSL3 led to a corresponding increase in cAMP production. Recombinant pINSL3 achieved a maximum response (E_{max}) of 97.0 ± 5.6% (*n* = 12), which is similar to the synthetic hINSL3's E_{max} of 106.8 ± 3.5% (*n* = 7). However, the EC_{50} value for recombinant pINSL3 was 24.5 ± 1.2 nM (pEC_{50} = 7.61), slightly higher than the standard hINSL3 containing the B- and A-domains (16.7 ± 1.1 nM, pEC_{50} = 7.78). These findings indicate that the recombinant pINSL3 peptide is properly folded and capable of binding to and activating its receptor, thereby inducing cAMP production. Nonetheless, recombinant INSL3 is approximately 30% less active than the synthetic version (Fig. 5.7), likely due to reduced receptor affinity as indicated by the pEC_{50} value. The cause of this reduced activity is unclear but may be associated with the presence of isomers.

5.2.5 Discussion

Unlike traditional chemical synthesis or *E. coli*-based expression systems, the silkworm system offers the advantage of post-translational modifications, such as those found in mammalian cells, resulting in a product with correct folding and bioactivity. Structural and functional analyses confirmed that the recombinant pINSL3 possesses proper domain organization and disulfide bond formation, despite the presence of some disulfide isomers. Retention of the C-domain does not hinder receptor binding or biological activity, suggesting that its role may be more closely related to peptide stability. Furthermore, the bioactivity assays demonstrated that the recombinant protein can effectively stimulate cAMP production, supporting its potential for clinical and research applications.

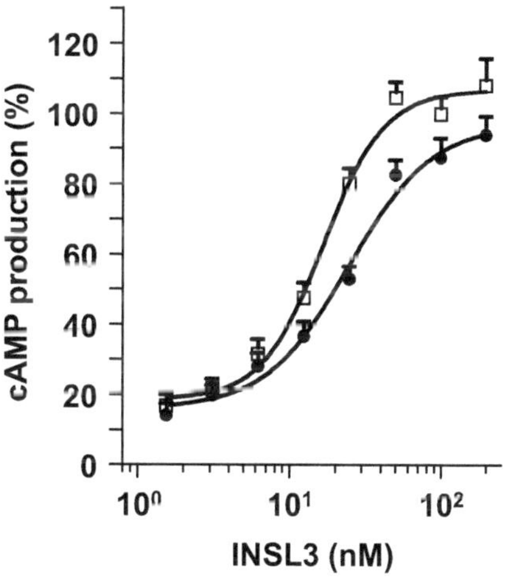

Fig. 5.7 cAMP activity of recombinant pINSL3 (Miyazaki et al. 2017). The ability of recombinant pINSL3 (●) to stimulate cAMP production was evaluated and compared with that of synthetic hINSL3 (□) in HEK293 cells expressing RXFP2. cAMP levels were measured as a percentage of the maximal response elicited by synthetic hINSL3 on RXFP2. Data are mean ± SEM (*n* = 3)

Importantly, INSL3 is recognized for its broad physiological and clinical relevance, participating in reproductive, bone, ophthalmic, and oncological processes. The ability to reliably produce active recombinant INSL3 opens new possibilities for therapeutic interventions and research in these fields. Overall, the silkworm-based expression system not only streamlines the production of complex peptide hormones but also paves the way for future studies on structure–function relationships and clinical applications of INSL3 and related proteins.

5.3 Phosphorylation

The recombinant protein produced in silkworms undergoes proper post-translational modifications, which are crucial for its catalytic activity and regulation. Acetyl-CoA carboxylases (ACCs) and malonyl-CoA decarboxylase (MCD) are key enzymes in lipid metabolism. There are two ACC isoforms: ACC1, mainly found in the cytosol of lipogenic organs like adipose tissue and liver, where it supports fatty acid synthesis, and ACC2, located on the mitochondrial outer membrane in tissues such as heart, liver, and skeletal muscle, where it regulates fatty acid oxidation (Abu-Elheiga et al. 1997; Abu-Elheiga et al. 1995). ACC plays a central role in controlling energy balance and maintaining lipid homeostasis (Folmes et al. 2013; Fullerton et al. 2013; Knobloch et al. 2013). Additionally, ACC2 knockout has shown anti-obesity effects and protection against cardiac remodeling (Abu-Elheiga et al. 2001; Abu-Elheiga et al. 2003; Kolwicz Jr. et al. 2012). Therefore, regulating ACC activity is considered a promising therapeutic strategy for metabolic disorders like obesity, diabetes, cancer, and cardiovascular diseases (Tong and Harwood Jr. 2006).

Malonyl-CoA decarboxylase (MCD, E.C.4.1.1.9), encoded by *MLYCD*, is an essential metabolic enzyme that maintains lipid homeostasis by regulating fatty acid metabolism (Dyck et al. 2006; Folmes et al. 2013; Saggerson 2008). In humans, MCD deficiency (OMIM 248360) leads to various disorders such as cardiomyopathy, hypoglycemia, hypotonia, mild intellectual disability, metabolic acidosis, malonic aciduria, seizures, and vomiting (Brown et al. 1984; Krawinkel et al. 1994; Xue et al. 2012). Post-translational modifications, including acetylation and deacetylation, modulate MCD enzymatic activity; deacetylation inhibits decarboxylase function and promotes de novo lipogenesis, whereas acetylation enhances fatty acid oxidation (Dyck et al. 2000; Laurent et al. 2013; Sambandam et al. 2004). In contrast, the functional significance of MCD phosphorylation remains unclear, as phosphorylation sites have yet to be identified and authentic MCD with appropriate post-translational modifications is difficult to obtain (Dyck et al. 2000; Saha et al. 2000; Sambandam et al. 2004).

Here, ACC and MCD produced by silkworms exhibited reliable catalytic activity, supported by appropriate post-translational modifications such as biotinylation and phosphorylation. Furthermore, their enzymatic functions were precisely regulated through allosteric modulation.

5.3.1 Expression and Purification of Human ACC and MCD

Silkworm pupae were used as bioreactors to express recombinant Δ148aa-hACC2. To produce the protein, 10 μg of BmNPV-Δ148aa-hACC2 bacmid DNA (Hwang et al. 2014) mixed with DMRIE-C reagent (Invitrogen) was injected into the dorsal side of the pupae. The pupae were maintained at 27 °C for 6–7 d and then stored at −80 °C until further analysis. For protein purification, five pupae were homogenized in 10 mL lysis buffer (50 mM Tris-HCl, 150 mM NaCl, pH 7.4, 0.1% TritonX-100) with an EDTA-free protease inhibitor (Roche) using a homogenizer (GLH-115, Yamato). After centrifugation at 12,000 *g* for 30 min to remove debris, the supernatant was filtered through a 0.45 μm syringe filter. The filtrate was loaded onto 500 μL of Anti-FLAG M2 antibody Affinity Gel (Sigma-Aldrich) pre-equilibrated with buffer (50 mM Tris-HCl, 150 mM NaCl, pH 7.4, 0.02% TritonX-100). The column was washed with 2.5 mL of equilibration buffer, and the protein was eluted with 100 μg/mL FLAG peptide in 50 mM Tris-HCl, 150 mM NaCl, pH 7.4. The eluted Δ148aa-hACC2 was then concentrated with a 100 K Amicon Ultra centrifugal filter (Millipore).

Human ACC2 is a large protein consisting of several domains: a mitochondrial attachment domain, a mitochondrial targeting sequence, a biotin carboxylase domain, a biotin carboxyl carrier protein (BCCP) domain, and a carboxyltransferase domain (Fig. 5.8a(I)) (Bianchi et al. 1990; Tong 2013). Biotin is covalently linked to a lysine residue in the BCCP domain, and several serine residues are phosphorylated by protein kinases (Beaty and Lane 1983; Brownsey et al. 2006; Ha et al. 1994). To increase protein solubility, we removed the N-terminal 148 hydrophobic amino acids, preserving the core functional domains (Fig. 5.8a(II)). The recombinant Δ148aa-hACC2 with a C-terminal FLAG tag was purified using an anti-FLAG affinity column. The protein migrated at ~260 kDa on SDS-PAGE, consistent with the expected size (Fig. 5.8b, upper panel). Western blotting with a FLAG antibody confirmed hACC2 expression (Fig. 5.8b lower panel). The yield of purified Δ148aa-hACC2 was 495 μg per pupa (Hwang et al. 2014). The purified hACC2 was highly homogeneous on SDS-PAGE and Western blot and then this protein for further functional analysis.

Human MCD (hMCD) has a 39-amino-acid mitochondrial targeting sequence at the N-terminus and a peroxisomal targeting signal (SKL) at the C-terminus. The SKL signal does not affect protein expression, but the mitochondrial targeting sequence reduces MCD expression (Voilley et al. 1999). To improve recombinant hMCD production, a version lacking the mitochondrial targeting sequence but containing the SKL signal was created (Fig. 5.9a). The recombinant Δ39aa-hMCD with an N-terminal FLAG tag was expressed in silkworm larvae and pupae and then purified using an anti-FLAG affinity column. The purified protein was over 95% pure and had an estimated molecular weight of ~50 kDa by SDS-PAGE (Fig. 5.9b). Western blot analysis confirmed specific expression of Δ39aa-hMCD, while no hMCD was detected in mock-injected silkworm samples (Fig. 5.9c). The average

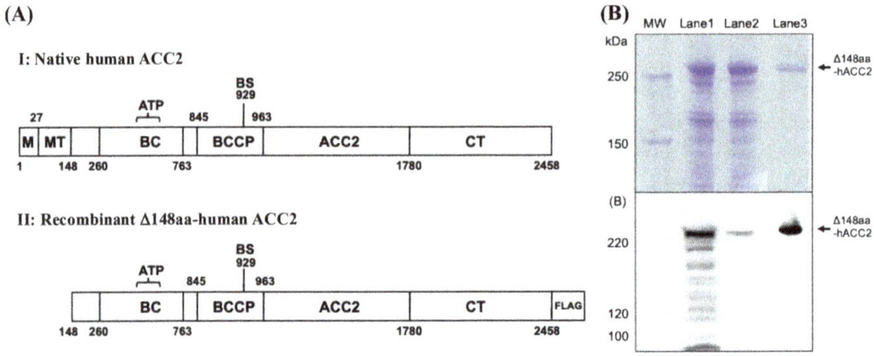

Fig. 5.8 (**a**) A schematic of human ACC2, showing its domain structure, including (I) ATP-grasp, biotin carboxylase, biotin carboxyl carrier protein, biotinylation site, carboxyltransferase, membrane attachment, and mitochondrial targeting sequence. (II) Recombinant Δ148aa-hACC2 variant lacks the N-terminal 148 amino acids, including a C-terminal FLAG tag for affinity purification. (**b**) Δ148aa-hACC2 expression was analyzed by SDS-PAGE and Western blotting, using anti-FLAG M2 and anti-mouse IgG-HRP antibodies to detect the purified protein. (Adapted with permission (Hwang et al. 2014). Copyright 2014 Springer Nature)

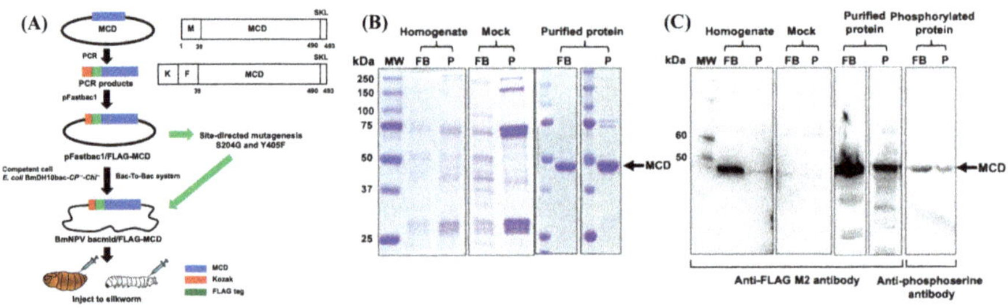

Fig. 5.9 (**a**) Schematic of recombinant Δ39aa-hMCD bacmid creation and expression in silkworms. The Δ39aa-hMCD gene was amplified by PCR and inserted into the pFastbac1 vector. The resulting pFastbac-Δ39aa-hMCD was transformed into *E. coli* BmDH10Bac containing a BmNPV-*CP*⁻-*Chi*⁻. The recombinant bacmid was injected into the dorsal side of fifth-instar larvae and pupae. M, mitochondrial-targeting sequence; K, Kozak sequence; F, FLAG tag; and SKL, peroxisomal-targeting sequence. Black and green arrows indicate bacmid construction and mutants, respectively. SDS-PAGE (**b**) and Western blot (**c**) analyses of recombinant Δ39aa-hMCD expressed in silkworm fat body and pupae. MW, molecular weight markers; FB, Δ39aa-hMCD from fat body; P, Δ39aa-hMCD from pupae. Δ39aa-hMCD was detected using anti-FLAG M2 and anti-mouse IgG-HRP antibodies, while phosphorylation was identified with an anti-phosphoserine antibody. (Adapted with permission (Hwang et al. 2015). Copyright 2015 Springer Nature)

yields from ten larvae and five pupae were 119 μg/larva and 344 μg/pupa, respectively (Hwang et al. 2015). Pupae produced a higher yield than larvae, likely due to differences in organ size and protease activity. Decarboxylase activity was 59.54 ± 7.68 nmol/mg/min from larvae and 480.16 ± 7.89 nmol/mg/min from pupae

($n = 6$ for each) (Hwang et al. 2015). These activities were not significantly different ($p > 0.05$), but both were higher than that of hMCD purified from *E. coli* (Zhou et al. 2004), indicating that post-translational modifications in silkworms enhance catalytic activity.

5.3.2 *Biotinylation of Δ148aa-hACC2*

Post-translational biotinylation of ACC2 is critical for its catalytic activity. Although the specific biotin-binding residue in human ACC2 has not been definitively determined, nuclear magnetic resonance structural studies suggest that biotin is covalently attached to lysine 929 within the BCCP domain (Lee et al. 2008). Biotinylation of Δ148aa-hACC2 expressed in silkworms was evaluated by Western blotting with an anti-biotin antibody. To further confirm the presence of biotin, a streptavidin–HRP conjugate was used, as streptavidin binds biotin with exceptionally high affinity. The anti-biotin antibody produced a stronger signal for hACC2 than the streptavidin probe (Fig. 5.10), but both detection methods verified biotinylation of hACC2. These findings demonstrate that silkworms can produce biotinylated heterologous proteins without additional biotin supplementation, similar to that observed in *T. ni* cells (Kim et al. 2007).

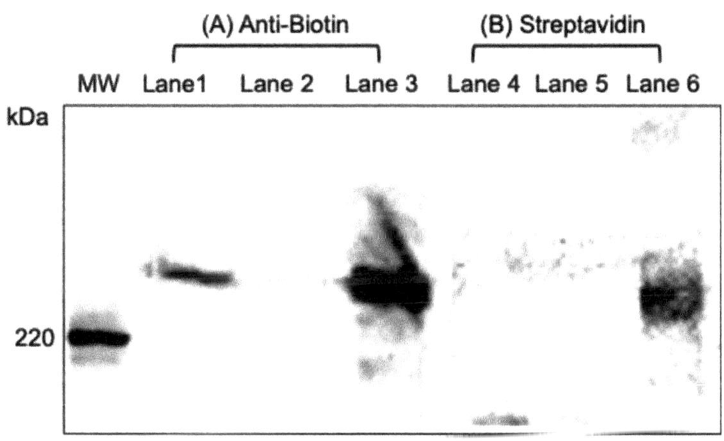

Fig. 5.10 Western blot analysis confirmed that the purified Δ148aa-hACC2 is post-translationally biotinylated, using anti-biotin antibody (**a**) and streptavidin–HRP conjugate (**b**). MW indicates molecular weight markers. Lanes 1 and 4 show protein extracts after infection, lanes 2 and 5 show flow-through from FLAG-tag purification, and lanes 3 and 6 show the purified and concentrated Δ148aa-hACC2. Anti-biotin antibody and streptavidin–HRP conjugate served as primary antibodies, while rabbit anti-goat IgG-HRP and anti-mouse IgG-HRP were used as secondary antibodies. (Adopted with permission (Hwang et al. 2014). Copyright 2017 Springer Nature)

5.3.3 Regulation of Enzyme Activity by Phosphorylation and Dephosphorylation of Δ148aa-hACC2

Phosphorylation by adenosine monophosphate-activated protein kinase (AMPK) serves as a key post-translational modification that allosterically regulates the catalytic activity of ACC. This modification inactivates ACC, while dephosphorylation restores its enzymatic activity. Phosphorylation of Ser222 in human ACC2 (corresponding to Ser212 in mouse ACC2) is particularly important for maintaining lipid metabolism homeostasis (Fullerton et al. 2013; Wakil and Abu-Elheiga 2009). Structural studies of the biotin carboxylase domain in human ACC2 have shown that phosphorylation at Ser222 impairs ACC2 polymerization, a crucial mechanism for regulating catalytic function (Cho et al. 2010; Lee et al. 2008). The post-translational phosphorylation status of recombinant Δ148aa-hACC2 was assessed to determine whether dephosphorylation would affect its catalytic activity. Western blot analysis using a monoclonal anti-phosphoserine antibody confirmed phosphorylation of Δ148aa-hACC2 purified from silkworm pupae. Treatment with Lambda protein phosphatase, which requires Mn^{2+}, gradually reduced phosphorylation compared to the control (Fig. 5.11a), but complete dephosphorylation was not achieved even after 6 h. This suggests that certain phosphorylation sites in full-length hACC2 are less accessible to Lambda protein phosphatase than those in isolated domains, such as the biotin carboxylase domain. The impact of dephosphorylation on Δ148aa-hACC2 was evaluated by measuring its catalytic activity (Fig. 5.11b). Δ148aa-hACC2 was incubated with or without Lambda protein phosphatase for 2 h, and activity was monitored over time. The phosphorylated form exhibited a specific activity of 0.786 ± 0.229 nmol Pi/mg/min ($n = 6$), while the phosphatase-treated protein showed a specific activity of 1.336 ± 0.441 nmol Pi/mg/min ($n = 6$), representing a roughly twofold increase. These results align with observations from knock-in mice, in which substitution of Ser212 with alanine abolishes this key phosphorylation event (Fullerton et al. 2013).

To analyze the allosteric activation of Δ148aa-hACC2 by citrate, changes in both structure and function were measured. When Δ148aa-hACC2 was treated with citrate, high-molecular-weight polymers formed, as shown on Native-PAGE, and their abundance increased with higher citrate concentrations (Fig. 5.11c). These polymerization results suggest that Δ148aa-hACC2 from pupae mainly exists as dimers and tetramers, based on molecular weight marker comparisons, and that more tetramers are produced as citrate concentration increases, with a corresponding decrease in dimers (Fig. 5.11c). Matching the increase in polymerization, the enzyme's catalytic activity also rose in a dose-dependent manner as citrate concentration increased (Fig. 5.11d). Specifically, incubation with 5, 10, and 20 mM citrate resulted in specific activities of 1.363 ± 0.279, 2.246 ± 0.870, and 4.186 ± 0.200 nmol P_i/mg/min ($n = 4$), respectively. These results show that doubling the citrate concentration roughly doubled the enzyme's activity, in direct proportion (Fig. 5.11d).

Moreover, when citrate concentrations exceeded 20 mM, the activation curve became sigmoidal, which agrees with earlier studies (Cheng et al. 2007). Structural

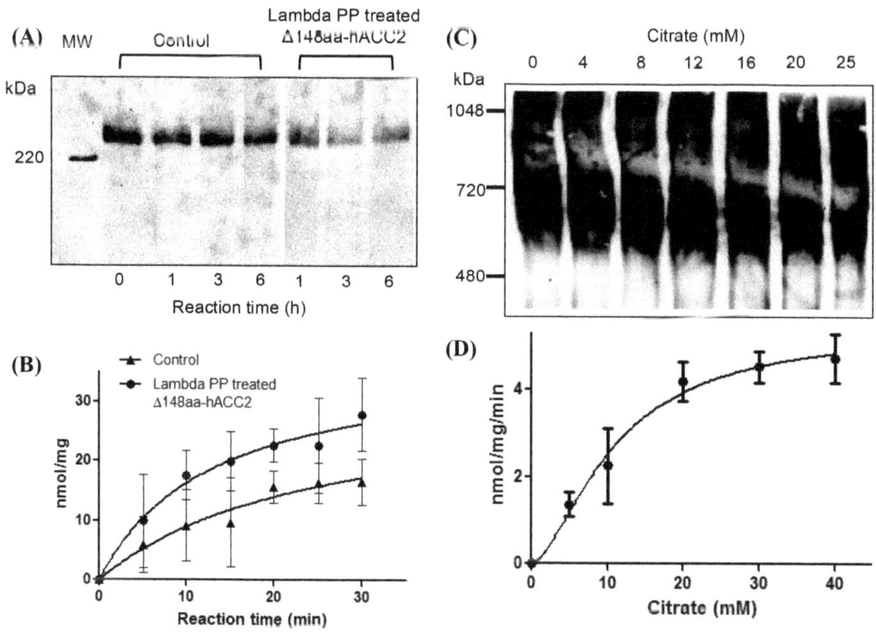

Fig. 5.11 Regulation of enzyme activity by dephosphorylation and phosphorylation of purified Δ148aa-hACC2. (**a**) Dephosphorylation of Δ148aa-hACC2 treated with Lambda protein phosphatase (PP) was assessed by Western blotting using a monoclonal anti-phosphoserine antibody produced in mouse and an anti-mouse IgG-HRP. (**b**) Lambda PP-treated Δ148aa-hACC2 enhanced catalytic activity. The polymerization and enzymatic activity of Δ148aa-hACC2 were influenced by the concentration of citrate. (**c**) The extent of Δ148aa-hACC2 polymerization was assessed at various citrate concentrations (0, 4, 8, 12, 16, 20, and 25 mM), with polymerization verified by Native-PAGE analysis. (**d**) Effect of citrate concentration on enzymatic activity. Data are mean ± SD ($n = 3$). (Adapted with permission (Hwang et al. 2014). Copyright 2014 Springer Nature)

changes in Δ148aa-hACC2 caused by citrate were further examined using high-resolution AFM imaging. Purified Δ148aa-hACC2 without citrate appeared as a nearly uniform distribution of particles (Fig. 5.12a). In contrast, adding citrate led to the appearance of large, filamentous polymer structures, much bigger than the protein alone (Fig. 5.12b). These observations confirm the formation of high molecular weight polymers seen in Native-PAGE and demonstrate that Δ148aa-hACC2 produced in silkworm is fully functional and responds properly to allosteric modulation.

5.3.4 Phosphorylation-Induced Catalytic Function

Previous studies evaluated the catalytic activity of MCD in relation to its phosphorylation and dephosphorylation status (Dyck et al. 1998; Park et al. 2002; Saha et al. 2000), but these investigations yielded conflicting results, showing either a

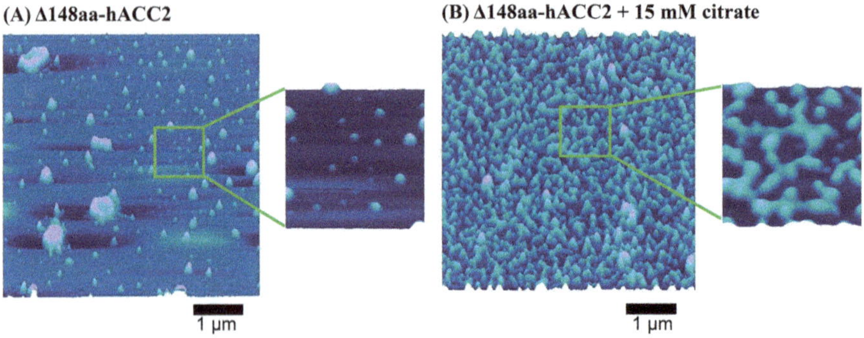

Fig. 5.12 Atomic force microscopy (AFM) images at the nanoscale of Δ148aa-hACC2. (**a**) Δ148aa-hACC2 by itself. (**b**) Δ148aa-hACC2 after being incubated with 15 mM 37 °C. The presence of citrate triggered the formation of filaments. (Adopted with permission (Hwang et al. 2014). Copyright 2014 Springer Nature)

reduction or an increase in MCD decarboxylase activity. To better understand how phosphorylation affects MCD's enzymatic function, potential phosphorylation sites in hMCD was analyzed. Using NetPhos 2.0, a neural network-based tool that predicts serine, threonine, and tyrosine phosphorylation sites in eukaryotic proteins (http://www.cbs.dtu.dk/services/NetPhos/), phosphorylation sites in hMCD were predicted (Blom et al. 1999) (Fig. 5.13), revealing four serine (Ser-204, Ser-275, Ser-326, Ser-380), four threonine (Thr-9, Thr-60, Thr-245, Thr-396), and one tyrosine (Tyr-468) residues as putative phosphorylation sites, all of which are highly conserved (Fig. 5.13a, b). Since each phosphorylation prediction program may identify slightly different amino acid residues based on their underlying bioinformatics algorithms, which suggested nine possible phosphorylation sites, a focused proteomic search using mass spectrometry was conducted. This approach led us to exclusively detect phosphorylation at Ser-204 and Tyr-405 in the recombinant Δ39aa-hMCD, thereby excluding the other predicted sites from further analysis. Consequently, we substituted these residues with glycine (S204G) and phenylalanine (Y405F) by site-directed mutagenesis to remove the hydroxyl groups necessary for phosphorylation (Fig. 5.13c).

The S204G and Y405F point mutants were expressed from silkworm fat body and pupae and then examined by Western blotting with anti-phosphoserine and anti-phosphotyrosine antibodies (Fig. 5.14). Compared to wild-type Δ39aa-hMCD, the mutants showed a significant decrease in phosphorylation signals on Western blot, confirming that Ser-204 and Tyr-405 are indeed phosphorylation sites. However, the phosphorylated MCD band was not completely absent, suggesting the possible presence of other phosphorylated amino acids besides Ser-204 and Tyr-405. To assess the impact of phosphorylation on hMCD's biological activity, the decarboxylase activity of Δ39aa-hMCD mutants was measured. The S204G mutant purified from fat body and pupae had specific activities of 30.36 ± 2.25 (n = 6) and 24.37 ± 1.99 nmol/mg/min (n = 6), respectively, which is about 50% less than the wild-type Δ39aa-hMCD. Additionally, the Y405F mutants had specific activities of

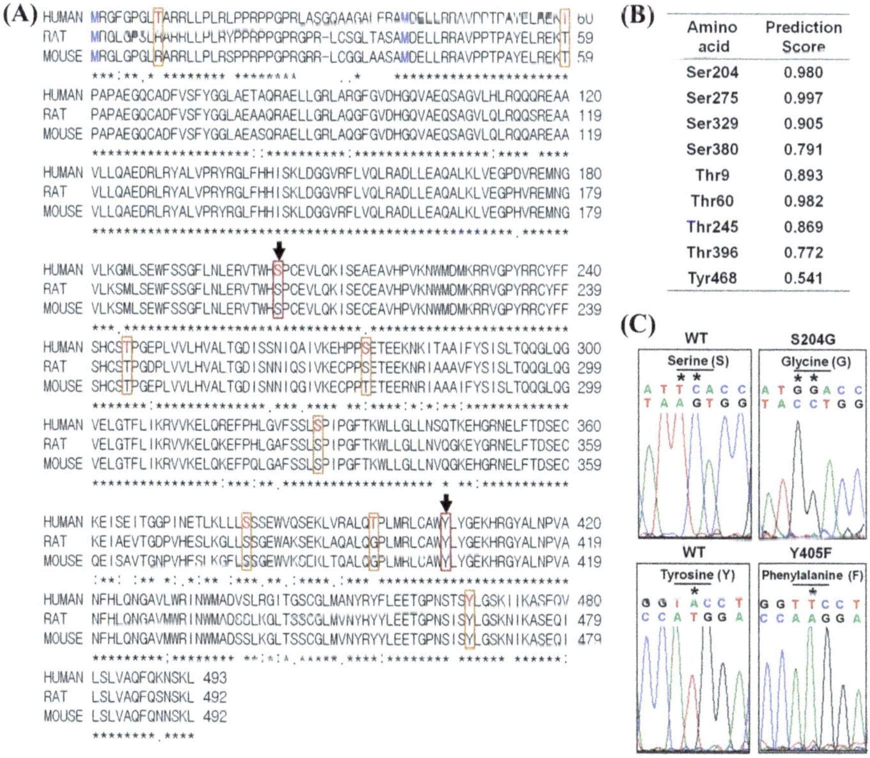

Fig. 5.13 Alignment of amino acid sequences for human, rat, and mouse MCD. (**a**) The two starting methionines are highlighted in blue; residues predicted to be phosphorylation sites in hMCD by the Netphos 2.0 server are marked in red; arrows indicate the specific mutation sites used to prevent phosphorylation of hMCD. (**b**) Predicted phosphorylation sites of hMCD. The prediction score reflects the likelihood of phosphorylation (maximum value = 1). (**c**) In hMCD, a point mutation replaced serine (*S*) at position 204 (codon TCA) with glycine (*G*) (codon GGA), and tyrosine (*Y*) at position 405 (codon TAC) was changed to phenylalanine (*F*) (codon TTC). (Adopted with permission (Hwang et al. 2015). Copyright 2015 Springer Nature)

33.45 ± 3.56 ($n = 6$) and 31.24 ± 1.69 nmol/mg/min ($n = 6$) from fat body and pupae, respectively. Taken together, these results show that dephosphorylation of hMCD reduces decarboxylase activity, highlighting the regulatory role of phosphorylation in its catalytic function.

5.3.5 Structural Effect of hMCD Phosphorylation

The crystal structure of human MCD (hMCD) shows that it forms a molecular tetramer, comprising two structural heterodimers, with each subunit displaying a distinct conformation (Fig. 5.15) (Aparicio et al. 2013). Each MCD monomer consists

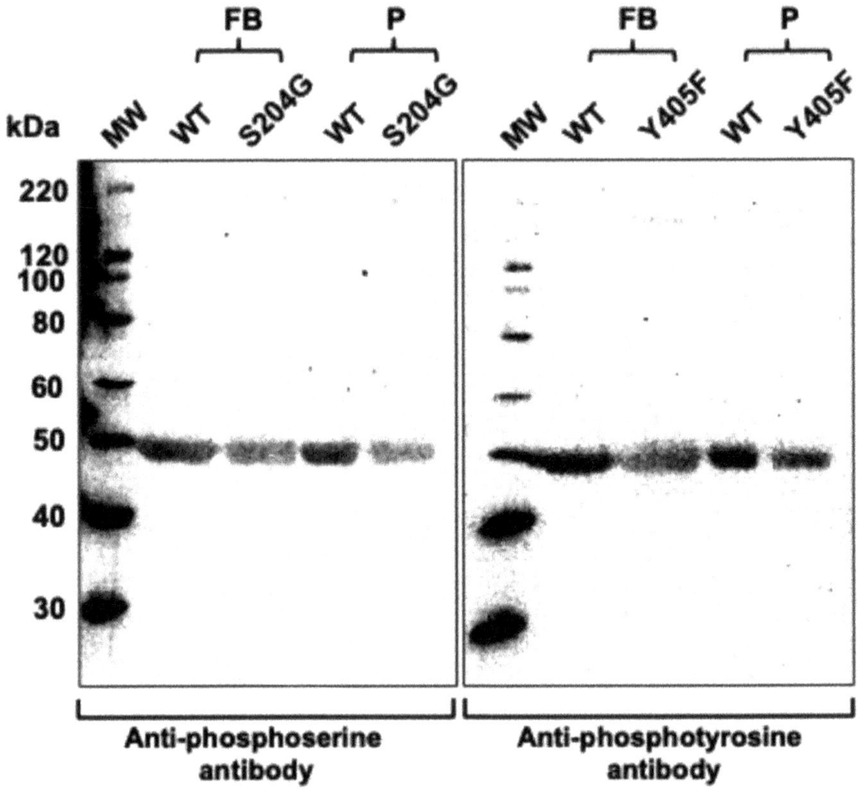

Fig. 5.14 Western blot of purified Δ39aa-hMCD and mutants: WT (wild-type), S204G (serine 204 mutation), and Y405F (tyrosine 405 mutation). FB and P indicate fat body and pupae proteins. Mouse monoclonal anti-phosphoserine and anti-phosphotyrosine antibodies were used with horseradish peroxidase (HRP)-conjugated secondary antibody. Each well held 5 μg protein. (Adopted with permission (Hwang et al. 2015). Copyright 2017 Springer Nature)

of an N-terminal helical domain that enables oligomerization and a C-terminal domain responsible for catalysis. The enzyme's active site is situated within a prominent groove containing highly conserved residues across evolution (Fig. 5.15a) (Aparicio et al. 2013; Froese et al. 2013). The four subunits of the tetramer are linked by inter-subunit disulfide bonds between Cys-206 and Cys-243 residues (Fig. 5.15b), which support positive cooperativity in the enzyme's decarboxylase activity (Aparicio et al. 2013).

The phosphorylated residue Ser-204 is found at the start of the catalytic domain. In the monomeric structure, the side chain of Ser-204 is completely exposed to the solvent (Fig. 5.15a, c). It is approximately 20 Å from the active site, indicating that its phosphorylation likely does not directly affect the structural integrity of the catalytic site. Nevertheless, the quaternary structure of hMCD places Ser-204 close to Cys-206, a critical residue for forming inter-subunit disulfide bonds (Fig. 5.15c). Therefore, phosphorylation or dephosphorylation of Ser-204 may influence the

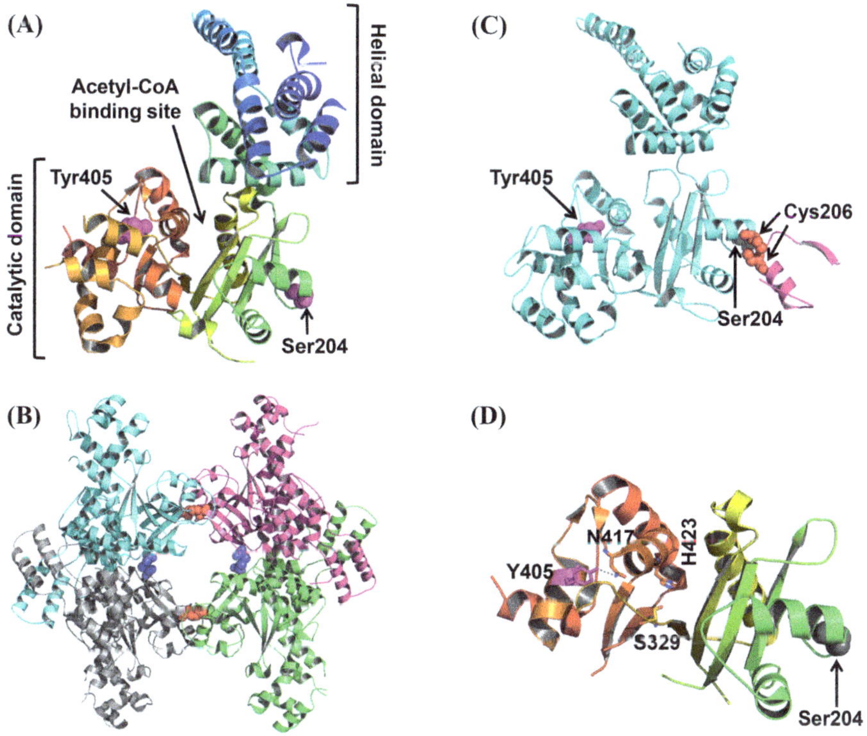

Fig. 5.15 Molecular structure of human peroxisomal MCD (Protein Data Bank accession: 2YGW). (**a**) An MCD monomer consisting of an N-terminal helical domain and a C-terminal catalytic domain containing the acetyl-CoA binding site. (**b**) The structural organization of the MCD tetramer. Disulfide bonds between subunits—Cys-206 to Cys-206 (marked in red) and Cys-243 to Cys-243 (marked in blue)—connect the four subunits. (**c**) MCD structure with phosphorylated Ser-204 and Tyr-405 residues, where Cys-206 forms a disulfide bond linking two subunits. (**d**) Tyr-405 is positioned near the catalytic dyad of Ser-329 and His-423, and creates a hydrogen bond with Asn-417. The cartoon models of MCD were created using PyMol. (Adopted with permission (Hwang et al. 2015). Copyright 2017 Springer Nature)

formation of these disulfide bridges, thereby affecting the enzyme's catalytic function. Additionally, Lys-210 in rats (Lys-211 in humans) has been shown to be essential for MCD enzyme activity via acetylation (Nam et al. 2006), suggesting that post-translational modifications near Cys-206 may play a significant role in regulating MCD function.

Unlike Ser-204, Tyr-405 is not located near any inter-subunit interactions. Instead, this residue is positioned close to the catalytic dyad formed by His-423 and Ser-329. The hydroxyl group of Tyr-405 creates a hydrogen bond with the side chain of Asn-417, which is part of the same α-helix that contains catalytic His-423 (Fig. 5.15d). Based on this structural arrangement, phosphorylation of Tyr-405 could potentially trigger a conformational change in the catalytic dyad, thereby fine-tuning the decarboxylase activity.

5.3.6 Discussion

ACC is a key enzyme in lipid regulation, influencing both fatty acid synthesis and oxidation. Its activity—mediated by a biotin group—is allosterically controlled by phosphorylation, dephosphorylation, and interactions with proteins such as Spot14 and Mig12. Given ACC2's role in metabolic disorders, it is considered a promising therapeutic target and requires authentic recombinant proteins for research. Human ACC2 produced in silkworms exhibited proper post-translational biotinylation and phosphorylation, indicating functional integrity. The recombinant ACC2 also demonstrated citrate-induced catalytic activity and polymer formation, validating its utility for structural and drug discovery studies. Biotin acts as a vital cofactor for five human carboxylases, including hACC2, where it attaches to lysine 929 in the BCCP domain to form carboxybiotin. Silkworm-expressed hACC2 showed effective biotinylation without extra biotin supplementation, marking a first for this expression system. The ACC is regulated by kinases and phosphatases; phosphorylation inhibits activity, while dephosphorylation restores it. Specifically, mutating the key serine residue increases the ACC2 activity, and the dephosphorylated hACC2 showed double the activity of the phosphorylated form. Citrate further activates the ACC but requires concentrations above physiological levels. The silkworm-produced ACC2 was properly folded, modified, catalytically active, and ideal for large-scale protein production.

Malonyl-CoA decarboxylase (MCD) converts malonyl-CoA to acetyl-CoA, impacting fatty acid and glucose metabolism. Modifying MCD expression alters metabolic outcomes, making it a therapeutic target. The silkworm expression system-derived human MCD has superior activity compared to *E. coli*-derived MCD, due to appropriate post-translational modifications. Two phosphorylation sites, Ser204 and Tyr405, were identified as critical regulators of its activity. AMPK phosphorylation activates MCD, lowering malonyl-CoA levels, while dephosphorylation inhibits it. Mutations at Ser204 and Tyr405 reduce MCD activity, indicating their regulatory significance. The silkworm-based protein expression system reliably produces hMCD with the necessary modifications for functional studies and potential drug development.

5.4 Glycosylation

Glycosylation is a critical factor affecting the functionality of various therapeutic proteins (Werner et al. 2007) and is one of the predominant post-translational modifications in eukaryotic cells (Jenkins and Curling 1994). The carbohydrate component of glycoproteins modulates solubility, structural integrity, enzymatic activity, secretion, and cellular processing (Varki 1993). Glycan structures are determined by the repertoire of glycosylating enzymes present in the endoplasmic reticulum and Golgi apparatus of the eukaryotic host cell (Werner et al. 2007). Insect cell and larval platforms have been utilized as hosts to produce a diverse array of proteins. These systems are capable of generating heterologous proteins at elevated levels and cost-effectively, while also facilitating post-translational modifications. Typically, the baculovirus–insect cell/larvae system enables transient expression of the target gene during the early phase following infection, after which the host cell enters a lytic phase, releasing intracellular contents into the culture medium. During this process, secreted proteins may exhibit incomplete post-translational modifications such as glycosylation. Notably, *N*-linked glycans produced in insect tissue cells generally lack terminal *N*-acetylglucosamine (GlcNAc), galactose (Gal), or sialic acid residues (Jenkins et al. 1996), and instead predominantly display high mannosidic and paucimannosidic glycan structures.

This chapter confirms glycosylation in silkworms and insect cells by expressing the model protein, β1,3-*N*-acetylglucosaminyltransferase 2 (β3GnT2) in both, and compares it with that from mammalian cells. Based on these findings, future prospects and possibilities for glycosylation in silkworms were discussed.

5.4.1 β3GnT2 Expression in Insect Cells and Silkworms

Tn-pXme11CNX6 cells that were stably transformed were grown for 5 days, after which the culture supernatant was harvested. For the silkworm larvae, hemolymph was collected 6 days following infection. The β3GnT2 enzyme activities measured in the hemolymph and supernatant were 32 mU/mL and 11 mU/mL, respectively (Dojima et al. 2009). The amount of β3GnT2 expressed in the silkworm hemolymph was nearly three times higher than that found in the supernatant from the stably transformed cell culture. SDS-PAGE analysis was used to examine the degradation products of GFP_{uv}-fused β3GnT2 (GFP_{uv}-β3GnT2) in both the culture supernatant and silkworm hemolymph (lane 1 of Fig. 5.16a, b). While low-molecular-weight fluorescent protein bands appeared in the cell culture supernatant, there was no evidence of fragmentation of GFP_{uv}-β3GnT2, consistent with results from the baculovirus expression system (Kato et al. 2004).

The culture supernatant (lane 1 of Fig. 5.16a) was diluted twofold and applied to Ni^{2+}-affinity chromatography. Approximately 46% of β3GnT2 activity was bound to the column, and the bound GFP_{uv}-β3GnT2 was eluted using a linear imidazole gradient. The purified sample displayed two prominent bands: GFP_{uv}-β3GnT2 (75 kDa) and a contaminant (lane 3 of Fig. 5.16a). The purity of the GFP_{uv}-β3GnT2 pool of *N*-glycoproteins was evaluated by lectin blotting. ConA detected a single GFP_{uv}-β3GnT2 band (lane 4 of Fig. 5.16a), indicating no contamination by Man-terminated *N*-glycans. The purified sample was dialyzed against distilled water overnight and then lyophilized for *N*-glycan analysis. One milligram of β3GnT2 fusion protein was obtained from 280 mL of culture supernatant, yielding a recovery rate of 8.8%. The specific β3GnT2 activity was increased 52-fold compared to the original supernatant (Dojima et al. 2009).

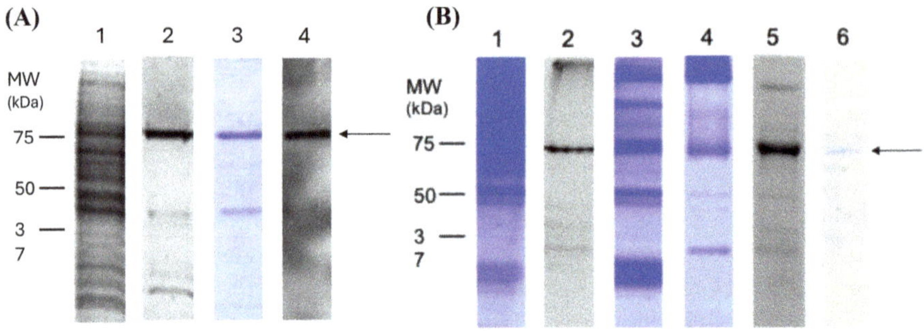

Fig. 5.16 Validation of GFP_{uv}-β3GnT2 in Tn-pXme11CNX6 cells (**a**) and silkworm larval hemolymph (**b**). In **a**, lanes 1–4 show: cell culture supernatant stained with Coomassie Blue, supernatant imaged with Molecular Image FX, purified GFP_{uv}-β3GnT2 from Ni^{2+} affinity chromatography detected by Coomassie Blue, and lectin blot of purified GFP_{uv}-β3GnT2 (Ni^{2+} affinity) detected with ConA. In **b**, lanes 1–4 show: hemolymph stained with Coomassie Blue, hemolymph imaged with Molecular Image FX, 50% ammonium sulfate precipitate of GFP_{uv}-β3GnT2 detected by Coomassie Blue, Ni^{2+} affinity chromatography eluent analyzed by Coomassie Blue, lectin blot of Ni^{2+} affinity eluent detected with ConA, and purified GFP_{uv}-β3GnT2 from size exclusion chromatography detected by Coomassie Blue. (Adapted with permission (Dojima et al. 2009). Copyright 2009 Elsevier)

Twenty-five milliliters of hemolymph was collected and diluted fivefold with 20 mM phosphate buffer at pH 7.4 and then subjected to precipitation using 40–50% saturated ammonium sulfate (lane 3 of Fig. 5.16b). The GFP_{uv}-β3GnT2-containing fraction was dialyzed against 20 mM phosphate buffer (pH 7.4) and subsequently applied to immobilized Ni^{2+} affinity chromatography. Elution was carried out with a stepwise increase in imidazole concentrations from 100 to 500 mM. The fractions that exhibited fluorescence were then analyzed for β3GnT2 enzyme activity and by SDS-PAGE (lane 4 of Fig. 5.16b). *N*-linked glycoprotein contamination was evaluated by lectin blotting, which revealed several bands detected by ConA (lane 5 of Fig. 5.16b). To further purify the sample, size-exclusion chromatography was performed using a Sephacryl S-200 HR column. The fractions eluted with 20 mM phosphate buffer were analyzed by SDS-PAGE (lane 6 of Fig. 5.16b). Fractions displaying a single GFP_{uv}-β3GnT2 band were dialyzed against pure water and then lyophilized for *N*-glycan analysis. Through these purification steps, 0.5 mg of the β3GnT2 fusion protein was isolated from 25 ml of silkworm hemolymph, yielding a 5% recovery. The specific activity of β3GnT2 showed a 155-fold increase compared to the starting hemolymph (Dojima et al. 2009).

5.4.2 *Characterization of* N-*Glycans*

The reducing ends of *N*-glycans, which were released from purified GFP_{uv}-β3GnT2 by glucoamidase A, were subjected to reductive amination with 2-aminopyridine (PA). The resulting PA-labeled oligosaccharides were separated on an ODS column, with their profiles shown in Fig. 5.17.

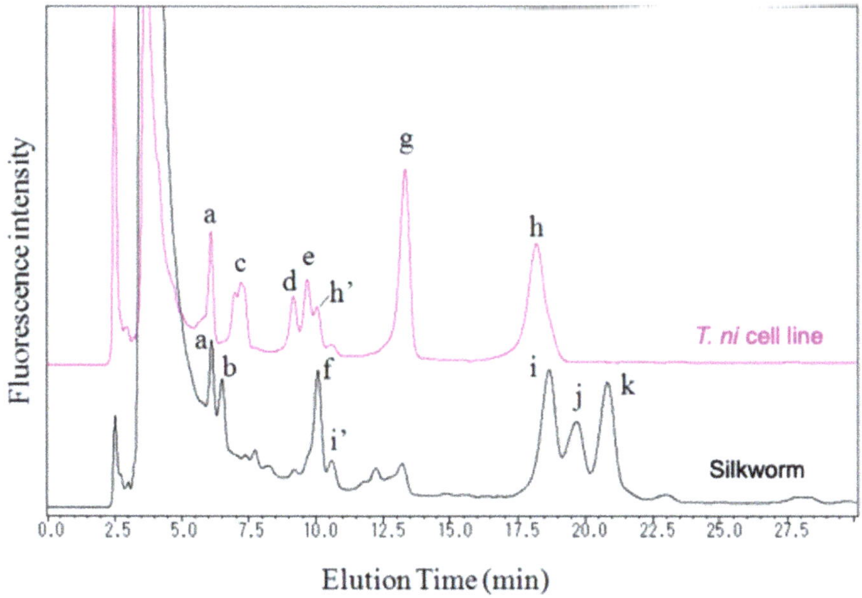

Fig. 5.17 *N* glycosylation patterns of recombinant β3GnT2 produced in *T. ni* cell lines (black) and silkworm larvae (pink), analyzed using an ODS column. The purified protein was treated with glycoamidase A to release *N*-glycans, whose reducing ends were then labeled with 2-aminopyridine (PA). By-products generated during the pyridylamination process are marked with a prime symbol, such as h' and i'. (Adopted with permission (Dojima et al. 2009). Copyright 2009 Elsevier)

Each isolated fraction, labeled a through k, was further analyzed using an amide column. The structures were determined by HPLC mapping, as detailed in Table 5.1. The β3GnT2 sample purified from silkworm larvae included GlcNAcβ1,2Manα1,3(GlcNAcβ1,4)(Manα1,6)Manβ1,4GlcNAcβ1,4GlcNAc (16.2%) and Galβ1,4GlcNAcβ1,2Manα1,3(GlcNAcβ1,4)(Manα1,6) Manβ1,4GlcNAcβ1,4GlcNAc (21.3%). These structures were not detected in samples derived from Tn-pXme11CNX6 cells. This suggests that the enzymes β1,2 *N*-acetylglucosaminyltransferase I (GlcNAcT I), β1,4 *N*-galactosyltransferase (GalT), and β1,4 *N*-acetylglucosaminyltransferase III (GlcNAcT III) may be present in silkworm larvae but are absent in Tn-pXme11CNX6 cells. The *N*-glycan structures are summarized in Table 5.2. The various glycoforms were categorized as high mannose (containing more than three mannose residues), paucimannose (with two core GlcNAc residues and three or fewer mannose residues), and those with terminal sugars other than mannose, such as GlcNAc or galactose. High-mannose structures accounted for 19.4% in Tn-pXme11CNX6 and 9.3% in silkworm larvae samples. The Tn-pXme11CNX6 sample had higher amounts of glycans containing three to seven mannose residues without fucosylation (23.6%) compared to silkworm larvae (0.9%). Most paucimannosidic glycans differed, particularly regarding the presence or absence of fucose residues. The *N*-glycans expressed in Tn-pXme11CNX6 were

Table 5.1 Comparison of glycosylation between silkworm and insect cells

Structure	Relative quantity (%)[a]	
	Silkworm	*T. ni* cell
Manα1-2Manα1-6 Manα1-6 Manα1-2Manα1-3 Manβ1-4GlcNAcβ1-4GlcNAc Manα1-2Manα1-2Manα1-3	4.5	6.8
Manα1-2Manα1-6 Manα1-6 Manα1-3 Manβ1-4GlcNAcβ1-4GlcNAc Manα1-2Manα1-2Manα1-3	3.9	-
Manα1-6 Manα1-6 Manα1-3 Manβ1-4GlcNAcβ1-4GlcNAc Manα1-2Manα1-3 Manα1-6 Manα1-6 Manα1-3 Manβ1-4GlcNAcβ1-4GlcNAc Manα1-2Manα1-2Manα1-3	-	11.7
Manα1-6 Manα1-6 Manα1-3 Manβ1-4GlcNAcβ1-4GlcNAc Manα1-3	0.9	5.9
Manα1-6 Manβ1-4GlcNAcβ1-4GlcNAc Manα1-3	-	11.0
Manα1-6 Manβ1-4GlcNAcβ1-4GlcNAc	14.0	4.0
Fucα1-6 Manα1-6 Manβ1-4GlcNAcβ1-4GlcNAc Manα1-3 Fucα1-3	-	30.8
Fucα1-6 Manα1-6 Manβ1-4GlcNAcβ1-4GlcNAc	-	22.2
Fucα1-6 Manα1-6 Manβ1-4GlcNAcβ1-4GlcNAc Manα1-3	29.4	-
Fucα1-6 Manα1-6 Manβ1-4GlcNAcβ1-4GlcNAc GlcNAcβ1-2Manα1-3	16.2	-
Fucα1-6 Manα1-6 Manβ1-4GlcNAcβ1-4GlcNAc Galβ1-4GlcNAcβ1-2Manα1-3	21.3	-
Others	12.5	11.5

[a] Molar percentage of each glycan.

This table was adapted with permission (Dojima et al. 2009).

Table 5.2 Glycosylation of β3GnT2 by Tn-pXme11CNX6 cells and silkworm larvae

Glycan type (%)	Tn – pXme11CNX6	Silkworm larvae
High mannosidic *N*-glycan	19.4	9.3
Paucimannosidic *N*-glycan	73.2	43.4
GlcNAc residue on Manα(1,3) branch	ND	16.2
Gal residue on Manα(1,3) branch	ND	21.3
Fucosylated *N*-glycans	58.2	29.4

This table was adapted with permission (Dojima et al. 2009). Copyright 2009 Elsevier

fucosylated at the α(1,3) and α(1,6) positions (36.0%). In silkworm larvae, the *N*-glycan structure was primarily α(1,6) fucosylated (29.4%). The proportion of fucosylated *N*-glycans was 58.2% in Tn-pXme11CNX6 and 29.4% in silkworm larvae.

5.4.3 N-*Glycan Modification of a Recombinant Protein in Silkworms*

To enhance the *N*-glycan profiles of recombinant glycoproteins produced in insect cells, *Caenorhabditis elegans* β-1,2-*N*-acetylglucosaminyltransferase II (GnT II) and bovine β1,4-galactosyltransferase I (GalT I) have been coexpressed using the Multibac system to achieve mammalian-like glycosylation (Palmberger et al. 2012). An insect cell line capable of producing complex *N*-glycans containing sialic acid was developed by coexpressing five mammalian glycosyltransferases, such as human GnT I, human GnT II, bovine GalT I, rat α2,6-sialyltransferase I (ST6GalT I), and α2,3-sialyltransferase IV (ST3Gal IV) (Hollister et al. 2002). Additionally, insect cell lines that synthesize cytidine-5′-monophospho-*N*-acetylneuraminic acid (CMP-Neu5Ac) and efficiently transport it to the Golgi have been established to improve sialic acid incorporation (Geisler and Jarvis 2012; Mabashi-Asazuma et al. 2013; Viswanathan et al. 2015). Furthermore, *N*-acetylglucosaminidase (FDL) is essential for generating pauci-mannose-type *N*-glycans in insect cells (Geisler et al. 2008; Leonard et al. 2006). Suppressing FDL expression increases GlcNAc-terminated *N*-glycans in recombinant glycoproteins expressed in insect cells (Kim et al. 2009; Mabashi-Asazuma et al. 2015a).

Like other insect cells, most *N*-glycans on glycoproteins from silkworms are of the paucimannose type. In the posterior silk gland (PSG), up to 17% of *N*-glycans are terminally galactosylated in transgenic silkworm larvae coexpressing human GnT II and bovine GalT I under a PSG-specific promoter (Mabashi-Asazuma et al. 2015b). This indicates that mammalian-like *N*-glycans can be produced in the PSG by expressing mammalian glycosyltransferases. Here, human GnT II (hGnT II) and human GalT I (hGalT I) were coexpressed with human immunoglobulin G (hIgG) in silkworm pupae to generate galactosylated *N*-glycans on hIgG.

5.4.3.1 Modification of *N*-Glycan by hGnT II Coexpression in Silkworms

Human glycosyltransferases were expressed in silkworm pupae using either the AcMNPV *polh* promoter or the *B. mori* actin A3 promoter. The hGnT II and hGalT I genes were inserted downstream of their respective promoters in their respective vectors. Recombinant BmNPV (BmNPV CP^{--}-Chi^{-}/P_{Act}-hGnT II, BmNPV CP^{--}-Chi^{-}/P_{Pol}-hGnT II, BmNPV CP^{--}-Chi^{-}/P_{Act}-hGalT I, and BmNPV CP^{--}-Chi^{-}/P_{Pol}-hGalT I) bacmids were used for coexpression. For hIgG expression, the BmNPV CP^{--}-Chi^{-}/29IJ6 hIgG bacmid was used (Park et al. 2009). Each recombinant bacmid was injected into silkworm larvae, BmNPV titers in hemolymph were measured, and the hemolymph was collected as recombinant BmNPV stock. hGnT II was expressed in silkworm pupae using either BmNPV CP^{--}-Chi^{-}/P_{Act}-hGnT II or BmNPV CP^{--}-Chi^{-}/P_{Pol}-hGnT II, as confirmed by Western blotting (Fig. 5.18a).

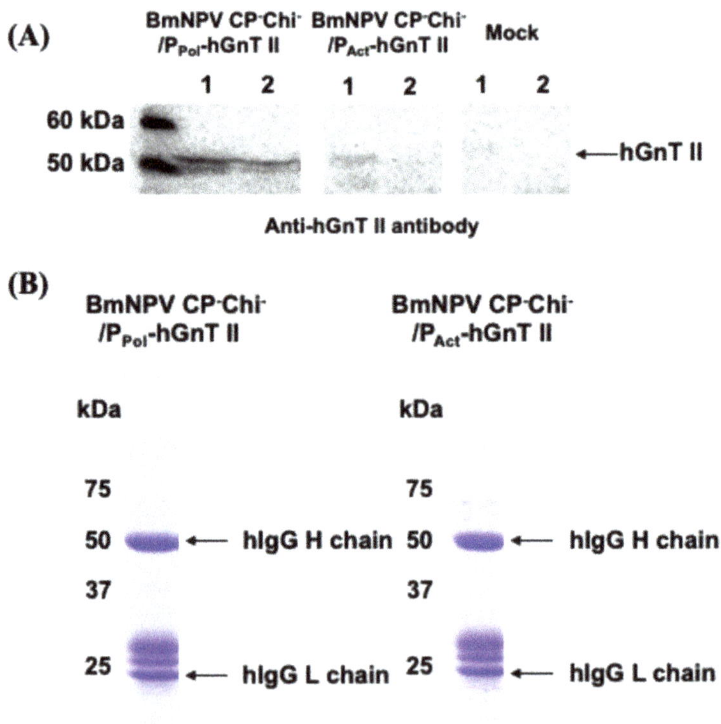

Fig. 5.18 Recombinant protein expression in silkworm pupae coexpressing hGnT II (Kato et al. 2017). (**a**) hGnT II was expressed in silkworm pupae using BmNPV CP^{--}-Chi^{-}/P_{Act}-hGnT II or BmNPV CP^{--}-Chi^{-}/P_{Pol}-hGnT II. Pupae were homogenized, centrifuged, and both supernatant and pellet fractions were analyzed by Western blot. Lanes 1 and 2 denote supernatant and pellet, respectively. (**b**) SDS-PAGE and CBB staining of hIgG purified from BmNPV CP^{--}-Chi^{-}/29IJ6 hIgG-injected pupae show protein coexpression with hGnT II under *polh* or actin A3 promoters

Specific bands at about 50 kDa were observed—higher than the predicted 48 kDa due to two potential *N*-glycosylation sites. No bands appeared in mock pupae. Expression was higher with the polyhedrin promoter than with the actin A3 promoter. To coexpress hIgG and hGnT II in silkworm pupae, BmNPV *CP*$^{--}$-*Chi*$^{-}$/29IJ6 IgG and either recombinant BmNPV (BmNPV *CP*$^{--}$-*Chi*$^{-}$/P_{Act}-hGnT II or BmNPV *CP*$^{--}$-*Chi*$^{-}$/P_{Pol}-hGnT II) were injected into pupae at a 1:1 pfu ratio (total 5×10^5 pfu/pupa). The resulting hIgG, coexpressed with hGnT II, was purified by Protein A Sepharose chromatography. Both heavy and light chains of hIgG were detected, as well as additional bands between them (Fig. 5.18b), likely degraded heavy chains, since the antibody reacted with some hemolymph proteins. This purified hIgG was then used for *N*-glycan structure analysis.

N-glycans were enzymatically extracted from purified human IgG, labeled with 2-aminopyridine (PA), and separated by ODS and amide column chromatography. The chromatographic profiles from the ODS column are presented in Fig. 5.19a. Notably, when human IgG was coexpressed with human GnT II, an additional peak (N8) was detected, absent in samples lacking hGnT II coexpression. Figure 5.19b details the specific *N*-glycan structures isolated from the purified IgG. For hIgG obtained from silkworm pupae, terminal GlcNAc residues were not identified on the *N*-glycans. However, coexpression of hIgG with hGnT II driven by the actin A3 promoter or the *polh* promoter resulted in 14.2% and 15.4% of *N*-glycans, respectively, being terminated with GlcNAc residues (N8: $GlcNAc_2Man_3GlcNAc_2$; N11:$GlcNAc_2Man_3GlcNAc(Fuc)GlcNAc$). These findings demonstrate that the introduced GnT II was capable of modifying hIgG *N*-glycans, leading to the formation of GlcNAc-terminated *N*-glycans in silkworm pupae.

(A)

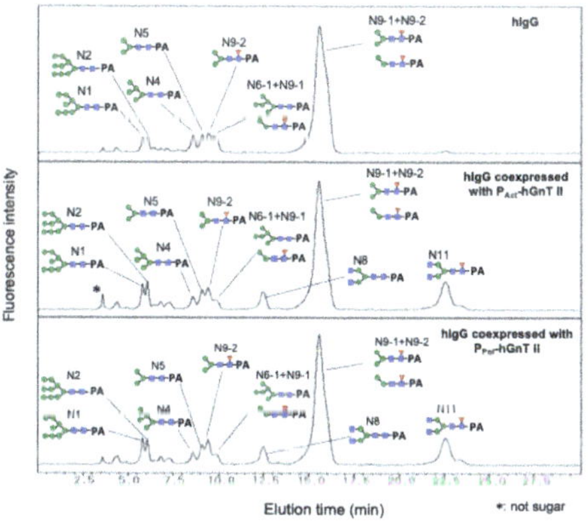

(B)

Peak No.	Structure	Mass Value	Without coexpression	Ratio (%) Coexpression with	
				P_{Act}-hGnT II	P_{Pol}-hGnT II
N1		1800	2.2	3.7	3.1
N2		1962	1.8	3.1	3.4
N4		1313	2.6	1.6	1.4
N5		989	2.8	3.1	3.0
N6-1		1475	2.4	0.5	0.4
N8		1395	-	4.3	4.5
N9-2		1135	65.2	61.0	58.5
N9-1		973	22.1	12.3	13.9
N11		1541	-	9.9	10.9
	Others		0.9	0.7	1.0

Fig. 5.19 (**a**) Chromatograms of PA-labeled *N*-glycans from recombinant IgG coexpressed with hGnT II in silkworms, analyzed using ODS column chromatography (Kato et al. 2017). N1–N11 refer to each *N*-glycan. (**b**) Structures of *N*-glycans from hIgG coexpressed with hGnT II in silkworm pupae, corresponding to N1–N11 identified in **a**

5.4.3.2 Modification of *N*-Glycan by Coexpression of hGnT II and hGalT I in Silkworms

The hGalT I was expressed in silkworm pupae using BmNPV *CP*$^{--}$-*Chi*$^{-}$/P_{Act}-hGalT I or BmNPV *CP*$^{--}$-*Chi*$^{-}$/P_{Pol}-hGalT I, and distinct bands at 40 kDa (expected 39 kDa) in Western blot were observed in both supernatant and pellet fractions from pupae extracts (Fig. 5.20a), indicating successful recombinant hGalT I expression. For terminal galactosylation of hIgG N-glycans in silkworm pupae, hIgG was coexpressed with hGnT II and hGalT I in three promoter combinations: (1) hGnT II under the actin A3 promoter (P_{Act}-hGnT II) with hGalT I under the *polh* promoter (P_{Pol}-hGalT I); (2) hGnT II under the *polh* promoter (P_{Pol}-hGnT II) with hGalT I under the actin A3 promoter (P_{Act}-hGalT I); and (3) both under the polyhedrin promoter (P_{Pol}-hGnT II and P_{Pol}-hGalT I). Recombinant BmNPVs were injected into pupae at a 1:3:3 ratio (IgG:hGnT II:hGalT I, total 5×10^{5} pfu/pupa). Purification of hIgG from pupae via Protein A Sepharose confirmed the presence of both heavy and light chains in all combinations (Fig. 5.20b).

Lectin blot analysis with FITC-labeled RCA120 confirmed the presence of terminal galactose on the *N*-glycans of hIgG. A specific band appeared when hIgG was coexpressed with either P_{Act}-hGnT II and P_{Pol}-hGalT I or P_{Pol}-hGnT II and P_{Pol}-hGalT I. No band was detected when P_{Pol}-hGnT II and P_{Act}-hGalT I were coexpressed, even though hGalT I was expressed (Kato et al. 2017). This shows that combining P_{Pol}-hGnT II with P_{Act}-hGalT I does not lead to galactosylation in silkworm pupae. For

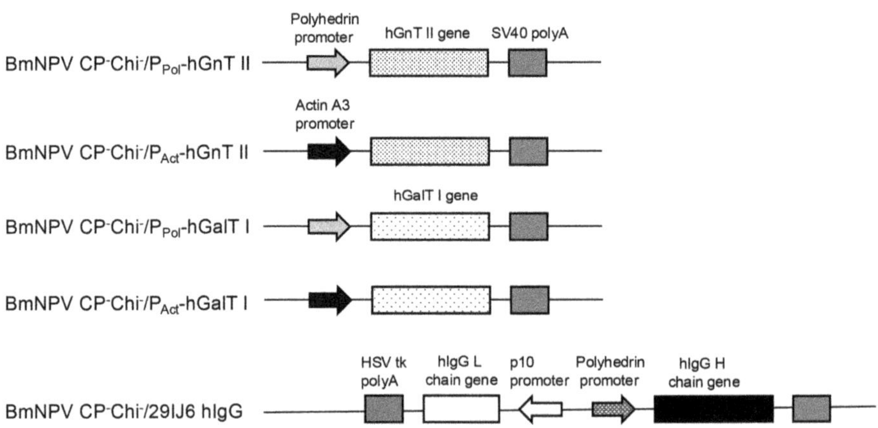

Fig. 5.20 Production of recombinant hIgG in silkworm pupae with coexpression of hGnT II and hGalT I (Kato et al. 2017). (**a**) hGalT I was expressed in silkworm pupae using either BmNPV *CP*$^{-}$-*Chi*$^{-}$/PAct-hGalT I or BmNPV *CP*$^{-}$-*Chi*$^{-}$/PPol-hGalT I. After homogenization with TBST and centrifugation, both the soluble supernatant and insoluble pellet fractions were analyzed by Western blotting. Lanes 1 and 2 indicate the supernatant and pellet, respectively. (**b**) SDS-PAGE stained with CBB results for purified hIgG produced with coexpression of PPol-hGnT II and PPol-hGal T I (left), PPol-hGnT II and PAct-hGal T I (center), and PAct-hGnT II and PPol-hGal T I (right)

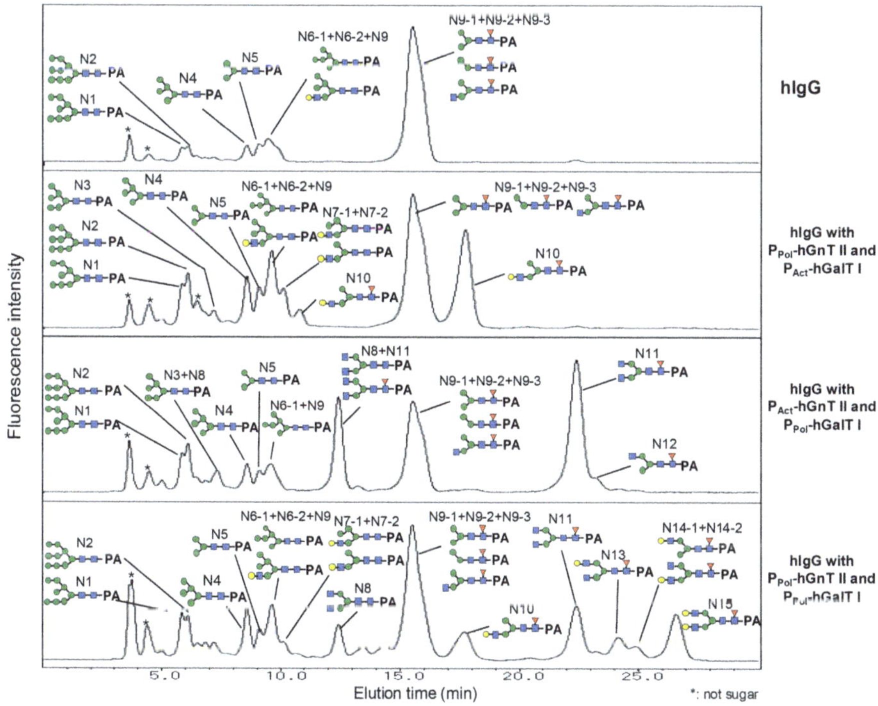

Fig. 5.21 Chromatograms of PA-derivatized *N*-glycans obtained from recombinant hIgG produced in silkworms coexpressing hGnT II and hGalT I, as analyzed by ODS column chromatography (Kato et al. 2017). N1 to N15 indicate individual *N*-glycans identified

detailed *N*-glycan analysis, PA-*N*-glycans from each recombinant hIgG were separated by ODS and amide column chromatography. In silkworm pupae coexpressing hIgG with hGnT II and hGalT I, new peaks (N10–N15) appeared at later retention times (Fig. 5.21), but the chromatograms differed depending on the specific coexpression combinations. Structural analysis (Fig. 5.22) showed that with P_{Act}-hGnT II and P_{Pol}-hGalT I, the mono-galactosylated *N*-glycan (N10: GalGlcNAcMan$_3$GlcNAc(Fuc)GlcNAc) was present at 26.7%, but no biantennary galactosylated *N*-glycans were found. In contrast, with P_{Pol}-hGnT II and P_{Pol}-hGalT I, a biantennary galactosylated *N*-glycan (N15: Gal$_2$GlcNAc$_2$Man$_3$GlcNAc(Fuc) GlcNAc) was detected at 11.4%. Several mono-galactosylated *N*-glycans (N6-2, N7-1, N10, N13, N14-1, N14-2) were also observed. No terminally galactosylated *N*-glycans were detected when P_{Pol}-hGnT II and P_{Act}-hGalT I were coexpressed, matching the lectin blot results. In this combination, 52.2% of the *N*-glycans had biantennary terminal GlcNAc residues (N11, N8). These findings suggest the *polh* promoter is effective for coexpressing both hGnT II and hGalT I to produce biantennary galactosylated *N*-glycans in silkworm pupae. However, pauci-mannose and high-mannose *N*-glycans were still found at 34.1% and 14.3%, respectively, with P_{pol}-hGnT II and P_{pol}-hGalT I. Additionally, a mono-terminally GlcNAcated

Peak No.	Structure	Mass Value	Ratio (%)			
			Without coexpression	Coexpression with		
				P_{Act}-hGnT II P_{Pol}-GalT I	P_{Pol}-hGnT II P_{Act}-GalT I	P_{Pol}-hGnT II P_{Pol}-hGalT I
N1	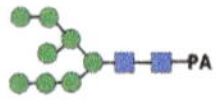	1800	2.2	3.7	2.8	3.7
N2	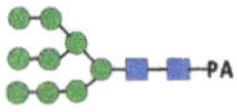	1962	1.9	3.6	3.6	2.9
N3	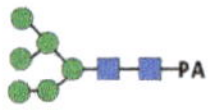	1475	-	0.8	0.7	-
N4	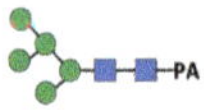	1313	2.6	4.5	2.3	4.9
N5	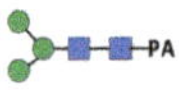	989	2.8	2.9	1.7	2.0
N6-1	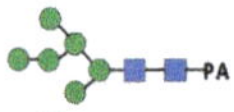	1475	2.4	2.7	2.0	2.8
N6-2	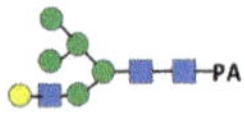	1679	-	4.3	-	2.0
N7-1	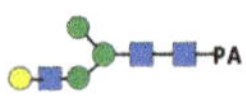	1354	-	2.0	-	0.5
N7-2	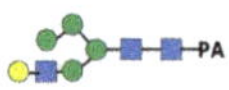	1516	-	1.7	-	0.8
N8	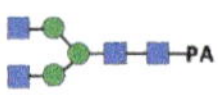	1395	-	-	13.5	4.5
N9-1	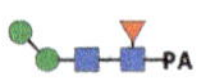	973	18.7	11.1	7.6	5.8
N9-2	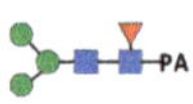	1135	49.4	25.7	18.2	26.3
N9-3	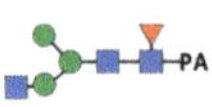	1338	19.1	10.1	6.2	7.6
N10	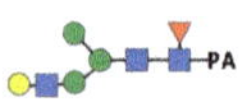	1500	-	26.7	-	7.0
N11	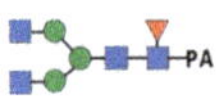	1541	-	-	38.7	11.0
N12	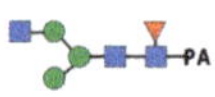	1338	-	-	2.3	-
N13	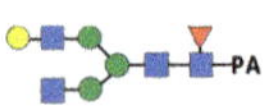	1704	-	-	-	3.9
N14-1	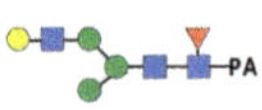	1500	-	-	-	0.9
N14-2	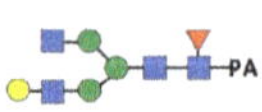	1704	-	-	-	1.3
N15	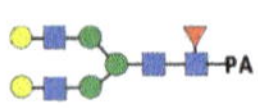	1866	-	-	-	11.4
	Others		0.9	0.7	0.4	0.7

Fig. 5.22 *N*-glycan profiles of hIgG coexpressed with both hGnT II and hGalT I in silkworm pupae (Kato et al. 2017)

N-glycan (N9-3: $GlcNAcMan_3GlcNAc(Fuc)GlcNAc$) was observed even when only BmNPV *CP*$^{--}$-*Chi*$^{-}$/29IJ6 hIgG was expressed. Pauci- and oligomannosidic glycans were often produced on recombinant hIgG in silkworm pupae without coexpressed human glycosyltransferases, a pattern also seen in other insect-expressed proteins (Farrell et al. 1998; Nomura et al. 2015; Tomiya et al. 2004). This indicates that silkworm pupae can generate complex *N*-glycans on recombinant hIgG, though at somewhat lower levels than in other systems.

5.4.4 Discussion

The baculovirus–insect cell system is widely used for recombinant protein production in insect cells, but its commercial applications are limited due to differences in glycosylation compared to mammalian cells. Insect cell *N*-glycans are typically modified to paucimannosidic or oligomannosidic forms (Tomiya et al. 2004) rather than terminally sialylated complex types, mainly because membrane-bound β1,2 *N*-acetylglucosaminidase (GlcNAcase) removes the terminal *N*-acetylglucosamine during intracellular transport (Altmann et al. 1995). Still, some reports show terminal galactose on insect-produced glycoproteins (Hsu et al. 1997; Joshi et al. 2000). Most *N*-glycans of a β3GnT2 fusion protein from stably transformed *T. ni* cells were paucimannosidic (73.2%), processed in the ER and Golgi but degraded by GlcNAcase (Altmann et al. 1999). These cells also produced allergenic α1,3-fucose in a fucosylated trimannosidic structure (36.0%). In silkworms, Gal and GlcNAc residues were found, indicating active GlcNAcT I and GalT enzymes. Previous studies show recombinant proteins in *B. mori* larvae are mainly fucosylated oligosaccharides. This study found a $Man_2GlcNAc(Fuc)GlcNAc$ structure on 29.4% of β3GnT2 from silkworms.

Manα(1,3) branches with bisecting GlcNAc, but no biantennary complex-type structures. GlcNAcT III may regulate glycosylation by adding bisecting GlcNAc, which blocks further modification. The *N*-glycosylation pathway in *T. ni* and silkworms resembles that in mammals, but alternative routes exist (Fig. 5.23). Insect-specific GlcNAcase removes terminal GlcNAc residues, and enzymes like α-mannosidase III convert $Man_5GlcNAc_2$ to $Man_2GlcNAc_2$ without GlcNAc addition. Fucosyltransferase requiring GlcNAc(1,2) on Man(1,3) has not been cloned from lepidopterans, suggesting different processing routes for fucosylated and non-fucosylated structures. Bisecting structures like Galβ1,4GlcNAcβ1,2Man are atypical for lepidopteran *N*-glycosylation. In silkworms, galactose terminal residues are found only in nonfucosylated glycans with bisecting GlcNAc, which is beneficial for therapeutic antibody production due to enhanced cytotoxicity (Shinkawa et al. 2003). Thus, silkworm larvae could be improved to produce human glycoproteins by modifying their glycosylation pathways.

An alternative method for producing biantennary, terminally galactosylated N-glycans in silkworms was investigated by coexpressing hIgG with the human glycosyltransferases hGnT II and hGalT I. The experiment utilized different

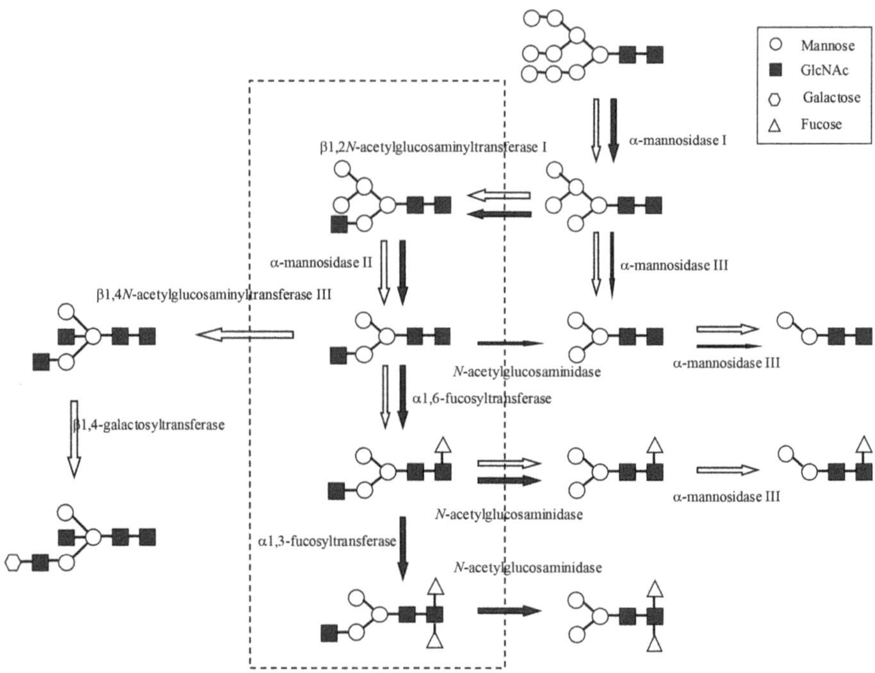

Fig. 5.23 Suggested *N*-glycan processing pathways in the *T. ni* cell line and silkworm larvae are shown, with open arrows for silkworm larvae and closed arrows for the *T. ni* cell line. The N-glycans outlined with dotted lines were not detected by HPLC mapping. (Adopted with permission (Dojima et al. 2009). Copyright 2009 Elsevier)

promoter combinations—actin A3 and *polh*—to drive the expression of these glycosyltransferases. The results demonstrated that using the *polh* promoter for both hGnT II and hGalT I yielded the highest proportion (up to 35%) of biantennary, terminally galactosylated *N*-glycans. In contrast, using the actin A3 promoter for either or both enzymes led to less efficient Golgi-targeted expression and lower yields of the desired glycan structures, often resulting in incomplete glycan processing. This indicates that promoter selection is critical for efficient glycosylation, as insufficient expression in the Golgi can hinder glycosylation. The discussion highlights the importance of proper localization and sequential activity of glycosyltransferases within the endoplasmic reticulum and Golgi for accurate *N*-glycan modification (Tu and Banfield 2010). Overexpression or mislocalization can disrupt this order, underscoring the need to optimize promoter selection and expression levels when coexpressing multiple enzymes (Hashimoto et al. 2002). Additionally, the *polh* promoter's activity during the late infection phase—when the secretory pathway is disrupted (Geisler et al. 2015)—may limit its effectiveness, suggesting further research is needed to refine promoter strategies for higher yields of galactosylated proteins.

The study also notes that, despite coexpression, a significant portion (46.4%) of hIgG *N*-glycans remained as paucimannose or high-mannose types. Prior research has shown that engineering glycosyltransferases, e.g., using bovine GalT I with specific mutations and human cytoplasmic/transmembrane/stem (CTS) domains, can improve Golgi localization and enhance the production of complex glycans (Geisler et al. 2015). The use of optimal CTS domains or orthologous genes may be necessary for efficient glycan processing, as demonstrated in yeast and suggested for silkworms.

Overall, the feasibility and challenges of engineering silkworm pupae to produce mammalian-like glycosylated proteins are demonstrated. The findings emphasize the critical role of promoter selection and glycosyltransferase localization for efficient Golgi modification and the production of complex *N*-glycans. The coinfection method using recombinant BmNPVs enables rapid production, but further optimization—especially regarding promoter choice and enzyme engineering—could enhance yield and glycan complexity. Future research should focus on refining these parameters, exploring alternative promoters and CTS domain configurations, and optimizing infection protocols to advance the production of therapeutic glycoproteins in silkworm systems.

References

Abu-Elheiga L, Jayakumar A, Baldini A, Chirala SS, Wakil SJ (1995) Human acetyl-CoA carboxylase: characterization, molecular cloning, and evidence for two isoforms. Proc Natl Acad Sci USA 92:4011–4015. https://doi.org/10.1073/pnas.92.9.4011

Abu-Elheiga L, Almarza-Ortega DB, Baldini A, Wakil SJ (1997) Human acetyl-CoA carboxylase 2. Molecular cloning, characterization, chromosomal mapping, and evidence for two isoforms. J Biol Chem 272:10669–10677. https://doi.org/10.1074/jbc.272.16.10669

Abu-Elheiga L, Matzuk MM, Abo-Hashema KA, Wakil SJ (2001) Continuous fatty acid oxidation and reduced fat storage in mice lacking acetyl-CoA carboxylase 2. Science 291:2613–2616. https://doi.org/10.1126/science.1056843

Abu-Elheiga L, Oh W, Kordari P, Wakil SJ (2003) Acetyl-CoA carboxylase 2 mutant mice are protected against obesity and diabetes induced by high-fat/high-carbohydrate diets. Proc Natl Acad Sci USA 100:10207–10212. https://doi.org/10.1073/pnas.1733877100

Adham IM, Burkhardt E, Benahmed M, Engel W (1993) Cloning of a cDNA for a novel insulin-like peptide of the testicular Leydig cells. J Biol Chem 268:26668–26672

Altmann F, Schwihla H, Staudacher E, Glossl J, Marz L (1995) Insect cells contain an unusual, membrane-bound beta-N-acetylglucosaminidase probably involved in the processing of protein N-glycans. J Biol Chem 270:17344–17349. https://doi.org/10.1074/jbc.270.29.17344

Altmann F, Staudacher E, Wilson IB, Marz L (1999) Insect cells as hosts for the expression of recombinant glycoproteins. Glycoconj J 16:109–123. https://doi.org/10.1023/a:1026488408951

Aparicio D, Perez-Luque R, Carpena X, Diaz M, Ferrer JC, Loewen PC, Fita I (2013) Structural asymmetry and disulfide bridges among subunits modulate the activity of human malonyl-CoA decarboxylase. J Biol Chem 288:11907–11919. https://doi.org/10.1074/jbc.M112.443846

Assis LH, Crespo D, Morais RD, Franca LR, Bogerd J, Schulz RW (2016) INSL3 stimulates spermatogonial differentiation in testis of adult zebrafish (Danio rerio). Cell Tissue Res 363:579–588. https://doi.org/10.1007/s00441-015-2213-9

Baneyx F, Mujacic M (2004) Recombinant protein folding and misfolding in Escherichia coli. Nat Biotechnol 22:1399–1408. https://doi.org/10.1038/nbt1029

Beaty NB, Lane MD (1983) Kinetics of activation of acetyl-CoA carboxylase by citrate. Relationship to the rate of polymerization of the enzyme. J Biol Chem 258:13043–13050

Bianchi A, Evans JL, Iverson AJ, Nordlund AC, Watts TD, Witters LA (1990) Identification of an isozymic form of acetyl-CoA carboxylase. J Biol Chem 265:1502–1509

Bogatcheva NV, Truong A, Feng S, Engel W, Adham IM, Agoulnik AI (2003) GREAT/LGR8 is the only receptor for insulin-like 3 peptide. Mol Endocrinol 17:2639–2646. https://doi.org/10.1210/me.2003-0096

Brown GK, Scholem RD, Bankier A, Danks DM (1984) Malonyl coenzyme A decarboxylase deficiency. J Inherit Metab Dis 7:21–26. https://doi.org/10.1007/BF01805615

Brownsey RW, Boone AN, Elliott JE, Kulpa JE, Lee WM (2006) Regulation of acetyl-CoA carboxylase. Biochem Soc Trans 34:223–227. https://doi.org/10.1042/BST20060223

Bullesbach EE, Schwabe C (2002) The primary structure and the disulfide links of the bovine relaxin-like factor (RLF). Biochemistry 41:274–281. https://doi.org/10.1021/bi0117302

Bullesbach EE, Schwabe C (2012) Replacement of disulfides by amide bonds in the relaxin-like factor (RLF/INSL3) reveals a role for the A11-B10 link in transmembrane signaling. Biochemistry 51:4198–4205. https://doi.org/10.1021/bi300352x

Cheng D, Chu CH, Chen L, Feder JN, Mintier GA, Wu Y, Cook JW, Harpel MR, Locke GA, An Y, Tamura JK (2007) Expression, purification, and characterization of human and rat acetyl coenzyme A carboxylase (ACC) isozymes. Protein Expr Purif 51:11–21. https://doi.org/10.1016/j.pep.2006.06.005

Cho YS, Lee JI, Shin D, Kim HT, Jung HY, Lee TG, Kang LW, Ahn YJ, Cho HS, Heo YS (2010) Molecular mechanism for the regulation of human ACC2 through phosphorylation by AMPK. Biochem Biophys Res Commun 391:187–192. https://doi.org/10.1016/j.bbrc.2009.11.029

Del Borgo MP, Hughes RA, Bathgate RD, Lin F, Kawamura K, Wade JD (2006) Analogs of insulin-like peptide 3 (INSL3) B-chain are LGR8 antagonists in vitro and in vivo. J Biol Chem 281:13068–13074. https://doi.org/10.1074/jbc.M600472200

Dojima T, Nishina T, Kato T, Uno T, Yagi H, Kato K, Park EY (2009) Comparison of the *N*-linked glycosylation of human beta1,3-N-acetylglucosaminyltransferase 2 expressed in insect cells and silkworm larvae. J Biotechnol 143:27–33. https://doi.org/10.1016/j.jbiotec.2009.06.013

Dyck JR, Barr AJ, Barr RL, Kolattukudy PE, Lopaschuk GD (1998) Characterization of cardiac malonyl-CoA decarboxylase and its putative role in regulating fatty acid oxidation. Am J Phys 275:H2122–H2129. https://doi.org/10.1152/ajpheart.1998.275.6.H2122

Dyck JR, Berthiaume LG, Thomas PD, Kantor PF, Barr AJ, Barr R, Singh D, Hopkins TA, Voilley N, Prentki M, Lopaschuk GD (2000) Characterization of rat liver malonyl-CoA decarboxylase and the study of its role in regulating fatty acid metabolism. Biochem J 350(Pt 2):599–608

Dyck JR, Hopkins TA, Bonnet S, Michelakis ED, Young ME, Watanabe M, Kawase Y, Jishage K, Lopaschuk GD (2006) Absence of malonyl coenzyme A decarboxylase in mice increases cardiac glucose oxidation and protects the heart from ischemic injury. Circulation 114:1721–1728. https://doi.org/10.1161/CIRCULATIONAHA.106.642009

Farrell PJ, Lu M, Prevost J, Brown C, Behie L, Iatrou K (1998) High-level expression of secreted glycoproteins in transformed lepidopteran insect cells using a novel expression vector. Biotechnol Bioeng 60:656–663. https://doi.org/10.1002/(sici)1097-0290(19981220)60:6<656::aid-bit2>3.0.co;2-9

Feugang JM, Rodríguez-Muñoz JC, Dillard DS, Crenshaw MA, Willard ST, Ryan PL (2015) Beneficial effects of relaxin on motility characteristics of stored boar spermatozoa. Reprod Biol Endocrinol 13:24. https://doi.org/10.1186/s12958-015-0021-4

Folmes CD, Park S, Terzic A (2013) Lipid metabolism greases the stem cell engine. Cell Metab 17:153–155. https://doi.org/10.1016/j.cmet.2013.01.010

Froese DS, Forouhar F, Tran TH, Vollmar M, Kim YS, Lew S, Neely H, Seetharaman J, Shen Y, Xiao R, Acton TB, Everett JK, Cannone G, Puranik S, Savitsky P, Krojer T, Pilka ES, Kiyani

W, Lee WH, Marsden BD, Von Delft F, Allerston CK, Spagnolo L, Gileadi O, Montelione GT, Oppermann U, Yue WW, Tong L (2013) Crystal structures of malonyl-coenzyme a decarboxylase provide insights into its catalytic mechanism and disease-causing mutations. Structure 21:1182–1192. https://doi.org/10.1016/j.str.2013.05.001

Fullerton MD, Galic S, Marcinko K, Sikkema S, Pulinilkunnil T, Chen ZP, O'Neill HM, Ford RJ, Palanivel R, O'brien M, Hardie DG, Macaulay SL, Schertzer JD, Dyck JR, Van Denderen BJ, Kemp BE, Steinberg GR (2013) Single phosphorylation sites in Acc1 and Acc2 regulate lipid homeostasis and the insulin-sensitizing effects of metformin. Nat Med 19:1649–1654. https://doi.org/10.1038/nm.3372

Geisler C, Jarvis DL (2012) Innovative use of a bacterial enzyme involved in sialic acid degradation to initiate sialic acid biosynthesis in glycoengineered insect cells. Metab Eng 14:642–652. https://doi.org/10.1016/j.ymben.2012.08.005

Geisler C, Aumiller JJ, Jarvis DL (2008) A fused lobes gene encodes the processing beta-N-acetylglucosaminidase in Sf9 cells. J Biol Chem 283:11330–11339. https://doi.org/10.1074/jbc.M710279200

Geisler C, Mabashi-Asazuma H, Kuo CW, Khoo KH, Jarvis DL (2015) Engineering beta1,4-galactosyltransferase I to reduce secretion and enhance N-glycan elongation in insect cells. J Biotechnol 193:52–65. https://doi.org/10.1016/j.jbiotec.2014.11.013

Granberg CF, Harrison SM, Dajusta D, Zhang S, Hajarnis S, Igarashi P, Baker LA (2012) Genetic basis of prune belly syndrome: screening for HNF1beta gene. J Urol 187:272–278. https://doi.org/10.1016/j.juro.2011.09.036

Ha J, Daniel S, Broyles SS, Kim KH (1994) Critical phosphorylation sites for acetyl-CoA carboxylase activity. J Biol Chem 269:22162–22168

Hashimoto H, Abe M, Hirata A, Noda Y, Adachi H, Yoda K (2002) Progression of the stacked Golgi compartments in the yeast *Saccharomyces cerevisiae* by overproduction of GDP-mannose transporter. Yeast 19:1413–1424. https://doi.org/10.1002/yea.925

Hollister J, Grabenhorst E, Nimtz M, Conradt H, Jarvis DL (2002) Engineering the protein N-glycosylation pathway in insect cells for production of biantennary, complex N-glycans. Biochemistry 41:15093–15104. https://doi.org/10.1021/bi026455d

Hsu TA, Takahashi N, Tsukamoto Y, Kato K, Shimada I, Masuda K, Whiteley EM, Fan JQ, Lee YC, Betenbaugh MJ (1997) Differential N-glycan patterns of secreted and intracellular IgG produced in Trichoplusia ni cells. J Biol Chem 272:9062–9070. https://doi.org/10.1074/jbc.272.14.9062

Hwang IW, Makishima Y, Kato T, Park S, Terzic A, Park EY (2014) Human acetyl-CoA carboxylase 2 expressed in silkworm Bombyx mori exhibits posttranslational biotinylation and phosphorylation. Appl Microbiol Biotechnol 98:8201–8209. https://doi.org/10.1007/s00253-014-5715-6

Hwang IW, Makishima Y, Suzuki T, Kato T, Park S, Terzic A, Chung SK, Park EY (2015) Phosphorylation of Ser-204 and Tyr-405 in human malonyl-CoA decarboxylase expressed in silkworm Bombyx mori regulates catalytic decarboxylase activity. Appl Microbiol Biotechnol 99:8977–8986. https://doi.org/10.1007/s00253-015-6687-x

Jenkins N, Curling EM (1994) Glycosylation of recombinant proteins: problems and prospects. Enzym Microb Technol 16:354–364. https://doi.org/10.1016/0141-0229(94)90149-x

Jenkins N, Parekh RB, James DC (1996) Getting the glycosylation right: implications for the biotechnology industry. Nat Biotechnol 14:975–981. https://doi.org/10.1038/nbt0896-975

Joshi L, Davis TR, Mattu TS, Rudd PM, Dwek RA, Shuler ML, Wood HA (2000) Influence of baculovirus-host cell interactions on complex N-linked glycosylation of a recombinant human protein. Biotechnol Prog 16:650–656. https://doi.org/10.1021/bp000057p

Kato T, Murata T, Usui T, Park EY (2004) Comparative analysis of GFP(UV)-beta1,3-N-acetylglucosaminyltransferase 2 production in two insect-cell-based expression systems. Protein Expr Purif 35:54–61. https://doi.org/10.1016/j.pep.2003.12.018

Kato T, Kajikawa M, Maenaka K, Park EY (2010) Silkworm expression system as a platform technology in life science. Appl Microbiol Biotechnol 85:459–470. https://doi.org/10.1007/s00253-009-2267-2

Kato T, Kato N, Kikuta K, Miyazaki T, Kondo S, Yagi H, Kato K, Park EY (2017) N-glycan modification of a recombinant protein via coexpression of human glycosyltransferases in silkworm pupae. Sci Rep 7:1409. https://doi.org/10.1038/s41598-017-01630-6

Kim KW, Yamane H, Zondlo J, Busby J, Wang M (2007) Expression, purification, and characterization of human acetyl-CoA carboxylase 2. Protein Expr Purif 53:16–23. https://doi.org/10.1016/j.pep.2006.11.021

Kim YK, Kim KR, Kang DG, Jang SY, Kim YH, Cha HJ (2009) Suppression of beta-N-acetylglucosaminidase in the N-glycosylation pathway for complex glycoprotein formation in Drosophila S2 cells. Glycobiology 19:301–308. https://doi.org/10.1093/glycob/cwn138

Knobloch M, Braun SM, Zurkirchen L, Von Schoultz C, Zamboni N, Arauzo-Bravo MJ, Kovacs WJ, Karalay O, Suter U, Machado RA, Roccio M, Lutolf MP, Semenkovich CF, Jessberger S (2013) Metabolic control of adult neural stem cell activity by Fasn-dependent lipogenesis. Nature 493:226–230. https://doi.org/10.1038/nature11689

Kolwicz SC Jr, Olson DP, Marney LC, Garcia-Menendez L, Synovec RE, Tian R (2012) Cardiac-specific deletion of acetyl CoA carboxylase 2 prevents metabolic remodeling during pressure-overload hypertrophy. Circ Res 111:728–738. https://doi.org/10.1161/CIRCRESAHA.112.268128

Krawinkel MB, Oldigs HD, Santer R, Lehnert W, Wendel U, Schaub J (1994) Association of malonyl-CoA decarboxylase deficiency and heterozygote state for haemoglobin C disease. J Inherit Metab Dis 17:636–637. https://doi.org/10.1007/BF00711609

Kumagai J, Hsu SY, Matsumi H, Roh JS, Fu P, Wade JD, Bathgate RA, Hsueh AJ (2002) INSL3/Leydig insulin-like peptide activates the LGR8 receptor important in testis descent. J Biol Chem 277:31283–31286. https://doi.org/10.1074/jbc.C200398200

Laurent G, German NJ, Saha AK, De Boer VC, Davies M, Koves TR, Dephoure N, Fischer F, Boanca G, Vaitheesvaran B, Lovitch SB, Sharpe AH, Kurland IJ, Steegborn C, Gygi SP, Muoio DM, Ruderman NB, Haigis MC (2013) SIRT4 coordinates the balance between lipid synthesis and catabolism by repressing malonyl CoA decarboxylase. Mol Cell 50:686–698. https://doi.org/10.1016/j.molcel.2013.05.012

Lee CK, Cheong HK, Ryu KS, Lee JI, Lee W, Jeon YH, Cheong C (2008) Biotinoyl domain of human acetyl-CoA carboxylase: structural insights into the carboxyl transfer mechanism. Proteins 72:613–624. https://doi.org/10.1002/prot.21952

Leonard R, Rendic D, Rabouille C, Wilson IB, Preat T, Altmann F (2006) The Drosophila fused lobes gene encodes an N-acetylglucosaminidase involved in N-glycan processing. J Biol Chem 281:4867–4875. https://doi.org/10.1074/jbc.M511023200

Luo X, Bathgate RA, Liu YL, Shao XX, Wade JD, Guo ZY (2009) Recombinant expression of an insulin-like peptide 3 (INSL3) precursor and its enzymatic conversion to mature human INSL3. FEBS J 276:5203–5211. https://doi.org/10.1111/j.1742-4658.2009.07216.x

Mabashi-Asazuma H, Shi X, Geisler C, Kuo CW, Khoo KH, Jarvis DL (2013) Impact of a human CMP-sialic acid transporter on recombinant glycoprotein sialylation in glycoengineered insect cells. Glycobiology 23:199–210. https://doi.org/10.1093/glycob/cws143

Mabashi-Asazuma H, Kuo CW, Khoo KH, Jarvis DL (2015a) Modifying an insect cell N-glycan processing pathway using CRISPR-Cas technology. ACS Chem Biol 10:2199–2208. https://doi.org/10.1021/acschembio.5b00340

Mabashi-Asazuma H, Sohn BH, Kim YS, Kuo CW, Khoo KH, Kucharski CA, Fraser MJ Jr, Jarvis DL (2015b) Targeted glycoengineering extends the protein N-glycosylation pathway in the silkworm silk gland. Insect Biochem Mol Biol 65:20–27. https://doi.org/10.1016/j.ibmb.2015.07.004

Minagawa I, Fukuda M, Ishige H, Kohriki H, Shibata M, Park EY, Kawarasaki T, Kohsaka T (2012) Relaxin-like factor (RLF)/insulin-like peptide 3 (INSL3) is secreted from testicular Leydig cells as a monomeric protein comprising three domains B-C-A with full biological activity in boars. Biochem J 441:265–273. https://doi.org/10.1042/BJ20111107

Minagawa I, Sagata D, Pitia AM, Kohriki H, Shibata M, Sasada H, Hasegawa Y, Kohsaka T (2014) Dynamics of insulin-like factor 3 and its receptor expression in boar testes. J Endocrinol 220:247–261. https://doi.org/10.1530/JOE-13-0430

Miyazaki T, Ishizaki M, Dohra H, Park S, Terzic A, Kato T, Kohsaka T, Park EY (2017) Insulin-like peptide 3 expressed in the silkworm possesses intrinsic disulfide bonds and full biological activity. Sci Rep 7:17339. https://doi.org/10.1038/s41598-017-17707-1

Nam HW, Lee GY, Kim YS (2006) Mass spectrometric identification of K210 essential for rat malonyl-CoA decarboxylase catalysis. J Proteome Res 5:1398–1406. https://doi.org/10.1021/pr050487g

Nef S, Parada LF (1999) Cryptorchidism in mice mutant for Insl3. Nat Genet 22:295–299. https://doi.org/10.1038/10364

Nomura T, Suganuma M, Higa Y, Kataoka Y, Funaguma S, Okazaki H, Suzuki T, Kobayashi I, Sezutsu H, Fujiyama K (2015) Improvement of glycosylation structure by suppression of beta-N-acetylglucosaminidases in silkworm. J Biosci Bioeng 119:131–136. https://doi.org/10.1016/j.jbiosc.2014.07.012

Palmberger D, Wilson IB, Berger I, Grabherr R, Rendic D (2012) SweetBac: a new approach for the production of mammalianised glycoproteins in insect cells. PLoS One 7:e34226. https://doi.org/10.1371/journal.pone.0034226

Park EY, Minkner R (2025) A systematic approach for scalable purification of virus-like particles. Protein Expr Purif 228:106664. https://doi.org/10.1016/j.pep.2025.106664

Park H, Kaushik VK, Constant S, Prentki M, Przybytkowski E, Ruderman NB, Saha AK (2002) Coordinate regulation of malonyl-CoA decarboxylase, sn-glycerol-3-phosphate acyltransferase, and acetyl CoA carboxylase by AMP activated protein kinase in rat tissues in response to exercise. J Biol Chem 277:32571–32577. https://doi.org/10.1074/jbc.M201692200

Park EY, Ishikiriyama M, Nishina T, Kato T, Yagi H, Kato K, Ueda H (2009) Human IgG1 expression in silkworm larval hemolymph using BmNPV bacmids and its N-linked glycan structure. J Biotechnol 139:108–114. https://doi.org/10.1016/j.jbiotec.2008.09.013

Pitia AM, Uchiyama K, Sano H, Kinukawa M, Minato Y, Sasada H, Kohsaka T (2017) Functional insulin-like factor 3 (INSL3) hormone-receptor system in the testes and spermatozoa of domestic ruminants and its potential as a predictor of sire fertility. Anim Sci J 88:678–690. https://doi.org/10.1111/asj.12694

Saggerson D (2008) Malonyl-CoA, a key signaling molecule in mammalian cells. Annu Rev Nutr 28:253–272. https://doi.org/10.1146/annurev.nutr.28.061807.155434

Saha AK, Schwarsin AJ, Roduit R, Masse F, Kaushik V, Tornheim K, Prentki M, Ruderman NB (2000) Activation of malonyl-CoA decarboxylase in rat skeletal muscle by contraction and the AMP-activated protein kinase activator 5-aminoimidazole-4-carboxamide-1-beta -D-ribofuranoside. J Biol Chem 275:24279–24283. https://doi.org/10.1074/jbc.C000291200

Sambandam N, Steinmetz M, Chu A, Altarejos JY, Dyck JR, Lopaschuk GD (2004) Malonyl-CoA decarboxylase (MCD) is differentially regulated in subcellular compartments by 5'AMP-activated protein kinase (AMPK). Studies using H9c2 cells overexpressing MCD and AMPK by adenoviral gene transfer technique. Eur J Biochem 271:2831–2840. https://doi.org/10.1111/j.1432-1033.2004.04218.x

Shinkawa T, Nakamura K, Yamane N, Shoji-Hosaka E, Kanda Y, Sakurada M, Uchida K, Anazawa H, Satoh M, Yamasaki M, Hanai N, Shitara K (2003) The absence of fucose but not the presence of galactose or bisecting N-acetylglucosamine of human IgG1 complex-type oligosaccharides shows the critical role of enhancing antibody-dependent cellular cytotoxicity. J Biol Chem 278:3466–3473. https://doi.org/10.1074/jbc.M210665200

Siqin MI, Okuno M, Yamada K, Sugawara Y, Nagura Y, Hamano K, Park EY, Sasada H, Kohsaka T (2013) The active form of goat insulin-like peptide 3 (INSL3) is a single-chain structure comprising three domains B-C-A, constitutively expressed and secreted by testicular Leydig cells. Biol Chem 394:1181–1194. https://doi.org/10.1515/hsz-2012-0357

Tomiya N, Narang S, Lee YC, Betenbaugh MJ (2004) Comparing N-glycan processing in mammalian cell lines to native and engineered lepidopteran insect cell lines. Glycoconj J 21:343–360. https://doi.org/10.1023/B:GLYC.0000046275.28315.87
Tong L (2013) Structure and function of biotin-dependent carboxylases. Cell Mol Life Sci 70:863–891. https://doi.org/10.1007/s00018-012-1096-0
Tong L, Harwood HJ Jr (2006) Acetyl-coenzyme A carboxylases: versatile targets for drug discovery. J Cell Biochem 99:1476–1488. https://doi.org/10.1002/jcb.21077
Tu L, Banfield DK (2010) Localization of Golgi-resident glycosyltransferases. Cell Mol Life Sci 67:29–41. https://doi.org/10.1007/s00018-009-0126-z
Varki A (1993) Biological roles of oligosaccharides: all of the theories are correct. Glycobiology 3:97–130. https://doi.org/10.1093/glycob/3.2.97
Viswanathan K, Narang S, Betenbaugh MJ (2015) Engineering sialic acid synthesis ability in insect cells. Methods Mol Biol 1321:171–178. https://doi.org/10.1007/978-1-4939-2760-9_12
Voilley N, Roduit R, Vicaretti R, Bonny C, Waeber G, Dyck JR, Lopaschuk GD, Prentki M (1999) Cloning and expression of rat pancreatic beta-cell malonyl-CoA decarboxylase. Biochem J 340(Pt 1):213–217
Wade JD, Lin F, Hossain MA, Shabanpoor F, Zhang S, Tregear GW (2009) The chemical synthesis of relaxin and related peptides. Ann N Y Acad Sci 1160:11–15. https://doi.org/10.1111/j.1749-6632.2009.03951.x
Wakil SJ, Abu-Elheiga LA (2009) Fatty acid metabolism: target for metabolic syndrome. J Lipid Res 50(Suppl):S138–S143. https://doi.org/10.1194/jlr.R800079-JLR200
Werner RG, Kopp K, Schlueter M (2007) Glycosylation of therapeutic proteins in different production systems. Acta Paediatr 96:17–22. https://doi.org/10.1111/j.1651-2227.2007.00199.x
Xue J, Peng J, Zhou M, Zhong L, Yin F, Liang D, Wu L (2012) Novel compound heterozygous mutation of MLYCD in a Chinese patient with malonic aciduria. Mol Genet Metab 105:79–83. https://doi.org/10.1016/j.ymgme.2011.09.007
Zhang S, Hughes RA, Bathgate RA, Shabanpoor F, Hossain MA, Lin F, Van Lierop B, Robinson AJ, Wade JD (2010) Role of the intra-A-chain disulfide bond of insulin-like peptide 3 in binding and activation of its receptor, RXFP2. Peptides 31:1730–1736. https://doi.org/10.1016/j.peptides.2010.05.021
Zhou D, Yuen P, Chu D, Thon V, Mcconnell S, Brown S, Tsang A, Pena M, Russell A, Cheng JF, Nadzan AM, Barbosa MS, Dyck JR, Lopaschuk GD, Yang G (2004) Expression, purification, and characterization of human malonyl-CoA decarboxylase. Protein Expr Purif 34:261–269. https://doi.org/10.1016/j.pep.2003.11.023

Chapter 6
Expression of Recombinant Proteins

Abstract This chapter explores the versatile application of the silkworm (*Bombyx mori*) expression system for producing a wide array of complex recombinant proteins. It details advanced strategies to enhance the yield and proper folding of secretory proteins, specifically highlighting the integration of the bombyxin signal sequence into the BmNPV bacmid and the coexpression of molecular chaperones, such as BiP and calnexin, to mitigate aggregation and improve cellular secretion. Furthermore, the chapter demonstrates the successful expression of challenging membrane proteins while maintaining their functional conformations. A major focus is placed on the generation of highly immunogenic virus-like particles (VLPs), detailing the production of both enveloped (eVLPs) and nonenveloped (neVLPs) structures suitable for vaccine development. To address downstream processing bottlenecks, the chapter presents systematic, scalable purification protocols tailored specifically for silkworm-derived eVLPs and neVLPs, ensuring high purity, structural integrity, and biological activity. Collectively, these technological advancements underscore the silkworm's efficacy as a robust, high-yield biofactory for the scalable production of therapeutically relevant glycoproteins, membrane receptors, and macromolecular viral assemblies.

Keywords Recombinant protein • Bombyxin signal sequence • Molecular chaperone • Virus-like particle • Enveloped VLP • Nonenveloped VLP • Silkworm biofactory

6.1 Introduction

Biomaterials sourced from living organisms have gained significant importance across both biomedical and industrial domains, with applications spanning pharmaceuticals, cosmetics, textiles, and sustainable materials. Since the initial development of recombinant protein expression in *E. coli* during the 1970s, a wide array of recombinant proteins has been produced using various host systems, profoundly influencing many facets of modern life and industry.

E. Y. Park, *Silkworm Biofactory*, https://doi.org/10.1007/978-981-92-0056-6_6

Regarding silkworms (*Bombyx mori*), the successful expression of interferon in 1985 (Maeda et al. 1985) marked a pioneering advance, establishing the first generation of recombinant protein production in this species. However, the initial technique required extensive construction of recombinant baculoviruses, which restricted its practical application and scalability. The introduction of the AcMNPV bacmid in 1993 (Luckow et al. 1993) and the subsequent development of the BmNPV bacmid in 2005 (Motohashi et al. 2005) led to the second generation of recombinant protein expression in silkworms, characterized by enhanced efficiency and broader applicability. The BmNPV bacmid system has enabled the effective production of a wide variety of proteins in silkworms, including secretory and membrane-associated proteins and, more recently, virus-like particle proteins. These technological advances have broadened the scope for functional protein synthesis and facilitated research and application of complex biomolecules. This chapter presents an extensive review of the expression profiles and biological functions of proteins produced in silkworms, systematically classifying them into three categories for in-depth analysis and discussion.

6.2 Secretory Proteins

Secreted proteins in silkworm larvae are fully modified after translation and accumulated in the larval hemolymph. Each larva typically provides about 0.5 mL of hemolymph. To enhance protein recovery efficiency, initial processing methods such as protein precipitation or membrane filtration are used. After this, affinity chromatography is the standard approach for purifying the proteins.

6.2.1 Expression of Glycosyltransferases in Silkworms

As an example, rat α2,6-sialyltransferase (ST6Gal1) was produced in the hemolymph of silkworm larvae using a BmNPV-*CP*$^{-}$-*Chi*$^{-}$ bacmid. To facilitate the secretion of ST6Gal1 into the hemolymph of silkworm larvae, the BmNPV-*CP*$^{-}$-*Chi*$^{-}$/ST6Gal1 construct was designed using bombyxin (bx) signal peptides (Fig. 6.1a). The BmNPV-*CP*$^{-}$-*Chi*$^{-}$/ST6Gal1 bacmid was injected into silkworms, and hemolymph was harvested. Initially, the hemolymph—diluted twofold and containing 9.0 units of FLAG-tagged ST6Gal1 (0.016 U/mg protein)—was treated with 25 to 70% ammonium sulfate to precipitate proteins. The precipitate was then dissolved in 50 mM MOPS buffer (pH 7.5), supplemented with 150 mM NaCl and 0.02% Triton X-100. This preparation yielded 7.2 units of FLAG-tagged ST6Gal1 (0.069 U/mg protein), reflecting an 80% recovery rate (Fig. 6.1b). After removing ammonium sulfate using a Sephadex G-25 M PD-10 column, the sample was loaded

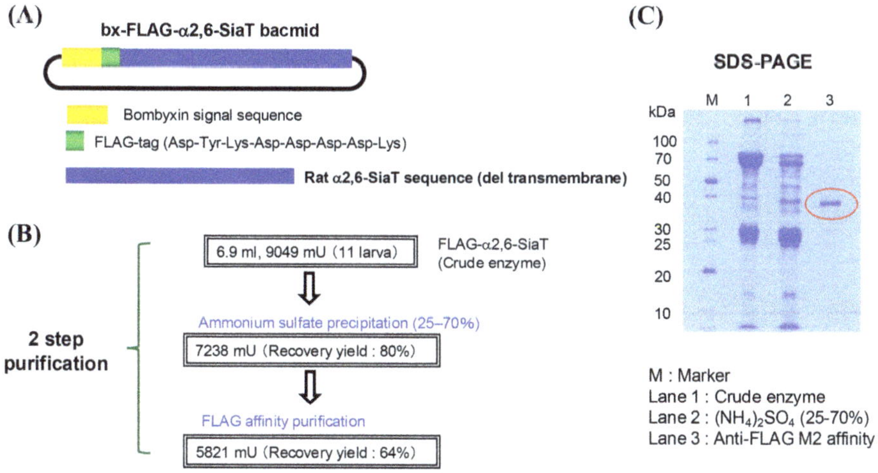

Fig. 6.1 Expression of FLAG-tagged ST6Gal1SDS-PAGE in silkworm larval hemolymph (Ogata et al. 2009). (**a**) Construction of recombinant bacmid encoding FLAG-tagged ST6Gal1 gene. (**b**) Purification of recombinant FLAG-tagged ST6Gal1 collected from silkworm larval hemolymph. (**c**) SDS-PAGE of FLAG-tagged ST6Gal1, and the gel was stained with Coomassie Brilliant Blue (CBB). Lanes M, 1, 2, and 3 indicate marker proteins, silkworm larval hemolymph (crude enzyme), ammonium sulphate precipitates, and elution from anti-FLAG M2 affinity (purified recombinant enzyme), respectively

onto an anti-FLAG M2 affinity column (Fig. 6.1b). The majority of ST6Gal1 activity bound to the column and was subsequently released by elution with the same buffer containing 100 μg/mL FLAG peptide. This final step produced 5.8 units of FLAG-tagged ST6Gal1 (2.6 U/mg protein), achieving a high yield of 64% and a purification factor greater than 150-fold. Each silkworm larva yields about 100–200 μg of purified glycosyltransferase enzymes, which exhibit high specific activity (Ogata et al. 2009). Analysis by SDS-PAGE and isoelectric focusing revealed a single band for FLAG-tagged ST6Gal1, with an estimated molecular weight of 40 kDa (Fig. 6.1c) and a pI of 6.3 (Ogata et al. 2009).

The substrate specificity of the enzyme matched that of ST6Gal1 found in rat liver. ST6Gal1 produced in silkworms was then used to sialylate asialoglycopolypeptide, converting it into α2,6-sialoglycopolypeptide (Fig. 6.2). This transformation was achieved using recombinant ST6Gal1 expressed in silkworm, following previously established procedures (Ogata et al. 2007) (Fig. 6.2). The desired glycopolypeptide was isolated using a Sephadex G-25 M PD-10 column. The molecular structures of the resulting α2,6-sialoglycopolypeptide were verified by 1H- and 13C-NMR spectroscopy and chemical analyses, according to the reference (Ogata et al. 2007). Analysis of the ^{1}H-NMR spectrum revealed complete sialylation, as indicated by the integration of proton signals. The substitution levels for both neutral sugar derivatives (NS) and sialylated sugar derivatives (Sia), calculated based

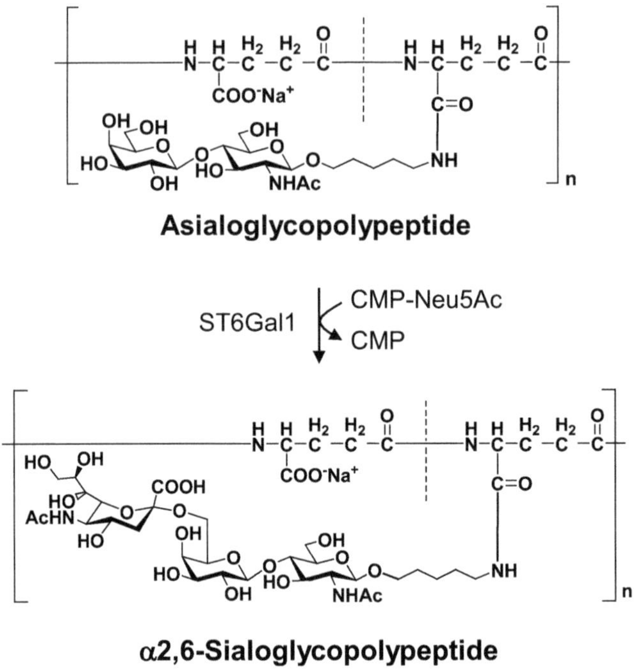

Fig. 6.2 Enzymatic synthesis of artificial sialoglycopolypeptide (Ogata et al. 2009). α2,6-Sialoglycopolypeptide was synthesized from asialoglycopolypeptide using recombinant ST6Gal1. The enzymatic synthesis was performed in a solution of 5.0 mg asialoglycopolypeptide, containing CMP-Neu5Ac, FLAG-tagged ST6Gal1, and calf intestine alkaline phosphatase in 50 mM MOPS buffer (pH 7.4) at 37 °C for 48 h

on the degree of polymerization (DP) of γ-PGA, reached 100% (Ogata et al. 2009). These findings demonstrate that the FLAG-tagged ST6Gal1 enzyme specifically transfers Neu5Ac to the nonreducing ends of type II sugar chains, demonstrating that ST6Gal1 expressed in silkworms is useful for sialylation of asialoglycopeptide.

Furthermore, glycosyltransferases such as human β1,3-*N*-acetylglucosaminyltransferase (β3GnT) (Dojima et al. 2009; Kato and Park 2007; Park et al. 2007) and human α1,4-*N*-acetylglucosaminyltransferase (Nakajima et al. 2009) are efficiently expressed in silkworms, exhibit high enzymatic activity, and are widely employed in diverse glycosylation synthetic processes. The antibody was also produced as a secreted protein, with each silkworm yielding several hundred micrograms. In addition to antibodies, human interleukin-18 (IL-18) (Tsutsumi et al. 2014) and the Spike protein from porcine epidemic diarrhea virus (Masuda et al. 2021) were both efficiently expressed and purified from silkworms, and their biological activities were successfully verified. Furthermore, a variety of receptor proteins were expressed in silkworms at yields ranging from tens to several hundred micrograms from a single silkworm (Kusaka et al. 2020; Muraki and Honda 2010; Nakaya et al. 2020). These results, including expression levels summarized in Table 6.1, demonstrate the versatility and effectiveness of the silkworm expression system for producing diverse proteins. This method could offer significant advantages for large-scale protein production, particularly for proteins that require complex folding or post-translational modifications.

Table 6.1 Secretory proteins expressed in silkworm

Secretory proteins	Purified protein amount	References
Human β3GnT2	92 μg/larva	Dojima et al. (2009), Park et al. (2007), Yagi et al. (2005)
Human α1,4-*N*-acetylglucosaminyltranferase	479 mU/mL 992 mU/mL (coexpressed with calnexin)	Nakajima et al. (2009)
Rat ST6Gal1	200 μg/larva	Ogata et al. (2009)
Huma 29IJ6 IgG	36 μg/larva 239 μg/larva (coexpressed with calreticulin)	Dojima et al. (2010a), Park et al. (2009)
Human single-chain variable fragment (scFV) 13CG2	132.4 μg/larva	Ishikiriyama et al. (2009)
Human interleukin-18	Crystal structure of IL-18 bound to two receptors	Tsutsumi et al. (2014)
Spike protein of porcine epidemic diarrhea virus	~100 μg/larva	Masuda et al. (2021)
Hemagglutinin of influenza virus A (H5N1)	~20 μg/larva	Dong et al. (2013)
Thyroid hormone receptor	25 μg/larva (fused with glutathione S-transferase)	Nakaya et al. (2020)
Human Fas receptor extracellular domain–human IgG1 heavy-chain Fc domain fusion protein	~433 μg/larva	Muraki and Honda (2010)
Human killer cell Ig-like receptor 2DL1 encoding extracellular region	200 μg/larva	Sasaki et al. (2009)
Human CD1d proteins	50 μg/larva	Kusaka et al. (2020)

This was adopted with permission (Park 2025), copyright 2025 Elsevier

6.2.2 Improved Secretion of Secretory Proteins in Silkworms

6.2.2.1 BmNPV Bacmid Integrated Signal Sequence

The bacmid expression system enables robust production of recombinant proteins in silkworm larvae and pupae. Nonetheless, complex proteins often fail to be expressed in a biologically active form, potentially due to inefficient recognition of signal peptides by the insect cell protein translocation machinery. Insect cells infected with baculovirus carrying the propapain gene fused to the honeybee melittin signal peptide secreted over five times more papain precursor compared to those with a plant signal peptide (Tessier et al. 1991). Similarly, human β1,3-*N*-acetylglucosaminyltransferase 2 (β3GnT2) required a signal sequence for functionality and became biologically active when fused with the honeybee melittin signal peptide (Kato et al. 2004). Thus, precise recognition and selection of signal peptides are crucial for efficient expression of functional proteins.

A variety of signal peptides were utilized, including those from cecropin B of *Hyalophora cecropia* (Van Hofsten et al. 1985), juvenile hormone esterase (Hirai et al. 2002), bombyxin (Adachi et al. 1989), diapause hormone precursor (Sato et al. 1993), and prophenoloxidase-activating enzyme (Satoh et al. 1999)—all derived from *Bombyx mori*—as well as the 64 K envelope protein from AcNPV (Volkman 1986). Additionally, the honeybee prepromelittin signal peptide (Vlasak et al. 1983), known to promote the secretion of propapain (Tessier et al. 1991) and β3GnT2 (Kato et al. 2004), was employed in the baculovirus expression system. This study presents the development of a practical engineering BmNPV bacmid system with various signal sequences, enabling efficient secretion of β3GnT2 into silkworm hemolymph.

β3GnT2 cDNA from Quick-Clone human fetal brain (Clontech) was inserted into the pBlueBacHis2-GFP_{uv} vector at the *BamHI* and *EcoRI* sites, producing pBlueBacHis2-GFP_{uv}-β3GnT2 (Kato et al. 2003). To add a signal peptide, the GFP_{uv}-β3GnT2 fragment was amplified by PCR using each primer (Table 6.2). Signal sequences used were: cecropin B (*ce*), silkworm juvenile hormone esterase (*jhe*), bombyxin (*bx*), diapause hormone precursor (*dh*), prophenoloxidase-activating enzyme (*ppae*), AcNPV 64 K envelope protein (*gp*), and honeybee prepromelittin (*me*). The PCR product was cloned into pENTR/D-TOPO (Invitrogen) and then recombined with pDEST8 to make pDEST/signal-GFP_{uv}-β3GnT2. This plasmid was transformed into *E. coli* BmDH10Bac cells, resulting in BmNPV bacmid/signal-GFP_{uv}-β3GnT2 that was isolated (Fig. 6.3).

To assess how signal sequences encoded by BmNPV influence protein secretion, recombinant BmNPV/signal-GFP_{uv}-β3GnT2 was introduced into Bm5 cells. The fusion protein remained intracellularly and was not secreted into the extracellular space when signal sequences from *gp*, *ce*, and *me* (Park et al. 2007). In contrast, using the *bx* and *ppae* signal sequences resulted in the fusion protein being secreted at 3 and 5 days after infection. Additionally, the *jhe* signal sequence produced only

Table 6.2 Primers set for PCR (Park et al. 2007)

Primer	Sequence
Forward primers	
me-GFP$_{uv}$-β3GnT2	5′-CACCATGAAATTCTTAGTCAACGTTGCCCTTGTTTTTATGGTCGTATACATTTCTTACATCTATGCCGGCCCGCGGGGTTCTCATCATC-3′
gp-GFP$_{uv}$-β3GnT2	5′-CACCATGCCCATCTTAAGCGCTATTGTTTTATATGTCCTTTTGGCGGCGGCGGCGCATTCTGCCTTTGCGCCCCCGCGGGGTTCTCATCATC-3′
ce-GFP$_{uv}$-β3GnT2	5′-CACCATGAATTTCTCAAGCATATTTTTCTTCGTGTTCGCTTTGGTTCTGGCTTTGTCAACCAGTTTCGGCTGCACCGGAGCCGAAACCGCGGGGTTCTCATCATC-3′
ppae-GFP$_{uv}$-β3GnT2	5′-CACCATGTTTTTAATTTGGACATTCATCGTGGCTGTTCTGGCGATCCAGACCAAAAGTGTTGTTCAACCGCGGGGTTCTCATCATC-3′
jhe-GFP$_{uv}$-β3GnT2	5′-CACCATGTGGTACCAAATTCGCTGTATTACTATATTACGTTATCGTGGGCTCAGTAAGGGCATACCCGCGGGGTTCTCATCATC-3′
dh-GFP$_{uv}$-β3GnT2	5′-CACCATGTATAAAACCAACATTGTTTTCAACGTTTTAGCTTTGGCATTGTTCAGTATTTTCTTCGCGAGTTGCACGCCGCGGGGTTCTCATCATC-3′
bx-GFP$_{uv}$-β3GnT2	5′-CACCATGAAGATACTCCTTGCTATTGCATTAATGTTGTCAACAGTAATGTGGGTGTCAACACAACCGCGGGGTTCTCATCATC-3′
Reverse primer	5′-CGGAATTCTGAAGGGTTTAGAGGCCCTCAAATGGG-3′

Thin and double underlines indicate signal sequences and the part of GFP$_{uv}$-β3GnT2 sequences, respectively

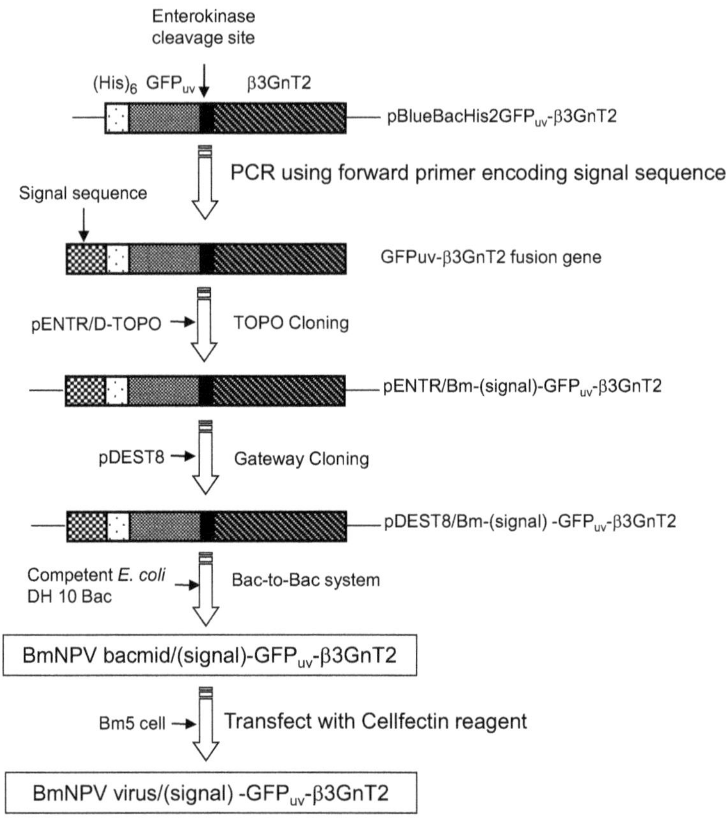

Fig. 6.3 Outline of the process for generating recombinant BmNPV and BmNPV bacmid encoding the signal-GFP_{uv}-β3GnT2. (Adopted with permission (Park et al. 2007). Copyright 2007 Elsevier)

a faint band. These findings demonstrate that the *bx* and *ppae* signal sequences effectively direct the fusion protein to be secreted into the silkworm hemolymph.

When the recombinant bacmid was used, expression was delayed by 1 day to compare with the recombinant baculovirus. In the case of recombinant BmNPV, the *bx* signal sequence greatly enhanced the secretion of the fusion protein, resulting in a β3GnT activity of 38.3 mU/mL at 4 d postinfection time (Table 6.3), which was the highest observed. The *ppae* signal sequence also promoted secretion, but it was less effective than *bx*. When the recombinant BmNPV bacmid was used, both *bx* and *ppae* demonstrated high secretory efficiency, yielding β3GnT activities of 56.1 and 46.7 mU/mL, respectively (Table 6.3). Notably, the *bx*-mediated β3GnT activity is, to our knowledge, the highest reported to date—approximately double the activity reported (Kato et al. 2005). Conversely, the *jhe* signal sequence was ineffective for secretion in silkworms (Table 6.3), achieving only 5% to 20% of the secretion levels seen with *bx* and *ppae*. However, the striking advantage of using recombinant bacmid is the expression of β3GnT2 in hemolymph, producing 91 μg/larva, which

Table 6.3 Effect of signal sequence on enzyme activity expressed in silkworm larval hemolymph

Signal sequence	Post-injection time (d)	β3GnT2 activity (mU/mL)[a]	Protein concentration (mg/mL)[a]	Specific activity (mU/mg protein)	β3GnT2 amount (μg/larva)[b]
BmNPV/signal-GFPuv-β3GnT2 infection					
none	3	0.0 ± 0.00	24.2 ± 0.18	0.0	0.00 ± 0.00
bx	3	3.6 ± 1.86	25.6 ± 0.97	0.1	5.25 ± 1.45
	4	38.3 ± 6.60	19.0 ± 0.73	2.0	56.30 ± 5.15
ppae	3	0.9 ± 1.23	21.2 ± 0.85	0.0	1.09 ± 081
	4	29.2 ± 0.60	19.8 ± 0.67	1.5	41.90 ± 0.43
jhe	3	6.1 ± 8.62	24.6 ± 1.10	0.3	7.82 ± 5.86
BmNPV bacmid/signal-GFPuv-β3GnT2 injection					
none	3	0.0 ± 0.00	24.2 ± 0.18	0.0	0.00 ± 0.00
bx	4	56.1 ± 0.86	44.7 ± 0.97	1.3	91.0 ± 0.74
	5	50.8 ± 4.06	37.1 ± 0.49	1.4	63.2 ± 2.68
ppae	4	46.7 ± 3.34	40.0 ± 1.34	1.2	74.20 ± 2.87
	5	51.5 ± 8.00	32.4 ± 2.50	1.6	52.50 ± 4.32
jhe	4	2.1 ± 1.46	41.4 ± 0.97	0.1	3.61 ± 1.34

This table was modified with permission (Park et al. 2007), Copyright 2007 Elsevier
[a]Collected hemolymph from 10 larvae. Data are ± standard deviation ($n = 3$)
[b]β3GnT2 assay was performed using five silkworm larvae, and data are ± standard deviation ($n = 3$). β3GnT2 amount was obtained from the correlation between enzyme activity and the amount of purified protein (Kato et al. 2004)

is almost twofold higher than that of the recombinant BmNPV (Table 6.3). This shows that 10 silkworm larvae produced almost 1 mg of the secreted biologically active protein.

To investigate the distribution of recombinant protein in the hemolymph, fat body, intestines, and silk gland of silkworm larvae, the *bx* and *ppae* signal sequences were used. Upon infection with recombinant BmNPV, the fusion protein was exclusively detected in the hemolymph (Fig. 6.4a(i)), accounting for 99% of total activity (Fig. 6.4a(ii)), where the degraded fusion protein appeared in the fat body (Fig. 6.4a(i)). In contrast, injection with recombinant BmNPV bacmid resulted in 88% of the fusion protein being secreted into the hemolymph (Fig. 6.4a(ii)), with minor degradation observed in both hemolymph and fat body (Fig. 6.4a(i)), and 12% of the total β3GnT activity remaining in the fat body. It indicates that proteolytic degradation of the fusion protein was more pronounced with BmNPV than with BmNPV bacmid. β3GnT2 is highly susceptible to cysteine proteases (Kato et al. 2003), making its expression challenging without protease inhibitors, which also leads to deactivation of β3GnT2 in both hemolymph and fat body. No degraded fusion proteins were detected in the hemolymph or fat body when using BmNPV bacmid.

The influence of signal sequences on the fusion protein's secretion into silkworm tissues was evaluated. The fusion protein was present in the fat body and intestines but absent from the silk gland (Fig. 6.4b(i)). Both the fat body and intestines

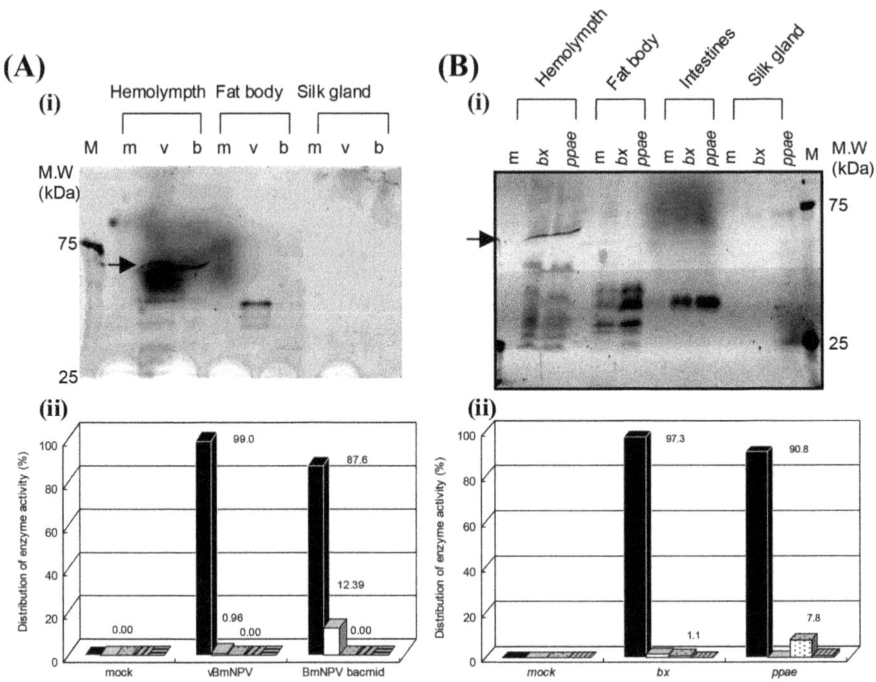

Fig. 6.4 (**a**) Comparative analysis of GFP$_{uv}$-β3GnT2 fusion protein secretion between recombinant BmNPV/*bx*-GFP$_{uv}$-β3GnT2 and BmNPV bacmid/*bx*-GFP$_{uv}$-β3GnT2 constructs. Both constructs were administered to silkworm larvae, after which the hemolymph, fat body, and silk gland were collected at 4 and 5 d postinfection. (i) Detection of fusion proteins was performed using SDS-PAGE, visualized by GFP$_{uv}$ fluorescence. The markers M, m, v, and b correspond to the protein molecular weight marker, mock control, BmNPV/*bx*-GFP$_{uv}$-β3GnT2, and BmNPV bacmid/*bx*-GFP$_{uv}$-β3GnT2, respectively. The arrow designates the GFP$_{uv}$-β3GnT2 fusion protein. (ii) Tissue-specific distribution of β3GnT activity secreted in silkworm tissues. The black, white, dotted, slashed, and horizontally striped bars represent the hemolymph, fat body, silk gland, intestines, and Malpighian tubules, respectively, from silkworm larvae. (**b**) Effects of various signal sequences on the secretion of GFP$_{uv}$-β3GnT2 fusion protein into silkworm larval tissues. BmNPV bacmid carrying either the *bx* or *ppae* signal sequence fused to GFP$_{uv}$-β3GnT2 was injected into silkworm larvae, and samples of hemolymph, fat body, intestines, and silk gland were collected 5 d after infection. (i) Fusion proteins isolated from larval tissues were analyzed for GFP$_{uv}$ fluorescence. The labels represent: M = molecular weight marker, m = mock control without recombinant bacmid, *bx* = bx signal sequence fused to GFP$_{uv}$-β3GnT2, and *ppae* = *ppae* signal sequence fused to GFP$_{uv}$-β3GnT2. The arrow identifies the GFP$_{uv}$-β3GnT2 fusion protein. (ii) Tissue distribution of β3GnT activity secreted in silkworm larvae. The black, gray, dotted, and slashed bars represent the hemolymph, fat body, intestines, and silk gland, respectively, of silkworm larvae. (Adapted with permission (Park et al. 2007). Copyright 2007 Elsevier)

accumulated a lower-molecular-weight 72 kDa fusion protein (Fig. 6.4b(i)), likely due to protease activity. For *ppae,* 7.8% of the fusion protein was detected in the intestines, which is 8 times greater than that observed for *bx*. Employing the *bx* signal sequence led to 97.3% of the fusion protein being secreted into the hemolymph, while the *ppae* sequence resulted in 90.8% secretion (Fig. 6.4b(ii)).

Assuming a hemolymph volume of 700 μL per larva, the concentration of secreted protein was 130 μg/mL, whereas the total protein concentration was 44.7 mg/mL (Table 6.3). Thus, the target recombinant protein constituted 3% of the total protein. To optimize recombinant protein production using silkworm larvae, it is critical to develop an efficient purification method for proteins secreted into the hemolymph.

6.2.2.2 Molecular Chaperone-Assisted Expression in Silkworms

Silkworm larvae serve as both effective and economical platforms for large-scale protein production. The use of potent *polh* and p10 promoters greatly enhances recombinant gene expression, compelling host cells to rapidly generate substantial amounts of foreign proteins. Nevertheless, baculovirus infection can inhibit the host's capacity for protein synthesis, thereby restricting the availability of factors necessary for secretion (Hsu et al. 1994; Jarvis and Garcia Jr. 1994). As the machinery responsible for post-translational modifications deteriorates a few days after infection (Ailor and Betenbaugh 1999), protein folding becomes defective, leading to pronounced aggregation and breakdown of the proteins produced. Considerable aggregation has been specifically documented during IgG expression in insect cells driven by the polyhedrin promoter (Hasemann and Capra 1990).

To counteract these issues, molecular chaperones play a critical role in ensuring the correct folding, assembly, and maturation of proteins in insect cells and larvae. During the transfer of glycoproteins from the endoplasmic reticulum (ER) to the Golgi apparatus (Kato and Kamiya 2007; Moremen and Molinari 2006; Ruddock and Molinari 2006), *N*-linked glycans are modified throughout the secretory pathway. Proteins newly synthesized in the ER interact with molecular chaperones in diverse ways (Molinari and Helenius 2000). Some initially associate with the immunoglobulin heavy-chain-binding protein (BiP, GRP78), followed by calnexin (CNX) and calreticulin (CRT) (Hammond and Helenius 1994), while others bind CNX and CRT without BiP (Braakman et al. 1991) or sequentially interact with BiP and GRP94 (Melnick et al. 1994). The formation of intramolecular disulfide bonds is facilitated by protein disulfide isomerase (PDI), ERp57, or both enzymes working together (Molinari and Helenius 1999).

This section focuses on the secretion of recombinant IgG1 and on the *N*-linked glycosylation patterns of IgG1 expressed in silkworm larvae and pupae using a BmNPV bacmid. IgG1 was produced through coexpression with five molecular chaperones—CNX, CRT, ERp57, BiP, and Hsp70—all of which are essential for successful IgG1 synthesis in silkworm larvae. Maintaining uniformity and reliability in protein synthesis—including amino acid sequence, biological activity, and post-translational modifications—is essential (Cabrera et al. 2005).

The 291J6 IgG protein was produced in both the hemolymph of larvae and the pupae of silkworms that were injected with 10 μg of rBmNPV bacmid per larva and incubated for 6 d. The concentration of 296IJ6 IgG produced in the larvae and

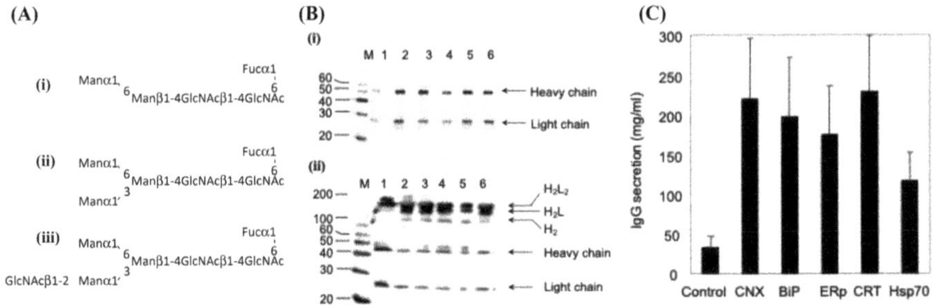

Fig. 6.5 (**a**) Structure of *N*-glycan. (i), (ii), and (iii) represent $Man_2GlcNAc(Fuc)GlcNAc$, $Man_3GlcNAc(Fuc)GlcNAc$, and $GlcNAcMan_3GlcNAc(Fuc)GlcNAc$, respectively. (**b**) Western blotting of the 29IJ6 IgG proteins secreted into the hemolymph of silkworm larvae, both with (i) and without reducing agents (ii). Lane 1, IgG expressed without any molecular chaperone; Lanes 2 through 6, IgG expression assisted by coexpression with CNX, BiP, ERp57, CRT, or Hsp70. In panel (ii), bands labeled as H_2L_2, H_2L, and H_2 correspond to fully assembled IgG, the heavy-chain dimer with a single light chain, and the heavy-chain dimer, respectively. (**c**) IgG secretion levels in the hemolymph of silkworm larvae. The control group represents IgG production without any coexpressed chaperones. Data are ± SD. (Adapted with permission (Dojima et al. 2010b). Copyright 2010 Elsevier)

pupae was 30.1 μg per larva, while the whole pupae had 78.0 μg per pupa (Dojima et al. 2010b) measured by ELISA.

Analysis of the *N*-glycan profile in the silkworm larval hemolymph revealed $Man_2GlcNAc(Fuc)GlcNAc$ (47.4%), $Man_3GlcNAc(Fuc)GlcNAc$ (28.5%), and a minor presence of $GlcNAcMan_3GlcNAc(Fuc)GlcNAc$ (3.4%) (Fig. 6.5b(i)) (Dojima et al. 2010b). In comparison, the *N*-linked glycan structures of 296IJ6 IgG in the whole pupae mainly consisted of paucimannosidic glycans (52.1%), similar to those found in hemolymph, and 14.6% featured a terminal *N*-acetylglucosamine (GlcNAc), indicating high mannose glycans made up 33.3% of the structures (Dojima et al. 2010b). These findings imply that the proteins from whole pupae underwent limited *N*-glycan processing and that pupae exhibited higher *N*-acetylglucosaminyltransferase activity than larvae. Additionally, α1,6-fucose groups were present on 296IJ6 IgG expressed in both larvae and pupae.

To assess whether 29IJ6 IgG secretion could be improved, five different molecular chaperones were coexpressed. Western blot analysis under reducing conditions confirmed the presence of both heavy-chain (HC) and light-chain (LC) proteins in the hemolymph (Fig. 6.5b(i)). Under nonreducing conditions (Fig. 6.5b(ii)), the fully assembled IgG (H_2L_2) and intermediate immunoglobulin oligomers (H_2L and H_2) were also detected.

Coexpression of CRT, CNX, and BiP with 296IJ6 IgG in silkworm larvae significantly boosted 296IJ6 IgG secretion in larval hemolymph to 229, 220, and 119 μg/mL, respectively (Fig. 6.5c), which is about five times higher than the levels without chaperone coexpression (43.0 μg/mL). When ERp57 or Hsp70 were coexpressed, the secretion levels reached 175 and 117 μg/mL, respectively, representing a three- to fourfold increase over controls. These findings indicate that CRT, CNX, and BiP substantially enhance 296IJ6 IgG folding and assembly during post-translational modification.

The coexpression of molecular chaperones such as CRT, CNX, and BiP, GRP78 significantly enhanced the secretion of 296IJ6 IgG in silkworm larvae, resulting in a fivefold increase. Specifically, the total concentration of recombinant 296IJ6 IgG reached 239 μg/mL when CRT was coexpressed. The efficiency of molecular chaperone coexpression was also demonstrated in human α1,4 *N*-acetylglucosaminyltransferase (α4GnT) expression in silkworm. The expression of GFP_{uv}-α4GnT in the silkworm larvae was 352 mU/mL within the larval haemolymph. Notably, when molecular chaperones such as CNX or BiP, GRP78 were coexpressed under the regulation of the ie-2 promoter, the α4GnT activity in larval haemolymph increased markedly to 922 mU/mL, representing a 2.6-fold enhancement compared to expression without chaperone coexpression (Nakajima et al. 2009).

6.2.3 *Discussion*

These findings suggest that the BmNPV bacmid-mediated, chaperone-assisted expression system in silkworm larvae offers significant potential to produce recombinant proteins, particularly for large-scale synthesis of structurally complex proteins. The analysis of the *N*-linked glycans present in the secreted 296IJ6 IgG within silkworm hemolymph revealed a predominance of paucimannose structures; however, the coexpression of molecular chaperones did not alter the composition of *N*-linked glycans in the IgG1 produced, as compared to those obtained without chaperones. These findings indicate that *N*-glycosylation is regulated by intrinsic mechanisms within the Golgi apparatus, independent of the translational enhancement provided by molecular chaperones. Consequently, further optimization of glycosylation pathways—such as coexpression of other glycosyltransferases, including N-acetylglucosaminyltransferase, galactosyltransferase, and sialyltransferase—has the potential to establish silkworm larvae as an effective platform for producing human antibodies.

6.3 Membrane Proteins

Investigating the structure and function of membrane proteins often necessitates large amounts of protein, typically in the milligram order and in an active form. However, producing such quantities presents a considerable challenge. For example, although mammalian cells produce proteins structurally and functionally similar to those found in their natural environment, their expression levels rarely exceed 5–10 pmol per mg of protein (Massotte 2003). Due to the complexities and high costs associated with boosting these expression levels, researchers often turn to alternative systems. Insect cells and silkworms have proven to be effective alternatives for this purpose. Since membrane proteins remain embedded in the cell membrane and are not secreted into the hemolymph, researchers harvest the fat body tissue from silkworm larvae and disrupt the cells using sonication. After the removal of cellular debris, the microsomal fraction is isolated, resuspended in an extraction buffer, and the membrane proteins are then solubilized and typically purified via affinity chromatography. Following this method, prorenin and (pro)renin receptors (PRR), which play a role in the renin–angiotensin system by nonproteolytically activating prorenin, have been expressed in silkworm larvae. Their presence in the fat body cells of silkworms has been verified, and after purification, their ability to bind renin has been successfully demonstrated (Du et al. 2008; Du et al. 2009a, 2009b).

In humans, guanine nucleotide-binding protein (G protein)-coupled receptors (GPCRs) comprise nearly 4% of all genes, making them the largest receptor superfamily in eukaryotic cells. GPCRs play crucial roles in numerous physiological functions and are considered key targets in pharmaceutical research and drug discovery. While many GPCR genes have been identified in the human genome, our understanding of their specific functions and the molecules that interact with them remains incomplete, largely due to insufficient structural and functional data. High-resolution three-dimensional structures of GPCRs are primarily obtained through X-ray crystallography. This technique requires large amounts of purified protein—typically on the milligram scale—which is often difficult to obtain. To address this challenge, several advanced heterologous protein expression systems have been established (Mccusker et al. 2007). The *E. coli* system is particularly attractive due to its affordability and ease of use (Bane et al. 2007). However, *E. coli* lacks native G-proteins, proteins that interact with GPCRs, and cannot perform post-translational modifications. By contrast, yeast expression systems can introduce post-translational modifications and create disulfide bonds. Some GPCRs have been successfully produced in high quantities using yeast (Niebauer and Robinson 2006; O'malley et al. 2007). Despite this, modifications in yeast are not always predictable, and inconsistent glycosylation can complicate crystallization. Insect cell systems offer another alternative, with reports of 16 GPCRs expressed at levels ranging from 1 to 250 pmol/mg protein (about 0.005 to 2 mg/L) in Sf9, Sf21, and *Trichoplusia ni* cell lines (Akermoun et al. 2005). Adrenergic receptors have been produced and crystallized using insect cells (Cherezov et al. 2007; Warne et al. 2008).

The silkworm-based GPCR expression remains challenging to purify while preserving biological activity, limiting its broader application. However, the use of silkworms has enabled high-yield GPCR expression; for instance, the human nociceptin receptor and its fusion with an inhibitory G protein alpha subunit have been produced in the fat bodies of silkworm larvae at much higher yields than in Sf9 cells (Kajikawa et al. 2009). The results of the silkworm genome project (International Silkworm Genome, 2008) have facilitated a comprehensive analysis of GPCRs in silkworms, shedding light on their regulatory roles in Lepidopteran development and in insect evolution and adaptation. To date, 90 potential GPCRs have been identified in *Bombyx mori*, including 33 previously unreported proteins (Fan et al. 2010). Recent literature has thoroughly summarized advances in GPCR signaling, feeding behavior, and other physiological processes in insects (Liu et al. 2021). It has highlighted ongoing research and new perspectives on GPCR signaling mechanisms (Calebiro et al. 2024).

6.4 Virus-Like Particles

Virus-like particles (VLPs) are nanoscale structures that assemble themselves into icosahedral shapes, consisting of viral capsid or coat proteins. These particles are non-infectious because they do not contain any genetic material. VLPs can originate from animal viruses, bacteriophages, or plant viruses and are widely researched for a range of uses. The process begins when the viral genome enters host cells and is transcribed in the nucleus. This leads to the creation of mRNAs, which are then translated into capsid or coat proteins (1 in Fig. 6.6). Membrane proteins are also generated through this pathway. After further post-translational modifications, these proteins are transported to the cell membrane (2 in Fig. 6.6). Inside the cell membrane, capsids self-assemble and bud outward, forming VLPs. During this budding process, VLPs become wrapped in the cell membrane (3 in Fig. 6.6), resulting in

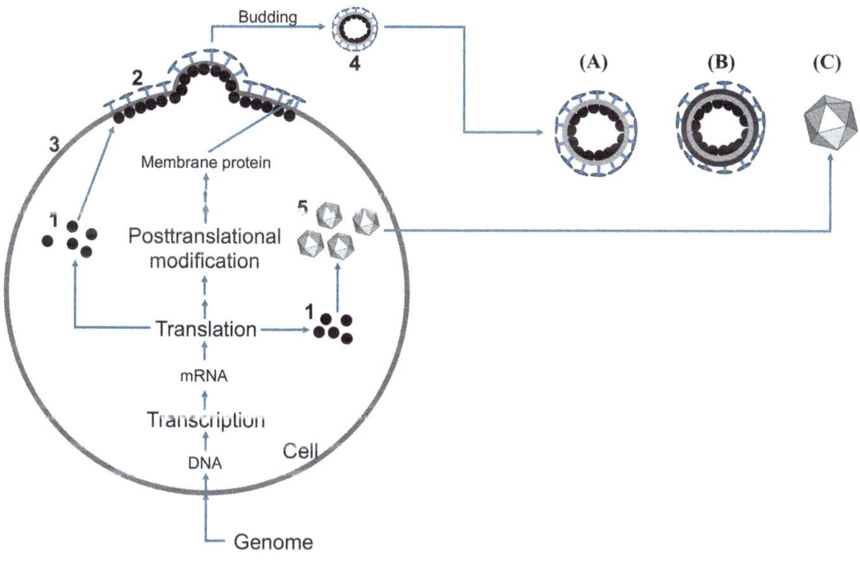

Fig. 6.6 Diagram illustrating the formation of eVLPs (**a**), multilayered VLPs (**b**), and neVLPs (**c**). 1 and 2 represent the capsid and membrane proteins; 3 indicates the cell membrane; 4 and 5 correspond to eVLP and neVLP, respectively. (Adopted with permission (Park 2025). Copyright 2025 Elsevier)

enveloped VLPs (eVLPs, Fig. 6.6a). Sometimes, eVLPs acquire several membrane layers, producing multilayered VLPs (Fig. 6.6b). Alternatively, capsid proteins can self-assemble into icosahedral forms without a membrane, creating nonenveloped VLPs (neVLPs, Fig. 6.6c).

The earliest recombinant VLPs were generated using cloned genes for the hepatitis B virus coat proteins, specifically the core antigen (HBcAg) (Burrell et al. 1979) and the surface antigen (HBsAg) (Valenzuela et al. 1982). As these particles exit the host cell, they acquire a membrane envelope, forming eVLPs. Since then, scientists have produced eVLPs from a wide range of viral families (Zeltins 2013). For example, human immunodeficiency virus-like particles (HIV-LPs) are also categorized as eVLPs. Additionally, during or after their assembly, it's feasible to incorporate one or more new proteins into the VLP envelope (Greenstone et al. 1998; Peacey et al. 2007; Rudolf et al. 2001; Wagner et al. 1996).

On the other hand, the neVLPs consist of multiple protein subunits arranged in icosahedral symmetry, giving them a more straightforward structure than eVLPs. Some VLPs can develop multiple layers by coexpressing and assembling distinct structural proteins within cells, forming multilayered VLPs. An example is the rotavirus-like particle (RV-LP) (Rodriguez-Limas et al. 2011), which consists of three distinct layers—inner, middle, and outer—constructed by the expression of RV proteins VP2, VP6, and VP7.

Various expression platforms are utilized to generate VLPs. Bacterial hosts, particularly *E. coli*, are frequently employed due to established protocols and widespread use (Tan et al. 2004). However, these systems are incapable of performing post-translational modifications, which often leads to the formation of inclusion bodies. In contrast, yeast species such as *Pichia* do not produce inclusion bodies and can execute some post-translational modifications. Still, yeast-based systems may require additional chaperonins to aid in proper protein folding and can sometimes introduce unsuitable glycosylation patterns (Morikawa et al. 2007). Insect cell expression systems, such as Sf9 and High Five, offer benefits including post-translational modification, rapid proliferation, and suitability for industrial-scale manufacturing. The development of baculovirus expression vectors has enabled the production of numerous eVLPs and neVLPs (Possee et al. 2020). Nevertheless, these systems require the creation of recombinant baculovirus vectors carrying the desired gene, as well as managing the infection process and operating bioreactors with sophisticated control mechanisms and expensive serum-free media (Krammer et al. 2010). Plants are now seen as a promising, more scalable, and economical option for VLP production (Marsian and Lomonossoff 2016; Mathew et al. 2014; Zhang et al. 2006). Mammalian cell cultures are highly proficient in producing recombinant VLPs because they facilitate post-translational modifications and multimeric assembly. Although transient transfection in these cells allows VLP collection within 48 to 72 h, this approach is typically associated with higher costs and lower yields than other systems.

VLPs serve a variety of purposes, with drug delivery being a prominent example (Deo et al. 2015). By embedding specific targeting molecules or antibodies within their membrane or capsid, VLPs are capable of encapsulating a range of therapeutic agents (Garcea and Gissmann 2004; Kato et al. 2015; Molino and Wang 2014; Yamada et al. 2003). For instance, *Rous sarcoma* virus-like particles (RSV-LPs) engineered to display the single-chain variable fragment (scFV) of the humanized CC49 antibody on their surfaces have achieved targeted delivery of anticancer drugs directly to colon carcinoma cells (Kato et al. 2015). Additionally, these proteins can provoke an immune response (Deo et al. 2013; Layton et al. 1996; Villa et al. 2006; Wagner et al. 1998; Young et al. 2006), which positions them as strong candidates for vaccine development. A well-known example is nonenveloped human papillomavirus-like particles (HPV-LPs), which are widely studied for use in preventive vaccines (Wang and Roden 2013).

The silkworm offers a cost-efficient platform for producing VLPs in large quantities. *B. mori* cells can perform post-translational modifications comparable to those seen in mammalian cells, as elaborated in Chap. 5. When silkworm larvae or pupae are injected with recombinant BmNPV bacmids, VLP production accelerates compared to the traditional baculovirus expression vector system. The use of silkworms for VLP synthesis is advantageous because they are inexpensive; expensive bioreactors and complex equipment are unnecessary—only a basic temperature-controlled environment and mulberry leaves for nutrition are required. This method is extremely economical because silkworms can generate substantial amounts of protein. Recent progress in VLP manufacturing utilizing the silkworm expression system is discussed here.

6.4.1 Enveloped VLPs

Viral proteins such as Gag, envelope, and coat are produced within the cell membrane, where they assemble and are subsequently released as eVLPs. When the Gag protein from Rous sarcoma virus (RSV) is expressed, it forms round RSV-like particles (RSV-LPs). The RSV-gag protein is composed of three distinct proteins: the mature matrix (MA), the capsid (CA), and the nucleocapsid (NC) (Ma and Vogt 2002). These proteins are required to package the viral RNA genome and are subsequently cleaved by viral proteases. Here, RSV-gag-577 (lacking the RSV-gag viral protease). The RSV-gag cDNAs were amplified by PCR and cloned into the pENTR/D-TOPO entry vector (Invitrogen, Carlsbad, CA, USA) to generate pENTR/RSV-gag-577. To express RSV-gag in silkworm larvae, the *RSV-gag*-577 inserts were transferred from their entry clones into pDEST8 by LR recombination, resulting in pDEST8/RSV-gag-577. This plasmid was then used to transform DH10 BmNPV *CP*$^{-}$-*Chi*$^{-}$ competent cells to generate recombinant bacmids (Deo et al. 2006), naming BmNPV bacmid/RSV-gag-577.

Silkworms were injected with BmNPV bacmid/RSV-gag-577, after which hemolymph and fat body tissues were collected on 7 dpi (Fig. 6.7). The VLPs generated.

in silkworms were predominantly found in the hemolymph and were purified using ultracentrifugation combined with step sucrose gradients. This process

Fig. 6.7 Expression of RSV-LPs in silkworms. A solution containing 10 μg of BmNPV bacmid DNA and 10% (v/v) DMREI-C reagent (Invitrogen) in PBS was prepared, and 40 μL was injected into silkworms using 1-mL syringes. The silkworms were fed an artificial diet (Silkmate 2S) every day, and hemolymph was collected from the silkworms on 7 dpi into Falcon tubes containing 2 mM phenylthiourea to prevent melanization. The collected hemolymph was divided into 1-mL portions in Eppendorf tubes and stored at −80 °C. Additionally, the fat bodies from the silkworms were scraped with a spatula and stored at −80 °C

resulted in the formation of two distinct layers (Fig. 6.8a). Each layer was carefully separated and subsequently analyzed by Western blot and SDS-PAGE (Fig. 6.8b, c) to verify the presence of the RSV-gag protein and assess sample purity. The RSV-LPs tend to cluster at the plasma membrane, supporting the hypothesis that VLPs assemble at the plasma membrane (Dalton et al. 2005; Johnson et al. 2005; Krishna et al. 1998). The 75 kDa band indicates RSV-LPs purified from silkworm larval hemolymph. Once sufficient gag monomers (MA, p10, CA, NC, and PR) accumulate in the plasma membrane, the resulting reduction in surface tension facilitates the release of VLPs from the plasma membrane. This process allows for straightforward collection of VLPs from both the supernatant and larval hemolymph, eliminating the need for cell-disruptive extraction techniques. Except for RSV-gag-577, several bands are shown in the purified sample in the Western blot. Thick bands around 30 kDa and 40 to 50 kDa might be monomers or aggregates of gag proteins, such as CA, p10-CA-NC, or MA-p10-CA (lanes 2 and 3 in Fig. 6.8b, c). As shown by this ultracentrifugation, the VLPs are not sufficiently pure for pharmaceutical purposes. An additional chromatographic method is needed. In this RSV-LP, morphology is shown as a spherical shape with a diameter of 100 nm (Fig. 6.9a).

Another example is the expression of dengue virus-like particles (DENV-LPs) in silkworms. For dengue virus (DENV), the capsid-premembrane-envelope (CprME) or premembrane-envelope (prME) gene sequences are expressed under the regulation of the *polh* promoter, resulting in the formation of DENV-LPs and DENV-LPs that have a size of approximately 77 nm (Fig. 6.9b) (Utomo et al. 2020). The particles produced from the CprME construct are spherical and measure about 55 nm, while those from the prME construct are around 35 nm in diameter (Utomo et al. 2019; Utomo et al. 2020). Additionally, a hemagglutinin (HA)-fused M1 protein

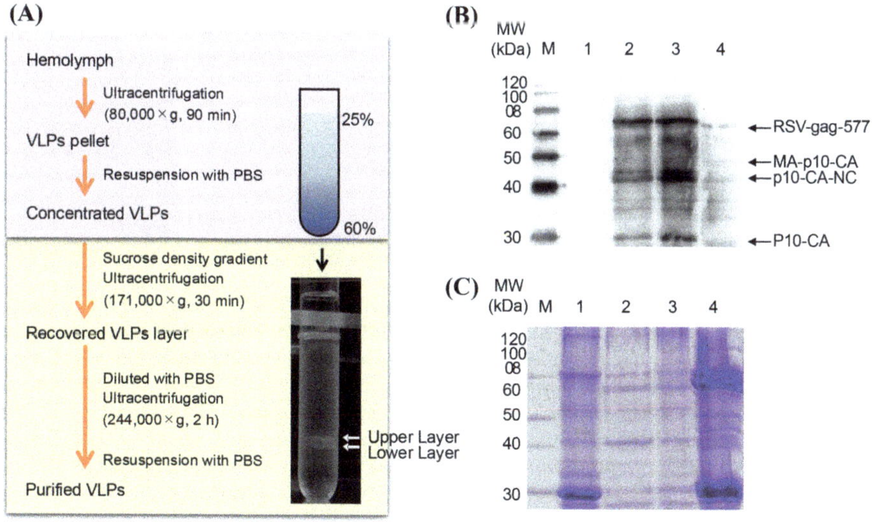

Fig. 6.8 Purification of RSV-LPs from silkworm larval hemolymph (**a**) and their analysis via Western blot (**b**) and SDS-PAGE (**c**). (**a**) The supernatant containing VLPs was harvested from silkworm larvae and first centrifuged at 1000 × g for 3 min using a swinging-bucket rotor to remove cells and debris. The clarified supernatant was then subjected to ultracentrifugation at 171,000 × g for 30 min at 4 °C, using a swinging-bucket rotor to precipitate the VLPs. The resulting pellets were dissolved in 200 μL of STE buffer (100 mM NaCl, 10 mM Tris-HCl at pH 7.5, 1 mM EDTA). This suspension was placed atop a 4.5 mL step sucrose gradient ranging from 10% to 60% (w/w) in STE. The gradients were then spun at 244,000 × g for 2 h at 4 °C with a swinging bucket rotor. Afterward, the gradients were divided into seven fractions (fractions 1–4: 500 μL each; fractions 5–7: 1 mL each). Each fraction underwent further concentration and buffer exchange using Microsep Spin 10 K columns (Pall, NY, USA). The fractions containing VLPs were combined and stored at 4 °C. Image in (**a**) captured after purifying silkworm larval hemolymph using a sucrose step gradient. Two distinct layers, indicated by white arrows, are visible. (**b**) Western blot analysis with the anti-RSV-gag antibody. (**c**) SDS-PAGE gels stained with Coomassie Blue. Lane M, molecular marker (Bio-Rad); lane 1, purified baculovirus (negative control); lane 2, lower density fraction; lane 3, higher density fraction; lane 4, larval hemolymph. (Adapted with permission (Deo et al. 2011). Copyright 2011 Elsevier)

from highly pathogenic avian influenza virus (HPAIV) is expressed to generate HPAIV-like particles (HPAIV-LPs) with a diameter of 150 nm. These HPAIV-LPs stimulate the production of HA-inhibiting antibodies, increase levels of splenic interferon-gamma (IFN-γ), elevate interleukin 4 (IL-4), and raise the $CD4^+/CD8^+$ ratio, supporting their potential use as vaccine candidates against HPAIV (Huang et al. 2021). For Japanese encephalitis virus (JEV), expressing the envelope (E) or prME protein leads to the formation of JEV-like particles (JEV-LPs), which are spherical and measure 80–100 nm (Fig. 6.9c). These JEV-LPs effectively inhibit JEV infection (Nerome et al. 2018). The presence of E proteins from DENV or JEV on the VLP surfaces has been verified through immuno-electron microscopy (IEM) using gold nanoparticle-tagged anti-E antibodies, where black dots indicate the protein locations around the VLPs (Fig. 6.9b, c).

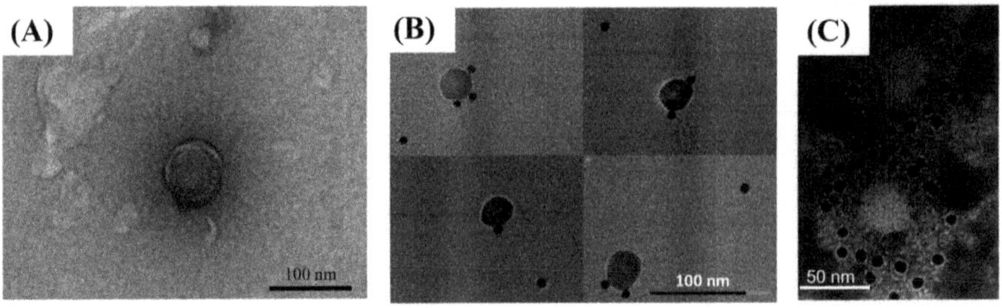

Fig. 6.9 Transmission electron microscopy (TEM) images of RSV-like particles (RSV-LPs) (**a**), DENV-like particles (DENV-LPs) (**b**), and JEV-like particles (JEV-LPs) (**c**). RSV-LPs were produced in silkworm larvae and isolated using sucrose gradient ultracentrifugation. DENV-LPs were produced in silkworm larvae and purified by affinity chromatography. JEV-LPs were produced in silkworm pupae and separated using sucrose gradient ultracentrifugation. The scale bars in (**a**), (**b**) denote 100 nm; in (**c**) 50 nm. In (**b**) and (**c**), the small dots represent gold-labeled antibodies specific for DENV or JEV envelope proteins, as observed by immuno-electron microscopy (IEM), respectively. (Adopted with permission (Park 2025). Copyright 2025 Elsevier)

While there are few examples of VLP expression in silkworms, the expression levels and properties of the VLPs are becoming clearer. The yield of VLPs was influenced by both the VLP type and the host used for expression. RSV-LP and DENV-LP were produced in larvae at levels of about 30–70 μg per larva, while JEV-LP was generated in pupae at roughly 300–700 μg per pupa, making its production more than ten times higher. This demonstrates that pupae have a much higher capacity for VLP production compared to larvae. Details regarding the matrix proteins used, as well as the sizes and production amounts of the expressed eVLPs, are provided in Table 6.4.

6.4.2 *Nonenveloped VLPs*

Protein subunits were produced in specific quantities and spontaneously assembled into particles with icosahedral shapes (processes 1 to 5 in Fig. 6.6), referred to as neVLPs, within the cellular membrane, as illustrated in Fig. 6.6c. Numerous examples of neVLP expression have been successfully generated in silkworm larvae or pupae.

Canine parvovirus-like particles (CPV-LPs) were generated utilizing the VP2 structural protein from Canine parvovirus (CPV) with a yield of 688 μg per silkworm pupa and about 300 μg per larva (Feng et al. 2011). Transmission electron microscopy (TEM) images demonstrated that the CPV-LPs isolated from the fat bodies of silkworm larvae exhibited a spherical structure, measuring approximately 20–22 nm in diameter (Lee et al. 2019) (Fig. 6.10a). Studies on the immunogenic properties of these CPV-LPs in mice and dogs showed that they could induce neutralizing antibodies against CPV without the need for an adjuvant. Additionally,

Table 6.4 eVLPs expressed in silkworm larvae or pupae

eVLP	Matrix protein	Diameter (nm)	Purified level	References
RSV-LP	*gag*-577 or 701	77	77 μg/larva	Deo et al. (2011)
DENV (serotype 2)-LP	C-PrME	35	28–40 μg/larva	Utomo et al. (2019)
DENV (serotype 3)-LP	C-PrME or prME	30–55	42 μg/larva 38 μg/larva	Utomo et al. (2020)
HPAIV (H5)-LP	HA-fused M1 protein	150	2.36 mg/mL pupa	Huang et al. (2021)
JEV (Nakayama)-LP	prME	–	300–500 μg/pupa	Matsuda et al. (2017)
JEV (Muar)-LP	E	80–100	725 μg/pupa	Nerome et al. (2018)

dogs that received oral administration of the CPV-LPs developed a comparable and durable immune response (Feng et al. 2014).

Papillomavirus-like particles (HPV-LPs) have already been developed and marketed as vaccines targeting cervical cancer in women. In 2012, researchers successfully expressed HPV-LPs in silkworm larvae (Palaniyandi et al. 2012). Bovine papillomavirus-like particles (BPV-LPs), which measure about 50 nm in diameter (Fig. 6.10b), were purified from a single silkworm pupa, yielding 700 μg. These particles induced specific antibody responses and demonstrated high sensitivity in detecting BPV (Watanabe et al. 2017). Capsid proteins from rotavirus (RV)—specifically VP2, VP6, and VP7—were also produced in larvae, resulting in a yield of 12.7 μg per larva after purification by ultracentrifugation (Yao et al. 2012). When only the VP6 protein was purified using affinity chromatography, the yield increased to 330 μg of RV-LP per larva (Fig. 6.10c) (Kato et al. 2023). The variation in yield may be attributed to differences in the purification procedures or the structural matrix proteins utilized. Rabbit hemorrhagic disease virus-like particles (RHDV-LPs) were produced at a rate of 10^7 hemagglutinin (HA) units per pupa. Without any adjuvant, administering a dose of 10^4 HA units provided rabbits with protection against RHDV for up to 180 d (Zheng et al. 2016). Foot-and-mouth disease virus (FMDV) outbreaks in livestock can cause severe economic damage in susceptible cloven-hoofed animals. FMDV-like particles (FMDV-LPs) were produced in silkworm larvae by expressing the FMDV polyprotein, yielding particles 30 nm in size (Fig. 6.10d). In immunization studies, two out of six vaccinated cattle were protected from FMDV infection (Li et al. 2007). Additional experiments using FMDV-LPs based on VP0, VP1, and VP3 proteins expressed in pupae led to complete protection in four out of five animals (Kumar et al. 2016; Li et al. 2012; Li et al. 2008). The expression level for FMDV-LPs using eri silkworm larva was quantified at 200 μg per single silkworm larva (Kumar et al. 2016). Porcine circovirus 2-like particles (PCV-LPs) were generated using the Cap protein from PCV, with purification yields reaching 400 μg from a single pupa (Masuda et al. 2018) and 927 μg

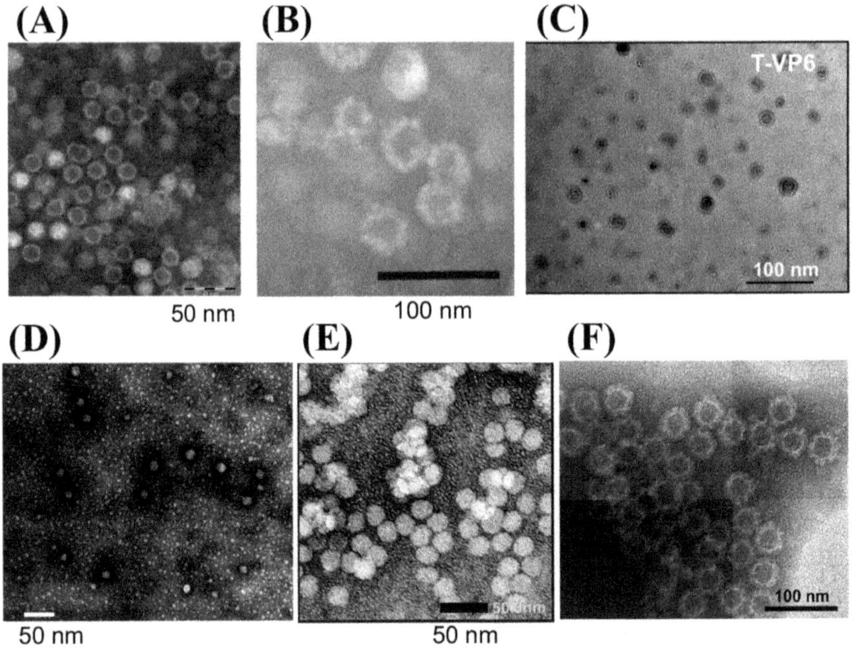

Fig. 6.10 TEM images of neVLPs produced in silkworms. Porcine parvovirus-like particles (PPV-LPs) (**a**) (Lee et al. 2019), BPV-LPs (**b**) (Watanabe et al. 2017), and PCV-LPs (**e**) (Masuda et al. 2018) were produced in the silkworm pupa, while porcine RV-LPs (**c**) (Kato et al. 2023), FMDV-LPs (**d**) (Li et al. 2007), and NoV-LPs (**f**) (Boonyakida et al. 2022) were purified from silkworm larvae. (Adapted with permission (Park 2025). Copyright Elsevier 2025)

from one larva (He et al. 2020). These particles range from 22 to 25 nm in diameter (Fig. 6.10e). Norovirus (NoV) is a primary cause of gastroenteritis worldwide, often known as stomach flu. The VP1 protein of NoV is used to produce NoV-like particles (NoV-LPs). Purification of these particles yielded more than 900 μg from either a single larva or a pupa (Masuda et al. 2023; Tsurumi et al. 2025), attributed to their stable structure. The resulting particles, roughly 36–40 nm in diameter, exhibit a rough but uniform surface (Boonyakida et al. 2022) (Fig. 6.10f). Sera from mice immunized with NoV-LPs bound to antigens in pig gastric mucin and effectively inhibited NoV infection in intestinal epithelial cells (Masuda et al. 2023). Furthermore, VP1 proteins from nine distinct NoV genotypes were expressed and purified as NoV-LPs in silkworm pupae, with each genotype successfully forming the desired particles (Tsurumi et al. 2025).

The matrix proteins utilized, along with the sizes and production yields of their corresponding neVLPs, are comprehensively detailed in Table 6.5.

Table 6.5 neVLPs expressed in silkworm larva or pupa

Names	Matrix protein	Diameter (nm)	Production level	References
Canine parvovirus-like particle (CPV-LP)	VP2 protein	22–25	688 μg/pupa 297 μg/larva	Feng et al. (2011)
CPV-LP	VP2 protein	25	–	Feng et al. (2014)
Porcine parvovirus-like particle	VP2 protein	20–22	–	Lee et al. (2019)
Human papillomavirus 6-like particle (HPV-LP)	L1 protein	~50	–	Palaniyandi et al. (2012)
Bovine papillomavirus 6-like particle (BPV-LP)	L1 protein	50–55	700 μg/pupa	Watanabe et al. (2017)
Rotavirus-like particle (RV-LP)	VP2, 6, 7 proteins	–	12.7 μg/larva	Yao et al. (2012)
Porcine RV-LP	VP6 protein	–	330 μg/larva	Kato et al. (2023)
Rabbit hemorrhagic disease virus-like particle (RHDV-LP)	VP60 protein	30–35	10^7 HA unit/pupa	Zheng et al. (2016)
Foot-and-mouth disease virus (Asia I/XJ)-like particle (FMDV-LP)	Polyprotein	30	- /larva	Li et al. (2007)
FMDV-LP	VP0, 1, 3 proteins	30	- /pupa	Li et al. (2012)
FMDV-LP	VP0, 1, 3 proteins	–	200 μg/eri silkworm larva	Kumar et al. (2016)
Porcine circovirus 2-like particle (PCV-LP)	Capsid protein	22	400 μg/pupa	Masuda et al. (2018)
PCV-LP	Capsid protein	25	917 μg/larva	He et al. (2020)
Norovirus-like particle (NoV-LP)	VP1 protein	34–37	930 μg/larva	Boonyakida et al. (2022)
NoV-LP	VP1 protein	~40	989 μg/pupa	Masuda et al. (2023)
NoV(GI, GII (2,3,4,6), GII (17))-LP	VP1 protein	~40	>900 μg/pupa	Tsurumi et al. (2025)

This table was adopted with permission (Park 2025). Copyright 2025 Elsevier

6.4.3 Systemic Purification of VLPs Expressed in Silkworms

neVLPs are not released from cells but instead accumulate within them, being released only when the cells break down. Both silkworm larvae and pupae serve as platforms for producing these neVLPs. For larvae, the fat body is typically collected before homogenization and then lysed in a buffer and solubilized for purification. In pupae, the whole organism is homogenized, lysed, and then solubilized to prepare it for purification. Although biotechnology has made significant advances, achieving efficient, cost-effective purification of VLPs with high purity—free of nucleic acids and viruses—remains a major hurdle. Silkworms are ideal hosts due to their capacity for large-scale VLP production. However, only a few studies have tackled the challenge of purifying VLPs from silkworms at scale, mainly because of the abundance of silkworm-derived proteins present. In particular, substantial concentrations of host cell proteins with molecular weights of 20–30 kDa and 60–80 kDa (Hou et al. 2010) persist in the solubilized extracts, making it difficult to separate the desired product. Traditional purification techniques such as filtration (FILT), precipitation (PPT), ultracentrifugation (UCEN), and affinity chromatography (AFC) have been used, but UCEN is unsuitable for large-scale applications, and AFC is costly. Here, a general scalable purification process is illustrated in Fig. 6.11.

6.4.3.1 eVLP Purification Method

After viruses or VLPs are introduced into host cells, the initial step is to remove cell debris and solid material from lysates or disrupted cells. This is typically accomplished through centrifugation or filtration as a preliminary treatment (Fig. 6.11). Following this, one-step purification techniques such as anion exchange

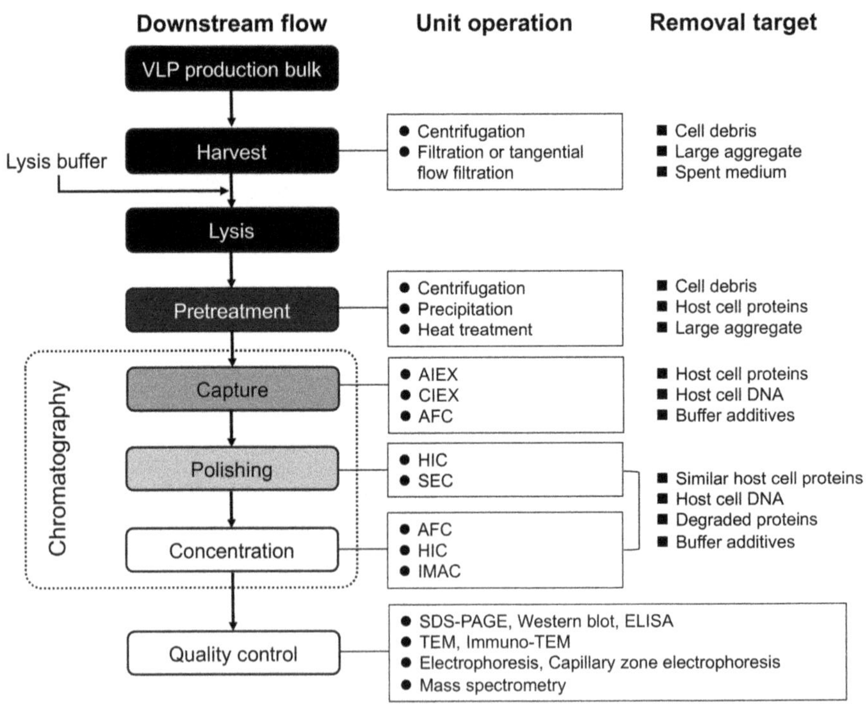

Fig. 6.11 General scalable purification process of VLPs. (Adopted with permission (Park and Minkner 2025). Copyright 2025 Elsevier)

chromatography (AIEX) (Pereira Aguilar et al. 2019; Steppert et al. 2016), ceramic hydroxyapatite chromatography (CHT) (Kurosawa et al. 2012), modern membrane-based separation (B Carvalho et al. 2018), or tangential flow filtration (TFF) are employed for the purification of dengue virus particles.

Because eVLPs possess a lipid membrane, certain buffers or pH adjustments may be limited during purification. In the initial step, a molecular sieve is used so that larger molecules pass through while smaller impurities are captured by ligands. Capto Core 700 has been employed at this stage and also serves as a polishing step for purifying large biomolecules, such as viruses, in scalable workflows (Lagoutte et al. 2016; Pereira Aguilar et al. 2020; Reiter et al. 2019). This technology features beads with an inert exterior and a ligand-functionalized interior; the outer layer has a pore size exclusion limit of about 700 kDa, preventing larger molecules, such as viruses or VLPs, from entering, while smaller molecules can be trapped by the core's ligand. There are a few examples, but size-exclusion chromatography (SEC) or AFC has been applied as a second step to concentrate VLPs after the first step, which dilutes them. While sequential SEC processes lead to low recovery, a workflow that uses flow-through (FT) in the first step, followed by AIEX in the second step, achieves a 51% recovery rate (Reiter et al. 2019). A variety of envelope-type viruses have been expressed and purified using diverse methods (Park and Minkner 2025); however, the eVLP purification methods, except for AFC, are very rare because of the difficulty of purification due to unstable buffers and the easy collapse of the particles.

6.4.3.2 neVLP Purification Method

Recently, there have been advances in scalable and cost-effective purification methods, such as the use of ion-exchange resins and novel chromatographic materials. Typically, a protein precipitation step is recommended before chromatography to eliminate a large proportion of host cell proteins. For example, after precipitation with ammonium sulfate, ion exchange chromatography was used to purify PCV-LPs from silkworm pupae, yielding 4 mg of PCV-LPs from just 10 pupae (Reiter et al. 2019).

Typically, achieving higher purity requires a two-step purification process. UCEN offers a traditional, straightforward, and user-friendly technique for purifying VLPs, though it is not suitable for large-scale applications. Despite this limitation, UCEN effectively separates VLPs from substantial host cell contaminants. For the purification of NoV-LPs produced in silkworm larvae, the UCEN was used as the initial step, followed by AFC as the second step (Fig. 6.12a, b). In the first phase, starting with lysate from the silkworm larval fat body, 61.2% of NoV-LPs were recovered at a purity level of 78%. The second phase yielded an overall recovery rate of 48.7%, with purity exceeding 90%, and eliminated double-stranded DNA by 99.6% (Fig. 6.12c). A total of 1 mg of NoV-LPs was obtained from seven silkworm larvae (Boonyakida et al. 2022). While this method is efficient and well-structured

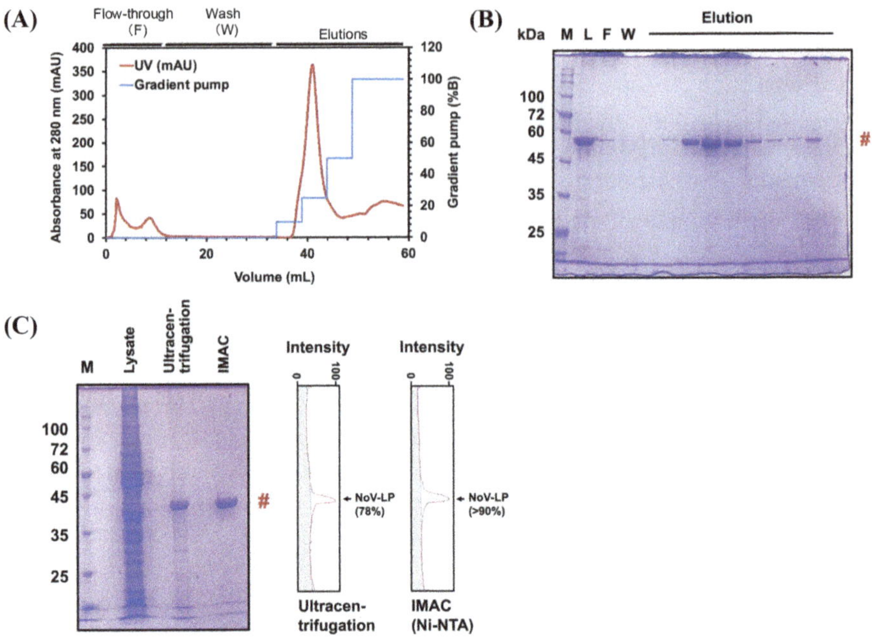

Fig. 6.12 Purification of NoV-LPs expressed in silkworm larva using the UCEN-AFC sequential purification. (**a**) Chromatography trace of VLPs separated on a HisTrap column. (**b**) SDS-PAGE image displaying the load (L), flowthrough (FT), wash (W), and elution. (**c**) SDS-PAGE comparison of NoV-LPs before and after each purification stage. The purity of VP1 post-ultracentrifugation and after Ni-NTA purification was quantified using densitometric analysis, with gray shading denoting background lane intensity. IMAC refers to Ni-NTA purification. (Adapted with permission (Boonyakida et al. 2022). Copyright 2022 Elsevier)

for producing substantial amounts of NoV-LPs from silkworm larvae, it presents significant challenges for scaling up to larger production volumes.

Because viruses and VLPs share many characteristics, achieving full separation through scalable chromatography can be challenging. For HPV-LPs produced in silkworm larvae, a multistep purification strategy was implemented to avoid DNA contamination: the first stage focuses on eliminating dsDNA, followed by the removal of baculovirus in the second stage, and finally, the third phase concentrates and refines the HPV-LPs. In the initial step, AIEX is used to effectively clear host cell proteins and reduce dsDNA content by more than 80% (the purple traces in Fig. 6.13A). Nevertheless, both baculovirus (represented in blue) and HPV-LPs (shown in pink) are released in the same set of fractions (fraction Nos. 3–7, marked in purple in Fig. 6.13A), making it impossible to distinguish between the two at this point. Western blot analysis demonstrated that fraction Nos. 3 through 10 contain HPV-LPs (Fig. 6.13B(a)), while fraction Nos. 2 through 7 contain baculovirus particles (Fig. 6.13B(b)).

Since ceramic hydroxyapatite (CHT) was the most widely used versatile chromatography material, it was employed in the second purification step to eliminate

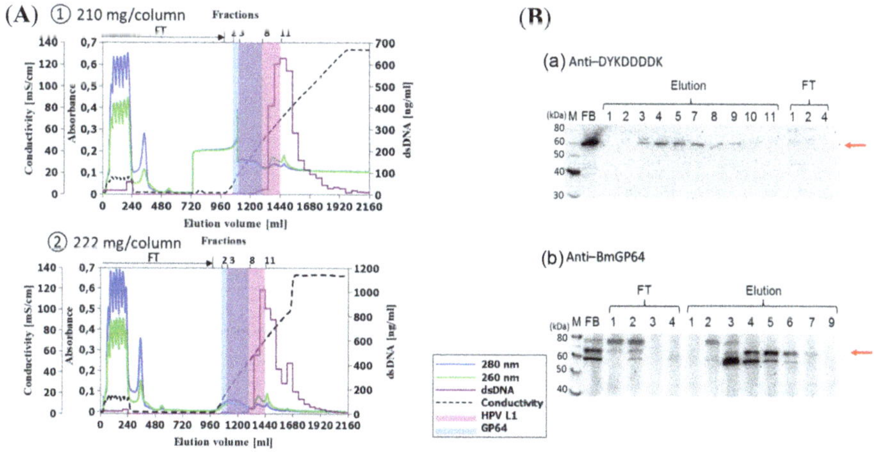

Fig. 6.13 First step purification of HPV-LPs. (**A**) The GigaCAP Q-650 M column (32 mm diameter × 35 mm length) operated at a flow rate of 1 mL/min. A linear salt gradient, ranging from 0 to 1.5 mol/L NaCl in a 150 mmol/L sodium phosphate buffer at pH 7.2. FT denotes the flow-through; blue lines represent absorbance at 280 nm, green lines indicate absorbance at 260 nm, dotted lines show conductivity (mS/cm), and purple lines reflect dsDNA concentration (ng/mL). The pink-shaded region marks where HPV-LP eluted, while the purple-shaded region shows the elution of BmGP64 from baculovirus. (**B**) Western blot analysis for each fraction collected from the GigaCAP Q-650 M column. Detection of HPV-LP and BmGP64 proteins was accomplished using anti-DYKDDDDK tag monoclonal antibodies and anti-BmGP64 antibodies, respectively. M, Magic Mark; XP and FB refer to HPV-LP-expressing silkworm fat bodies. Red arrows in panels (a) and (b) point to the molecular weights of HPV-LP and BmGP64, respectively. (Adapted with permission (Minkner et al. 2018). Copyright 2018 Elsevier)

the baculovirus. The HPV-LPs were separated using a sodium phosphate linear gradient at pH 7.2 (Fig. 6.14A). The majority of HPV-LP was found in elution fraction No. 5 through 17 (Fig. 6.14B(a)), whereas nearly all of the baculovirus was detected in the flow-through and in elution fraction Nos. 5 through 15 (Fig. 6.14B(b)). Although traces of baculovirus remained in the elution, the second-step purification reduced baculovirus particles by a order of 10^6 to 10^3.

During the third stage of the process, the samples underwent concentration and purification using a heparin affinity column. A large amount of sample was applied to the column, and HPV-LP was separated by gradually increasing the concentration of 10 mmol/L sodium phosphate (Fig. 6.15A(a), B). HPV-LP was primarily collected in fractions No. 25 through 30 (Fig. 6.15B(a)). Notably, neither baculovirus nor double-stranded DNA (dsDNA) was present in any of the eluted fractions (Fig. 6.15A, B(b)).

When the results of the three-step purification were compared with those of one-step methods, such as ultracentrifugation or affinity chromatography, HPV-LPs showed significantly less contamination. The three-step method yielded a total of 350 μg of HPV-LP from a 10 mL fat body sample (Minkner et al. 2018). Recovery rates for HPV-LP during the three-step purification were 31.4%, 22.7%, and 96.9%

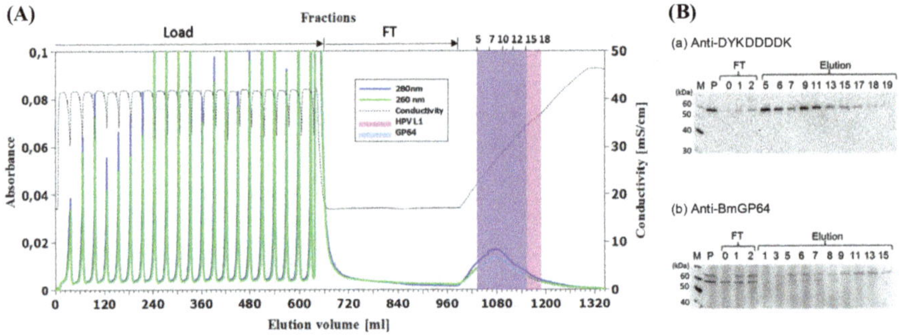

Fig. 6.14 The second purification step using CHT column chromatography. (**A**) The chromatographic separation was carried out at a flow rate of 1 mL/min. FT denotes the flowthrough; all other symbols and lines correspond to those in Fig. 6.13. (**B**) Western blot for each fraction collected during the CHT column purification. (a) Western blot of HPV-LP using anti-DYKDDDDK tag monoclonal antibody. (b) Western blot of BmGP64 using anti-BmGP64 antibody. M represents Magic Mark XP; P represents the pooled sample from column purification. Red arrows in panels (a) and (b) highlight the bands for HPV L1 and BmGP64, respectively. (Adapted with permission (Minkner et al. 2018). Copyright 2018 Elsevier)

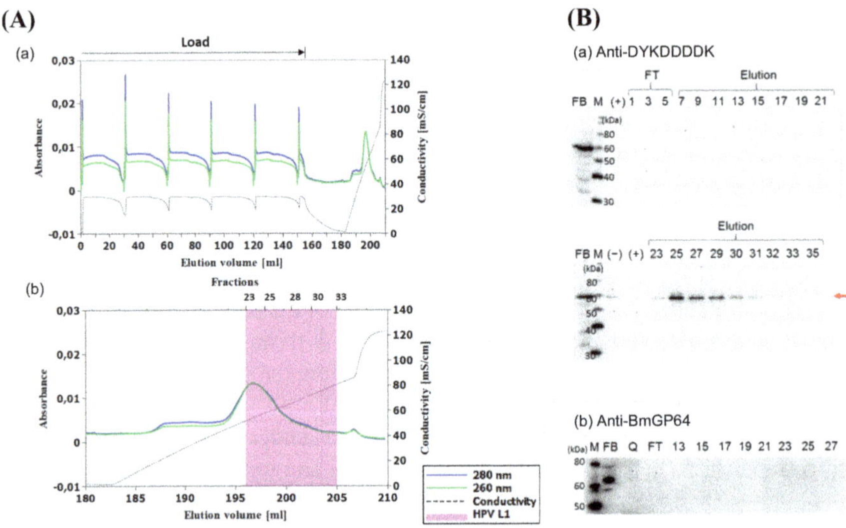

Fig. 6.15 The third stage of purification of HPV-LPs. (**A**) Heparin column chromatography. (a) Illustration of the sample loading process at a flow rate of 1 mL/min. (b) The elution process with a 0 to 1 mol/L sodium chloride gradient in 25 mL of a 10 mmol/L sodium phosphate buffer at pH 7.2. The symbols and line styles match those shown in Fig. 6.13. (**B**) Western blot analysis detects HPV-LP (a) and BmGP64 (b), using a monoclonal antibody against the DYKDDDDK tag and an anti-BmGP64 antibody, respectively. M for Magic Mark XP, FB for HPV-LP expressing fat body, Q for a large pool of CHT, (+) for the CHT fraction Nos. 4–17 pool, (−) for the unfiltered CHT fraction Nos. 4–17 pool, and 1 to 35 indicating fraction No. Red arrows in both (a) and (b) highlight the molecular bands corresponding to HPV L1 and BmGP64, respectively. (Adapted with permission (Minkner et al. 2018). Copyright 2018 Elsevier)

at each step, with a total recovery of 5.5% as determined (Minkner et al. 2018). Using one-step purification, the overall yield was 6.9%. The variation in yields was attributed to protein loss from repeated freezing and thawing during the multistep process. Although single-step ultracentrifugation or affinity chromatography yielded a slightly higher yield than the three-step approach, their final products contained higher levels of host cell proteins, baculovirus, and dsDNA. Notably, the three-step process achieved a 99.99% reduction in dsDNA, whereas the reductions for ultracentrifugation and affinity chromatography were 97.64% and 99.89%, respectively (Minkner et al. 2018).

In addition to the aforementioned findings, a significant number of neVLPs were expressed and purified in silkworms. While VLPs that are easy to purify can be processed using a one-step purification method, multistep purification methods are often employed to enhance purity or sufficiently prevent contamination. Table 6.6 summarizes the purification methods, recovery rates, and purities for each purification method.

6.4.3.3 Discussion

VLP stability is not fully understood, but neVLPs are generally more robust than eVLPs, reflecting trends in the virus. eVLPs are most stable at pH 7–8 and temperatures between 0 °C and 30 °C; they aggregate easily at lower pH or higher temperatures (Hu et al. 2011). Ionic strength and salt concentration are important, resulting

Table 6.6 Purification of neVLPs expressed in silkworm

Target	Pretreatment	Matrix	Recovery (%)	Purity (%)	Reference
One-step purification					
HPV-LP	CEN and FILT	Proteema 300 Å (PSS, Mainz, Germany)	–	–	Hahne et al. (2011)
PCV2-VLP	PPT	Hitrap Q HP (GE Healthcare)	4 mg/10 silkworm	Single band in SDS-PAGE	Zheng et al. (2016)
Two-step purification					
NoV-LP	CEN	1st: Sucrose gradient 2nd: Histrap HP (GE Healthcare)	67.4	>90 2.7 mg/10 larva	Boonyakida et al. (2022)
Multistep purification					
HPV-LP	FILT	1st: GigaCap Q-650 M (Tosoh Co. Japan) 2nd: CHT II (Bio-Rad, USA) 3rd: HiTrap Heparin HP (GE Healthcare)	6.2	DNA removal 99.997%	Minkner and Park (2018)

This table was modified with permission (Park and Minkner 2025). Copyright 2025 Elsevier

in a narrower stability range for eVLPs (Dalton et al. 2005). The lipid envelope is more fragile than the protein capsid, so precipitation steps require caution. PEG and ammonium sulfate are effective for neVLP purification (Lewis and Metcalf 1988), but only low concentrations of ammonium sulfate are tolerated by eVLPs without damaging the envelope (Steppert et al. 2017).

After purification, VLPs might contain trace host DNA or proteins (Lipin et al. 2008), which is a key concern for pharmaceutical applications. The main risk is from DNA with oncogene sequences; however, residual DNA below 100 pg per dose from continuous cell lines is considered safe (Grachev et al. 1998). Verifying the VLP structure and purity is critical, with SDS-PAGE, CBB staining, Western blotting, and ELISA as standard methods. TEM is essential for visual confirmation, especially of eVLP envelopes. Additional characterization methods include DLS, circular dichroism, fluorescence spectroscopy, and hemolysis assays. For VLPs used in humans or animals, high purity is crucial. HPLC, alone or combined with other methods, is recommended for purity testing, and capillary zone electrophoresis provides high-resolution analysis. Mass spectrometry can confirm surface modifications. Tracking the target protein during purification is challenging but necessary for yield estimation. Since a universal protocol is impractical due to differences among particle types, it is best to develop purification procedures specific to each particle type.

References

Adachi T, Takiya S, Suzuki Y, Iwami M, Kawakami A, Takahashi SY, Ishizaki H, Nagasawa H, Suzuki A (1989) cDNA structure and expression of bombyxin, an insulin-like brain secretory peptide of the silkmoth *Bombyx mori*. J Biol Chem 264:7681–7685

Ailor E, Betenbaugh MJ (1999) Modifying secretion and post-translational processing in insect cells. Curr Opin Biotechnol 10:142–145. https://doi.org/10.1016/s0958-1669(99)80024-x

Akermoun M, Koglin M, Zvalova-Iooss D, Folschweiller N, Dowell SJ, Gearing KL (2005) Characterization of 16 human G protein-coupled receptors expressed in baculovirus-infected insect cells. Protein Expr Purif 44:65–74. https://doi.org/10.1016/j.pep.2005.04.016

B Carvalho S, Fortuna AR, Wolff MW, Peixoto C, M Alves P, Reichl U, JT Carrondo M (2018) Purification of influenza virus-like particles using sulfated cellulose membrane adsorbers. J Chem Technol Biotechnol 93:1988–1996. https://doi.org/10.1002/jctb.5474

Bane SE, Velasquez JE, Robinson AS (2007) Expression and purification of milligram levels of inactive G-protein coupled receptors in *E. coli*. Protein Expr Purif 52:348–355. https://doi.org/10.1016/j.pep.2006.10.017

Boonyakida J, Utomo DIS, Soma FN, Park EY (2022) Two-step purification of tag-free norovirus-like particles from silkworm larvae (Bombyx mori). Protein Expr Purif 190:106010. https://doi.org/10.1016/j.pep.2021.106010

Braakman I, Hoover-Litty H, Wagner KR, Helenius A (1991) Folding of influenza hemagglutinin in the endoplasmic reticulum. J Cell Biol 114:401–411. https://doi.org/10.1083/jcb.114.3.401

Cabrera G, Cremata JA, Valdes R, Garcia R, Gonzalez Y, Montesino R, Gomez H, Gonzalez M (2005) Influence of culture conditions on the *N*-glycosylation of a monoclonal antibody specific for recombinant hepatitis B surface antigen. Biotechnol Appl Biochem 41:67–76. https://doi.org/10.1042/BA20040032

Calebiro D, Miljus T, O'brien S (2024) Endomembrane GPCR signaling: 15 years on, the quest continues. Trends Biochem Sci 50:46–60. https://doi.org/10.1016/j.tibs.2024.10.006

Cherezov V, Rosenbaum DM, Hanson MA, Rasmussen SG, Thian FS, Kobilka TS, Choi HJ, Kuhn P, Weis WI, Kobilka BK, Stevens RC (2007) High-resolution crystal structure of an engineered human beta2-adrenergic G protein-coupled receptor. Science 318:1258–1265. https://doi.org/10.1126/science.1150577

Dalton AK, Murray PS, Murray D, Vogt VM (2005) Biochemical characterization of rous sarcoma virus MA protein interaction with membranes. J Virol 79:6227–6238. https://doi.org/10.1128/JVI.79.10.6227-6238.2005

Deo VK, Hiyoshi M, Park EY (2006) Construction of hybrib Autographa californica nuclear polyhedrosis bacmid by modification of p143 helicase. J Virol Methods 134:212–216. https://doi.org/10.1016/j.jviromet.2006.01.007

Deo VK, Tsuji Y, Yasuda T, Kato T, Sakamoto N, Suzuki H, Park EY (2011) Expression of an RSV-gag virus-like particle in insect cell lines and silkworm larvae. J Virol Methods 177:147–152. https://doi.org/10.1016/j.jviromet.2011.07.012

Deo VK, Yoshimatsu K, Otsuki T, Dong J, Kato T, Park EY (2013) Display of *Neospora caninum* surface protein related sequence 2 on Rous sarcoma virus-derived gag protein virus-like particles. J Biotechnol 165:69–75. https://doi.org/10.1016/j.jbiotec.2013.02.013

Deo VK, Kato T, Park EY (2015) Chimeric virus-like particles made using GAG and M1 capsid proteins providing dual drug delivery and vaccination platform. Mol Pharm 12:839–845. https://doi.org/10.1021/mp500860x

Dojima T, Nishina T, Kato T, Uno T, Yagi H, Kato K, Park EY (2009) Comparison of the *N*-linked glycosylation of human beta1,3-N-acetylglucosaminyltransferase 2 expressed in insect cells and silkworm larvae. J Biotechnol 143:27–33. https://doi.org/10.1016/j.jbiotec.2009.06.013

Dojima T, Nishina T, Kato T, Ueda H, Park EY (2010a) Production of scFv-displaying BmNPV in silkworm larvae and its efficient purification. Biotechnol Appl Biochem 57:63–69. https://doi.org/10.1042/BA20100173

Dojima T, Nishina T, Kato T, Uno T, Yagi H, Kato K, Ueda H, Park EY (2010b) Improved secretion of molecular chaperone-assisted human IgG in silkworm, and no alterations in their *N*-linked glycan structures. Biotechnol Prog 26:232–238. https://doi.org/10.1002/btpr.313

Dong J, Harada M, Yoshida S, Kato Y, Murakawa A, Ogata M, Kato T, Usui T, Park EY (2013) Expression and purification of bioactive hemagglutinin protein of highly pathogenic avian influenza A (H5N1) in silkworm larvae. J Virol Methods 194:271–276. https://doi.org/10.1016/j.jviromet.2013.08.040

Du D, Kato T, Nabi AH, Suzuki F, Park EY (2008) Expression of functional human (pro)renin receptor in silkworm (*Bombyx mori*) larvae using BmMNPV bacmid. Biotechnol Appl Biochem 49:195–202. https://doi.org/10.1042/BA20070136

Du D, Kato T, Suzuki F, Park EY (2009a) Binding affinity of full-length and extracellular domains of recombinant human (pro)renin receptor to human renin when expressed in the fat body and hemolymph of silkworm larvae. J Biosci Bioeng 108:304–309. https://doi.org/10.1016/j.jbiosc.2009.04.018

Du D, Kato T, Suzuki F, Park EY (2009b) Expression of protein complex comprising the human prorenin and (pro)renin receptor in silkworm larvae using *Bombyx mori* nucleopolyhedrovirus (BmNPV) bacmids for improving biological function. Mol Biotechnol 43:154–161. https://doi.org/10.1007/s12033-009-9183-7

Fan Y, Sun P, Wang Y, He X, Deng X, Chen X, Zhang G, Chen X, Zhou N (2010) The G protein-coupled receptors in the silkworm, *Bombyx mori*. Insect Biochem Mol Biol 40:581–591. https://doi.org/10.1016/j.ibmb.2010.05.005

Feng H, Liang M, Wang HL, Zhang T, Zhao PS, Shen XJ, Zhang RZ, Hu GQ, Gao YQ, Wang CY, Wang TC, Zhang W, Yang ST, Xia XZ (2011) Recombinant canine parvovirus-like particles express foreign epitopes in silkworm pupae. Vet Microbiol 154:49–57. https://doi.org/10.1016/j.vetmic.2011.06.022

Feng H, Hu GQ, Wang HL, Liang M, Liang H, Guo H, Zhao P, Yang YJ, Zheng XX, Zhang ZF, Zhao YK, Gao YW, Yang ST, Xia XZ (2014) Canine parvovirus VP2 protein expressed in silkworm pupae self-assembles into virus-like particles with high immunogenicity. PLoS One 9:e79575. https://doi.org/10.1371/journal.pone.0079575

Garcea RL, Gissmann L (2004) Virus-like particles as vaccines and vessels for the delivery of small molecules. Curr Opin Biotechnol 15:513–517. https://doi.org/10.1016/j.copbio.2004.10.002

Grachev V, Magrath D, Griffiths E (1998) WHO requirements for the use of animal cells as in vitro substrates for the production of biologicals (Requirements for biological susbstances no. 50). Biologicals 26:175–193. https://doi.org/10.1006/biol.1998.0153

Greenstone HL, Nieland JD, De Visser KE, De Bruijn ML, Kirnbauer R, Roden RB, Lowy DR, Kast WM, Schiller JT (1998) Chimeric papillomavirus virus-like particles elicit antitumor immunity against the E7 oncoprotein in an HPV16 tumor model. Proc Natl Acad Sci USA 95:1800–1805. https://doi.org/10.1073/pnas.95.4.1800

Hahne T, Palaniyandi M, Kato T, Fleischmann P, Watzig H, Park EY (2014) Characterization of human papillomavirus 6b L1 virus-like particles isolated from silkworms using capillary zone electrophoresis. J Biosci Bioeng 118:311–314. https://doi.org/10.1016/j.jbiosc.2014.02.026

Hammond C, Helenius A (1994) Folding of VSV G protein: sequential interaction with BiP and calnexin. Science 266:456–458. https://doi.org/10.1126/science.7939687

Hasemann CA, Capra JD (1990) High-level production of a functional immunoglobulin heterodimer in a baculovirus expression system. Proc Natl Acad Sci USA 87:3942–3946. https://doi.org/10.1073/pnas.87.10.3942

He Q, Cao Z, Wang P, Lu Q, Zheng H, Sun J (2020) Efficient application of a baculovirus-silkworm larvae expression system for obtaining porcine circovirus type 2 virus-like particles for a vaccine. Arch Virol 165:2301–2309. https://doi.org/10.1007/s00705-020-04754-9

Hirai M, Kamimura M, Kikuchi K, Yasukochi Y, Kiuchi M, Shinoda T, Shiotsuki T (2002) cDNA cloning and characterization of *Bombyx mori* juvenile hormone esterase: an inducible gene by the imidazole insect growth regulator KK-42. Insect Biochem Mol Biol 32:627–635. https://doi.org/10.1016/s0965-1748(01)00141-2

Hou Y, Zou Y, Wang F, Gong J, Zhong X, Xia Q, Zhao P (2010) Comparative analysis of proteome maps of silkworm hemolymph during different developmental stages. Proteome Sci 8:45. https://doi.org/10.1186/1477-5956-8-45

Hsu TA, Eiden JJ, Bourgarel P, Meo T, Betenbaugh MJ (1994) Effects of co-expressing chaperone BiP on functional antibody production in the baculovirus system. Protein Expr Purif 5:595–603. https://doi.org/10.1006/prep.1994.1082

Hu L, Trefethen JM, Zeng Y, Yee L, Ohtake S, Lechuga-Ballesteros D, Warfield KL, Aman MJ, Shulenin S, Unfer R, Enterlein SG, Truong-Le V, Volkin DB, Joshi SB, Middaugh CR (2011) Biophysical characterization and conformational stability of Ebola and Marburg virus-like particles. J Pharm Sci 100:5156–5173. https://doi.org/10.1002/jps.22724

Huang D, Chao YC, Lv Z, Jan JT, Yang YC, Hsiao PW, Wu CY, Liao CH, Wu TH, Wang LC (2021) Comparison of chicken immune responses after inoculation with H5 avian influenza virus-like particles produced by insect cells or pupae. J Vet Res 65:139–145. https://doi.org/10.2478/jvetres-2021-0026

Ishikiriyama M, Nishina T, Kato T, Ueda H, Park EY (2009) Human single-chain antibody expression in the hemolymph and fat body of silkworm larvae and pupae using BmNPV bacmids. J Biosci Bioeng 107:67–72. https://doi.org/10.1016/j.jbiosc.2008.11.001

Jarvis DL, Garcia A Jr (1994) Biosynthesis and processing of the *Autographa californica* nuclear polyhedrosis virus gp64 protein. Virology 205:300–313. https://doi.org/10.1006/viro.1994.1646

Johnson MC, Spidel JL, Ako-Adjei D, Wills JW, Vogt VM (2005) The C-terminal half of TSG101 blocks Rous sarcoma virus budding and sequesters Gag into unique nonendosomal structures. J Virol 79:3775–3786. https://doi.org/10.1128/JVI.79.6.3775-3786.2005

Kajikawa M, Sasaki K, Wakimoto Y, Toyooka M, Motohashi T, Shimojima T, Takeda S, Park EY, Maenaka K (2009) Efficient silkworm expression of human GPCR (nociceptin receptor) by a

Bombyx mori bacmid DNA system. Biochem Biophys Res Commun 385:375–379. https://doi.org/10.1016/j.bbrc.2009.05.063376

Kato K, Kamiya Y (2007) Structural views of glycoprotein fate determination in cells. Glycobiology 17:1031–1044. https://doi.org/10.1093/glycob/cwm046

Kato T, Park EY (2007) Specific expression of GFPuv-beta1,3-*N*-acetylglucosaminyltransferase 2 fusion protein in fat body of *Bombyx mori* silkworm larvae using signal peptide. Biochem Biophys Res Commun 359:543–548. https://doi.org/10.1016/j.bbrc.2007.05.137

Kato T, Murata T, Usui T, Park EY (2003) Improvement of GFPuv-beta 3GnT2 fusion protein production by suppressing protease in baculovirus expression system. Biosci Biotechnol Biochem 67:2388–2395. https://doi.org/10.1271/bbb.67.2388

Kato T, Murata T, Usui T, Park EY (2004) Efficient production of human β-1,3-*N*-acetylglucosaminyltransferase-2 fused with green fluorescence protein in insect cell. Biochem Eng J 19:15–23. https://doi.org/10.1016/j.bej.2003.09.009

Kato T, Murata T, Usui T, Park EY (2005) Improvement of the production of GFPuv-beta1,3-*N*-acetylglucosaminyltransferase 2 fusion protein using a molecular chaperone-assisted insect-cell-based expression system. Biotechnol Bioeng 89:424–433. https://doi.org/10.1002/bit.20362

Kato T, Yui M, Deo VK, Park EY (2015) Development of Rous sarcoma virus-like particles displaying hCC49 scFv for specific targeted drug delivery to human colon carcinoma cells. Pharm Res 32:3699–3707. https://doi.org/10.1007/s11095-015-1730-2

Kato T, Kakuta T, Yonezuka A, Sekiguchi T, Machida Y, Xu J, Suzuki T, Park EY (2023) Expression and purification of porcine rotavirus structural proteins in silkworm larvae as a vaccine candidate. Mol Biotechnol 65:401–409. https://doi.org/10.1007/s12033-022-00548-3

Krammer F, Schinko T, Palmberger D, Tauer C, Messner P, Grabherr R (2010) *Trichoplusia ni* cells (high five) are highly efficient for the production of influenza A virus-like particles: a comparison of two insect cell lines as production platforms for influenza vaccines. Mol Biotechnol 45:226–234. https://doi.org/10.1007/s12033-010-9268-3

Krishna NK, Campbell S, Vogt VM, Wills JW (1998) Genetic determinants of Rous sarcoma virus particle size. J Virol 72:564–577. https://doi.org/10.1128/JVI.72.1.564-577.1998

Kumar M, Saravanan P, Jalali SK (2016) Expression and purification of virus like particles (VLPs) of foot-and-mouth disease virus in Eri silkworm (Samia cynthia ricini) larvae. Virus 27:84–90. https://doi.org/10.1007/s13337-015-0290-8

Kurosawa Y, Saito M, Kobayashi S, Okuyama T (2012) Purification of dengue virus particles by one-step ceramic hydroxyapatite chromatography. World J Vaccines 02. https://doi.org/10.4236/wjv.2012.23020

Kusaka H, Kita S, Tadokoro T, Yoshida K, Kasai Y, Niiyama H, Fujimoto Y, Hanashima S, Murata M, Sugiyama S, Ose T, Kuroki K, Maenaka K (2020) Efficient preparation of human and mouse CD1d proteins using silkworm baculovirus expression system. Protein Expr Purif 172. https://doi.org/10.1016/j.pep.2020.105631

Lagoutte P, Mignon C, Donnat S, Stadthagen G, Mast J, Sodoyer R, Lugari A, Werle B (2016) Scalable chromatography-based purification of virus-like particle carrier for epitope based influenza A vaccine produced in Escherichia coli. J Virol Methods 232:8–11. https://doi.org/10.1016/j.jviromet.2016.02.011

Layton GT, Harris SJ, Myhan J, West D, Gotch F, Hill-Perkins M, Cole JS, Meyers N, Woodrow S, French TJ, Adams SE, Kingsman AJ (1996) Induction of single and dual cytotoxic T-lymphocyte responses to viral proteins in mice using recombinant hybrid Ty-virus-like particles. Immunology 87:171–178. https://doi.org/10.1046/j.1365-2567.1996.464539.x

Lee SH, Bae SM, Gwak WS, Woo SD (2019) Production of porcine parvovirus virus-like particles using silkworm larvae. J Asia Pac Entomol 22:1167–1172. https://doi.org/10.1016/j.aspen.2019.10.002

Lewis GD, Metcalf TG (1988) Polyethylene glycol precipitation for recovery of pathogenic viruses, including hepatitis A virus and human rotavirus, from oyster, water, and sediment samples. Appl Environ Microbiol 54:1983–1988. https://doi.org/10.1128/aem.54.8.1983-1988.1988

Li Z, Yi Y, Yin X, Zhang Z, Liu J (2007) Immune response in cattle inoculated with the recombinant complete polyprotein of foot-and-mouth disease virus from Bombyx mori larvae. Chin Sci Bull 52:2805–2810. https://doi.org/10.1007/s11434-007-0436-1
Li Z, Yi Y, Yin X, Zhang Z, Liu J (2008) Expression of foot-and-mouth disease virus capsid proteins in silkworm-baculovirus expression system and its utilization as a subunit vaccine. PLoS One 3:e2273. https://doi.org/10.1371/journal.pone.0002273
Li Z, Yi Y, Yin X, Zhang Y, Liu M, Liu H, Li X, Li Y, Zhang Z, Liu J (2012) Development of a foot-and-mouth disease virus serotype A empty capsid subunit vaccine using silkworm (Bombyx mori) pupae. PLoS One 7:e43849. https://doi.org/10.1371/journal.pone.0043849
Lipin DI, Chuan YP, Lua LH, Middelberg AP (2008) Encapsulation of DNA and non-viral protein changes the structure of murine polyomavirus virus-like particles. Arch Virol 153:2027–2039. https://doi.org/10.1007/s00705-008-0220-9
Liu N, Wang Y, Li T, Feng X (2021) G-protein coupled receptors (GPCRs): signaling pathways, characterization, and functions in insect physiology and toxicology. Int J Mol Sci 22:5260. https://doi.org/10.3390/ijms22105260
Luckow VA, Lee SC, Barry GF, Olins PO (1993) Efficient generation of infectious recombinant baculoviruses by site-specific transposon-mediated insertion of foreign genes into a baculovirus genome propagated in *Escherichia coli*. J Virol 67:4566–4579. https://doi.org/10.1128/JVI.67.8.4566-4579.1993
Ma YM, Vogt VM (2002) Rous sarcoma virus Gag protein-oligonucleotide interaction suggests a critical role for protein dimer formation in assembly. J Virol 76:5452–5462. https://doi.org/10.1128/jvi.76.11.5452-5462.2002
Maeda S, Kawai T, Obinata M, Fujiwara H, Horiuchi T, Saeki Y, Sato Y, Furusawa M (1985) Production of human alpha-interferon in silkworm using a baculovirus vector. Nature 315:592–594. https://doi.org/10.1038/315592a0
Marsian J, Lomonossoff GP (2016) Molecular pharming - VLPs made in plants. Curr Opin Biotechnol 37:201–206. https://doi.org/10.1016/j.copbio.2015.12.007
Massotte D (2003) G protein-coupled receptor overexpression with the baculovirus-insect cell system: a tool for structural and functional studies. Biochim Biophys Acta 1610:77–89. https://doi.org/10.1016/s0005-2736(02)00720-4
Masuda A, Lee JM, Miyata T, Sato T, Hayashi S, Hino M, Morokuma D, Karasaki N, Mon H, Kusakabe T (2018) Purification and characterization of immunogenic recombinant virus-like particles of porcine circovirus type 2 expressed in silkworm pupae. J Gen Virol 99:917–926. https://doi.org/10.1099/jgv.0.001087
Masuda A, Lee JM, Miyata T, Ebihara T, Kakino K, Hino M, Fujita R, Mon H, Kusakabe T (2021) Stable trimer formation of spike protein from porcine epidemic diarrhea virus improves the efficiency of secretory production in silkworms and induces neutralizing antibodies in mice. Vet Res 52:102. https://doi.org/10.1186/s13567-021-00971-5
Masuda A, Man Lee J, Miyata T, Sato S, Masuda A, Taniguchi M, Fujita R, Ushijima H, Morimoto K, Ebihara T, Hino M, Kakino K, Mon H, Kusakabe T (2023) High yield production of norovirus GII.4 virus-like particles using silkworm pupae and evaluation of their protective immunogenicity. Vaccine 41:766–777. https://doi.org/10.1016/j.vaccine.2022.12.015
Mathew LG, Herbst-Kralovetz MM, Mason HS (2014) Norovirus Narita 104 virus-like particles expressed in Nicotiana benthamiana induce serum and mucosal immune responses. Biomed Res Int 2014:807539. https://doi.org/10.1155/2014/807539
Matsuda S, Nerome R, Maegawa K, Kotaki A, Sugita S, Kawasaki K, Kuroda K, Yamaguchi R, Takasaki T, Nerome K (2017) Development of a Japanese encephalitis virus-like particle vaccine in silkworms using codon-optimised prM and envelope genes. Heliyon 3:e00286. https://doi.org/10.1016/j.heliyon.2017.e00286
Mccusker EC, Bane SE, O'malley MA, Robinson AS (2007) Heterologous GPCR expression: a bottleneck to obtaining crystal structures. Biotechnol Prog 23:540–547. https://doi.org/10.1021/bp060349b

Melnick J, Dul JL, Argon Y (1994) Sequential interaction of the chaperones BiP and GRP94 with immunoglobulin chains in the endoplasmic reticulum. Nature 370:373–375. https://doi.org/10.1038/370373a0

Minkner R, Park EY (2018) Purification of virus-like particles (VLPs) expressed in the silkworm Bombyx mori. Biotechnol Lett 40:659–666. https://doi.org/10.1007/s10529-018-2516-5

Minkner R, Baba R, Kurosawa Y, Suzuki S, Kato T, Kobayashi S, Park EY (2018) Purification of human papillomavirus-like particles expressed in silkworm using a Bombyx mori nucleopolyhedrovirus bacmid expression system. J Chromatogr B Analyt Technol Biomed Life Sci 1096:39–47. https://doi.org/10.1016/j.jchromb.2018.08.007

Molinari M, Helenius A (1999) Glycoproteins form mixed disulphides with oxidoreductases during folding in living cells. Nature 402:90–93. https://doi.org/10.1038/47062

Molinari M, Helenius A (2000) Chaperone selection during glycoprotein translocation into the endoplasmic reticulum. Science 288:331–333. https://doi.org/10.1126/science.288.5464.331

Molino NM, Wang SW (2014) Caged protein nanoparticles for drug delivery. Curr Opin Biotechnol 28:75–82. https://doi.org/10.1016/j.copbio.2013.12.007

Moremen KW, Molinari M (2006) *N*-linked glycan recognition and processing: the molecular basis of endoplasmic reticulum quality control. Curr Opin Struct Biol 16:592–599. https://doi.org/10.1016/j.sbi.2006.08.005

Morikawa Y, Goto T, Yasuoka D, Momose F, Matano T (2007) Defect of human immunodeficiency virus type 2 Gag assembly in *Saccharomyces cerevisiae*. J Virol 81:9911–9921. https://doi.org/10.1128/JVI.00027-07

Motohashi T, Shimojima T, Fukagawa T, Maenaka K, Park EY (2005) Efficient large-scale protein production of larvae and pupae of silkworm by *Bombyx mori* nuclear polyhedrosis virus bacmid system. Biochem Biophys Res Commun 326:564–569. https://doi.org/10.1016/j.bbrc.2004.11.060

Muraki M, Honda S (2010) Efficient production of human Fas receptor extracellular domain-human IgG1 heavy chain Fc domain fusion protein using baculovirus/silkworm expression system. Protein Expr Purif 73:209–216. https://doi.org/10.1016/j.pep.2010.05.007

Nakajima M, Kato T, Kanamasa S, Park EY (2009) Molecular chaperone-assisted production of human alpha-1,4-*N*-acetylglucosaminyltransferase in silkworm larvae using recombinant BmNPV bacmids. Mol Biotechnol 43:67–75. https://doi.org/10.1007/s12033-009-9174-8

Nakaya H, Tatematsu KI, Sezutsu H, Kuwabara N, Koibuchi N, Takeda S (2020) Secretory expression of thyroid hormone receptor using transgenic silkworms and its DNA binding activity. Protein Expr Purif 176:105723. https://doi.org/10.1016/j.pep.2020.105723

Nerome K, Yamaguchi R, Fuke N, Izzati UZ, Maegawa K, Sugita S, Kawasaki K, Kuroda K, Nerome R (2018) Development of a Japanese encephalitis virus genotype V virus like particle vaccine in silkworms. J Gen Virol 99:897–907. https://doi.org/10.1099/jgv.0.001081

Niebauer RT, Robinson AS (2006) Exceptional total and functional yields of the human adenosine (A2a) receptor expressed in the yeast *Saccharomyces cerevisiae*. Protein Expr Purif 46:204–211. https://doi.org/10.1016/j.pep.2005.09.020

O'malley MA, Lazarova T, Britton ZT, Robinson AS (2007) High-level expression in *Saccharomyces cerevisiae* enables isolation and spectroscopic characterization of functional human adenosine A2a receptor. J Struct Biol 159:166–178. https://doi.org/10.1016/j.jsb.2007.05.001

Ogata M, Murata T, Murakami K, Suzuki T, Hidari KI, Suzuki Y, Usui T (2007) Chemoenzymatic synthesis of artificial glycopolypeptides containing multivalent sialyloligosaccharides with a gamma-polyglutamic acid backbone and their effect on inhibition of infection by influenza viruses. Bioorg Med Chem 15:1383–1393. https://doi.org/10.1016/j.bmc.2006.11.006

Ogata M, Nakajima M, Kato T, Obara T, Yagi H, Kato K, Usui T, Park EY (2009) Synthesis of sialoglycopolypeptide for potentially blocking influenza virus infection using a rat alpha2,6-sialyltransferase expressed in BmNPV bacmid-injected silkworm larvae. BMC Biotechnol 9:54. https://doi.org/10.1186/1472-6750-9-54

Palaniyandi M, Kato T, Park EY (2012) Expression of human papillomavirus 6b L1 protein in silkworm larvae and enhanced green fluorescent protein displaying on its virus-like particles. SpringPlus 1:29. https://doi.org/10.1186/2193-1801-1-29

Park EY (2025) From silkroad to bioroad-silkworm biotechnology. Protein Expr Purif 234:106762. https://doi.org/10.1016/j.pep.2025.106762

Park EY, Minkner R (2025) A systematic approach for scalable purification of virus-like particles. Protein Expr Purif 228:106664. https://doi.org/10.1016/j.pep.2025.106664

Park EY, Kageshima A, Kwon MS, Kato T (2007) Enhanced production of secretory beta1,3-*N*-acetylglucosaminyltransferase 2 fusion protein into hemolymph of *Bombyx mori* larvae using recombinant BmNPV bacmid integrated signal sequence. J Biotechnol 129:681–688. https://doi.org/10.1016/j.jbiotec.2007.01.028

Park EY, Ishikiriyama M, Nishina T, Kato T, Yagi H, Kato K, Ueda H (2009) Human IgG1 expression in silkworm larval hemolymph using BmNPV bacmids and its N-linked glycan structure. J Biotechnol 139:108–114. https://doi.org/10.1016/j.jbiotec.2008.09.013

Peacey M, Wilson S, Baird MA, Ward VK (2007) Versatile RHDV virus-like particles: incorporation of antigens by genetic modification and chemical conjugation. Biotechnol Bioeng 98:968–977. https://doi.org/10.1002/bit.21518

Pereira Aguilar P, Schneider TA, Wetter V, Maresch D, Ling WL, Tover A, Steppert P, Jungbauer A (2019) Polymer-grafted chromatography media for the purification of enveloped virus-like particles, exemplified with HIV-1 gag VLP. Vaccine 37:7070–7080. https://doi.org/10.1016/j.vaccine.2019.07.001

Pereira Aguilar P, Reiter K, Wetter V, Steppert P, Maresch D, Ling WL, Satzer P, Jungbauer A (2020) Capture and purification of human immunodeficiency Virus-1 virus-like particles: convective media vs porous beads. J Chromatogr A 1627:461378. https://doi.org/10.1016/j.chroma.2020.461378

Possee RD, Chambers AC, Graves LP, Aksular M, King LA (2020) Recent developments in the use of Baculovirus expression vectors. Curr Issues Mol Biol 34:215–230. https://doi.org/10.21775/cimb.034.215

Reiter K, Aguilar PP, Wetter V, Steppert P, Tover A, Jungbauer A (2019) Separation of virus-like particles and extracellular vesicles by flow-through and heparin affinity chromatography. J Chromatogr A 1588:77–84. https://doi.org/10.1016/j.chroma.2018.12.035

Rodriguez-Limas WA, Tyo KE, Nielsen J, Ramirez OT, Palomares LA (2011) Molecular and process design for rotavirus-like particle production in *Saccharomyces cerevisiae*. Microb Cell Factories 10:33. https://doi.org/10.1186/1475-2859-10-33

Ruddock LW, Molinari M (2006) *N*-glycan processing in ER quality control. J Cell Sci 119:4373–4380. https://doi.org/10.1242/jcs.03225

Rudolf MP, Fausch SC, Da Silva DM, Kast WM (2001) Human dendritic cells are activated by chimeric human papillomavirus type-16 virus-like particles and induce epitope-specific human T cell responses in vitro. J Immunol 166:5917–5924. https://doi.org/10.4049/jimmunol.166.10.5917

Sasaki K, Kajikawa M, Kuroki K, Motohashi T, Shimojima T, Park EY, Kondo S, Yagi H, Kato K, Maenaka K (2009) Silkworm expression and sugar profiling of human immune cell surface receptor, KIR2DL1. Biochem Biophys Res Commun 387:575–580. https://doi.org/10.1016/j.bbrc.2009.07.065

Sato Y, Oguchi M, Menjo N, Imai K, Saito H, Ikeda M, Isobe M, Yamashita O (1993) Precursor polyprotein for multiple neuropeptides secreted from the suboesophageal ganglion of the silkworm *Bombyx mori*: characterization of the cDNA encoding the diapause hormone precursor and identification of additional peptides. Proc Natl Acad Sci USA 90:3251–3255. https://doi.org/10.1073/pnas.90.8.3251

Satoh D, Horii A, Ochiai M, Ashida M (1999) Prophenoloxidase-activating enzyme of the silkworm, *Bombyx mori*. Purification, characterization, and cDNA cloning. J Biol Chem 274:7441–7453. https://doi.org/10.1074/jbc.274.11.7441

Steppert P, Burgstaller D, Klausberger M, Berger E, Aguilar PP, Schneider TA, Kramberger P, Tover A, Nobauer K, Razzazi-Fazeli E, Jungbauer A (2016) Purification of HIV-1 gag virus-like particles and separation of other extracellular particles. J Chromatogr A 1455:93–101. https://doi.org/10.1016/j.chroma.2016.05.053

Steppert P, Burgstaller D, Klausberger M, Tover A, Berger E, Jungbauer A (2017) Quantification and characterization of virus-like particles by size-exclusion chromatography and nanoparticle tracking analysis. J Chromatogr A 1487:89–99. https://doi.org/10.1016/j.chroma.2016.12.085

Tan M, Zhong W, Song D, Thornton S, Jiang X (2004) *E. coli*-expressed recombinant norovirus capsid proteins maintain authentic antigenicity and receptor binding capability. J Med Virol 74:641–649. https://doi.org/10.1002/jmv.20228

Tessier DC, Thomas DY, Khouri HE, Laliberte F, Vernet T (1991) Enhanced secretion from insect cells of a foreign protein fused to the honeybee melittin signal peptide. Gene 98:177–183. https://doi.org/10.1016/0378-1119(91)90171-7

Tsurumi Y, Morimoto K, Masuda A, Lee JM, Mon H, Kusakabe T (2025) Production of norovirus VLPs of the nine representative genotypes widely distributed in Japan using the silkworm-baculovirus expression vector system. J Virol Methods 331:115038. https://doi.org/10.1016/j.jviromet.2024.115038

Tsutsumi N, Kimura T, Arita K, Ariyoshi M, Ohnishi H, Yamamoto T, Zuo X, Maenaka K, Park EY, Kondo N, Shirakawa M, Tochio H, Kato Z (2014) The structural basis for receptor recognition of human interleukin-18. Nat Commun 5:5340. https://doi.org/10.1038/ncomms6340

Utomo DIS, Hirono I, Kato T, Park EY (2019) Formation of virus-like particles of the dengue virus serotype 2 expressed in silkworm larvae. Mol Biotechnol 61:852–859. https://doi.org/10.1007/s12033-019-00210-5

Utomo DIS, Pambudi S, Sjatha F, Kato T, Park EY (2020) Production of dengue virus-like particles serotype-3 in silkworm larvae and their ability to elicit a humoral immune response in mice. AMB Express 10:147. https://doi.org/10.1186/s13568-020-01087-3

Valenzuela P, Medina A, Rutter WJ, Ammerer G, Hall BD (1982) Synthesis and assembly of hepatitis B virus surface antigen particles in yeast. Nature 298:347–350. https://doi.org/10.1038/298347a0

Van Hofsten P, Faye I, Kockum K, Lee JY, Xanthopoulos KG, Boman IA, Boman HG, Engstrom A, Andreu D, Merrifield RB (1985) Molecular cloning, cDNA sequencing, and chemical synthesis of cecropin B from *Hyalophora cecropia*. Proc Natl Acad Sci USA 82:2240–2243. https://doi.org/10.1073/pnas.82.8.2240

Villa LL, Ault KA, Giuliano AR, Costa RL, Petta CA, Andrade RP, Brown DR, Ferenczy A, Harper DM, Koutsky LA, Kurman RJ, Lehtinen M, Malm C, Olsson SE, Ronnett BM, Skjeldestad FE, Steinwall M, Stoler MH, Wheeler CM, Taddeo FJ, Yu J, Lupinacci L, Railkar R, Marchese R, Esser MT, Bryan J, Jansen KU, Sings HL, Tamms GM, Saah AJ, Barr E (2006) Immunologic responses following administration of a vaccine targeting human papillomavirus Types 6, 11, 16, and 18. Vaccine 24:5571–5583. https://doi.org/10.1016/j.vaccine.2006.04.068

Vlasak R, Unger-Ullmann C, Kreil G, Frischauf AM (1983) Nucleotide sequence of cloned cDNA coding for honeybee prepromelittin. Eur J Biochem 135:123–126. https://doi.org/10.1111/j.1432-1033.1983.tb07626.x

Volkman LE (1986) The 64K envelope protein of budded *Autographa californica* nuclear polyhedrosis virus. Curr Top Microbiol Immunol 131:103–118. https://doi.org/10.1007/978-3-642-71589-1_6

Wagner R, Deml L, Schirmbeck R, Niedrig M, Reimann J, Wolf H (1996) Construction, expression, and immunogenicity of chimeric HIV-1 virus-like particles. Virology 220:128–140. https://doi.org/10.1006/viro.1996.0293

Wagner R, Teeuwsen VJ, Deml L, Notka F, Haaksma AG, Jhagjhoorsingh SS, Niphuis H, Wolf H, Heeney JL (1998) Cytotoxic T cells and neutralizing antibodies induced in rhesus monkeys by virus-like particle HIV vaccines in the absence of protection from SHIV infection. Virology 245:65–74. https://doi.org/10.1006/viro.1998.9104

Wang JW, Roden RB (2013) Virus-like particles for the prevention of human papillomavirus-associated malignancies. Expert Rev Vaccines 12:129–141. https://doi.org/10.1586/erv.12.151
Warne T, Serrano-Vega MJ, Baker JG, Moukhametzianov R, Edwards PC, Henderson R, Leslie AG, Tate CG, Schertler GF (2008) Structure of a beta1-adrenergic G-protein-coupled receptor. Nature 454:486–491. https://doi.org/10.1038/nature07101
Watanabe S, Iizuka T, Hatama S, Kanno T, Mase M, Shibahara T (2017) Production of highly immunogenic virus-like particles of bovine papillomavirus type 6 in silkworm pupae. Vaccine 35:5878–5882. https://doi.org/10.1016/j.vaccine.2017.08.079
Yagi H, Takahashi N, Yamaguchi Y, Kimura N, Uchimura K, Kannagi R, Kato K (2005) Development of structural analysis of sulfated *N*-glycans by multidimensional high performance liquid chromatography mapping methods. Glycobiology 15:1051–1060. https://doi.org/10.1093/glycob/cwi092
Yamada T, Iwasaki Y, Tada H, Iwabuki H, Chuah MK, Vandendriessche T, Fukuda H, Kondo A, Ueda M, Seno M, Tanizawa K, Kuroda S (2003) Nanoparticles for the delivery of genes and drugs to human hepatocytes. Nat Biotechnol 21:885–890. https://doi.org/10.1038/nbt843
Yao L, Wang S, Su S, Yao N, He J, Peng L, Sun J (2012) Construction of a baculovirus-silkworm multigene expression system and its application on producing virus-like particles. PLoS One 7:e32510. https://doi.org/10.1371/journal.pone.0032510
Young SL, Wilson M, Wilson S, Beagley KW, Ward V, Baird MA (2006) Transcutaneous vaccination with virus-like particles. Vaccine 24:5406–5412. https://doi.org/10.1016/j.vaccine.2006.03.052
Zeltins A (2013) Construction and characterization of virus-like particles: a review. Mol Biotechnol 53:92–107. https://doi.org/10.1007/s12033-012-9598-4
Zhang X, Buehner NA, Hutson AM, Estes MK, Mason HS (2006) Tomato is a highly effective vehicle for expression and oral immunization with Norwalk virus capsid protein. Plant Biotechnol J 4:419–432. https://doi.org/10.1111/j.1467-7652.2006.00191.x
Zheng X, Wang S, Zhang W, Liu X, Yi Y, Yang S, Xia X, Li Y, Zhang Z (2016) Development of a VLP-based vaccine in silkworm pupae against rabbit hemorrhagic disease virus. Int Immunopharmacol 40:164–169. https://doi.org/10.1016/j.intimp.2016.08.016

Chapter 7
Engineering of Virus-Like Particles (VLPs)

Abstract This chapter comprehensively details the advanced engineering and application of virus-like particles (VLPs) utilizing the silkworm expression system. It systematically covers the design and production of both enveloped VLPs (eVLPs) and nonenveloped VLPs (neVLPs). For eVLPs, strategies including genetic fusion and coexpression are explored to develop bivalent and multivalent antigen-displaying platforms, highlighting their potential as targeted delivery vehicles. For neVLPs, the focus shifts to advanced conjugation techniques, particularly SpyTag/SpyCatcher (SpT/SpC) and SnoopTag/SnoopCatcher (SnT/SnC) systems for in vivo and in vitro covalent bonding. The chapter elaborates on rational design approaches to optimize insertion sites within viral structural proteins to enhance the display efficiency. Significant applications are showcased, including the development of bivalent VLPs displaying *Plasmodium yoelii* malaria antigens and the complex engineering of tetravalent VLPs displaying envelope domain IIIs (EDIIIs) from all four dengue virus serotypes. These engineered VLPs, with their high immunogenicity and precise antigen display, demonstrate the immense potential of silkworm-derived platforms in developing next-generation multivalent vaccines and advanced therapeutics.

Keywords Silkworm expression system • Antigen-display • Virus-like particle • SpyTag/SpyCatcher • SnoopTag/SnoopCatcher • Covalent bonding • *Plasmodium yoelii* • Dengue virus • VLP-derived vaccine

7.1 Introduction

VLPs have attracted interest as promising vaccine platforms. However, a key challenge is that they cannot inherently serve as multivalent vaccines because they cannot naturally display multiple antigens simultaneously. To overcome this, a variety of modification strategies have been developed to display multiple antigens on the VLP surface, thereby achieving multivalency.

E. Y. Park, *Silkworm Biofactory*, https://doi.org/10.1007/978-981-92-0056-6_7

This chapter explores approaches to modifying both enveloped VLPs (eVLPs) and nonenveloped VLPs (neVLPs) and examines their resulting immunogenic properties.

7.2 Enveloped VLPs

The eVLPs have an envelope protein on their surface and are generally more effective at displaying membrane proteins. Target proteins, such as antigens, can be displayed by genetically fusing them to viral membrane proteins or transmembrane domains.

7.2.1 *Genetic Fusion*

This approach represents the most straightforward technique for displaying functional proteins on the surface of eVLPs. In this method, the viral envelope protein is genetically fused with a signal sequence, expressed under the control of the *polh* promoter, and secreted into the hemolymph of silkworm larvae. The outline is illustrated in Fig. 7.1a.

For example, premembrane envelope (prME) proteins from dengue virus (DENV) serotypes 2 or 3 have been successfully expressed, resulting in the assembly of DENV-like particles (DENV-LPs) with diameters of approximately 35 nm (Utomo et al. 2019) and 30–55 nm (Utomo et al. 2020), respectively. Similarly, capsid premembrane envelope (CprME) proteins have produced DENV-LPs of comparable size. Furthermore, various E proteins can be coexpressed and functionally displayed on the VLP surface (Fig. 7.1b). Coexpression of prMEs or CprMEs derived from DENV serotypes 1 and 4 led to the formation of DENV-LPs with a diameter of 30–55 nm (Utomo et al. 2022). The successful display of dengue immunogenic antigen, specifically envelope domain III (EDIII), on VLP surfaces was validated using heparin-binding assays, ELISA, and isothermal titration calorimetry. This antigen display was found to induce a humoral immune response in murine models (Utomo et al. 2022).

7.2.2 *Coexpression*

By coexpressing multiple proteins, functional proteins or several antigens can be displayed on the surface of eVLPs via host cell envelopes (Fig. 7.1c). VLPs displaying these functional proteins have been applied in ligand–receptor binding studies. For instance, the Rous sarcoma virus (RSV)-Gag protein and the human prorenin receptor (hPRR) were coexpressed in silkworms, and the resulting particles were

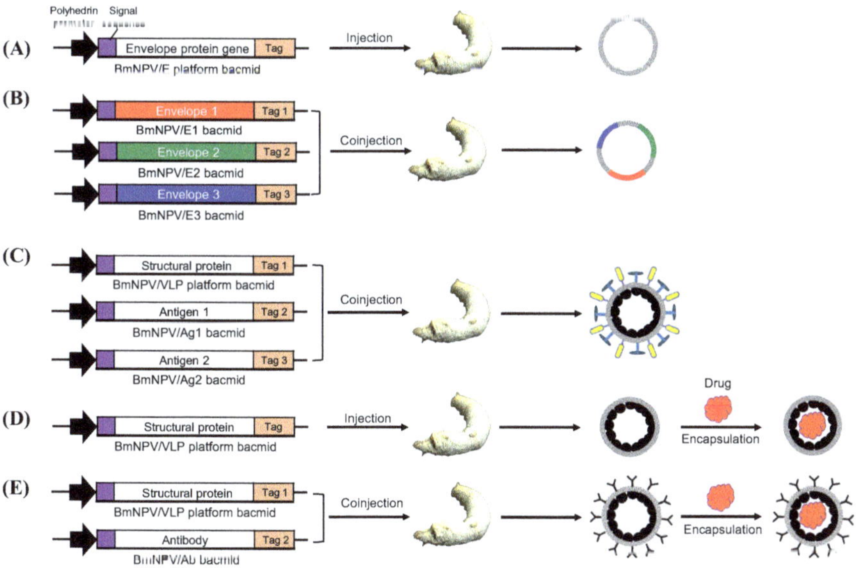

Fig. 7.1 Illustrations of eVLP expression in silkworms (**a**); multivalent eVLPs produced by coexpression of multiple envelope proteins (**b**); tagged antigens displayed on the eVLP surface (**c**); drug-encapsulated eVLPs (**d**); and drug-encapsulated eVLPs targeted to specific cells (**e**). (Adopted with permission (Park 2025). Copyright 2025 Elsevier)

purified, confirming that hPRR was displayed on the surface of RSV-like particles (RSV-LPs), as shown by immunoelectron microscopy (IEM). RSV-LPs displaying hPRR were subsequently employed in ELISA to assess the binding of hPRR to the human prorenin protein (Tsuji et al. 2011).

7.2.2.1 Bivalent Antigen-Displaying eVLPs

Similarly, the RSV-LP scaffold was used to display either *Neospora caninum* surface protein-related sequence 2 (NcSRS2) or *N. caninum* surface antigen 1 (NcSAG1), generating monovalent (Deo et al. 2013) and both displaying bivalent (Xu et al. 2019a) vaccines, respectively. For surface display on RSV-LP, the N-terminus of NcSRS2 and NcSAG1 was fused with the GP64 transmembrane domain from the *B. mori* envelope, and a FLAG tag was added to the C-terminus for affinity purification (Fig. 7.2a). As illustrated in Fig. 7.2b, antigens containing the baculovirus GP64 transmembrane anchor (Fig. 7.2a) were successfully expressed as membrane proteins within silkworm cells. When immature enveloped viruses, such as BmNPV and RSV, exit the cell into the extracellular space, they use the host cell membrane to form their viral envelope. Consequently, these viral particles incorporate numerous natural host components, including membrane proteins and lipids. This approach is commonly used as the principal method for displaying subunit

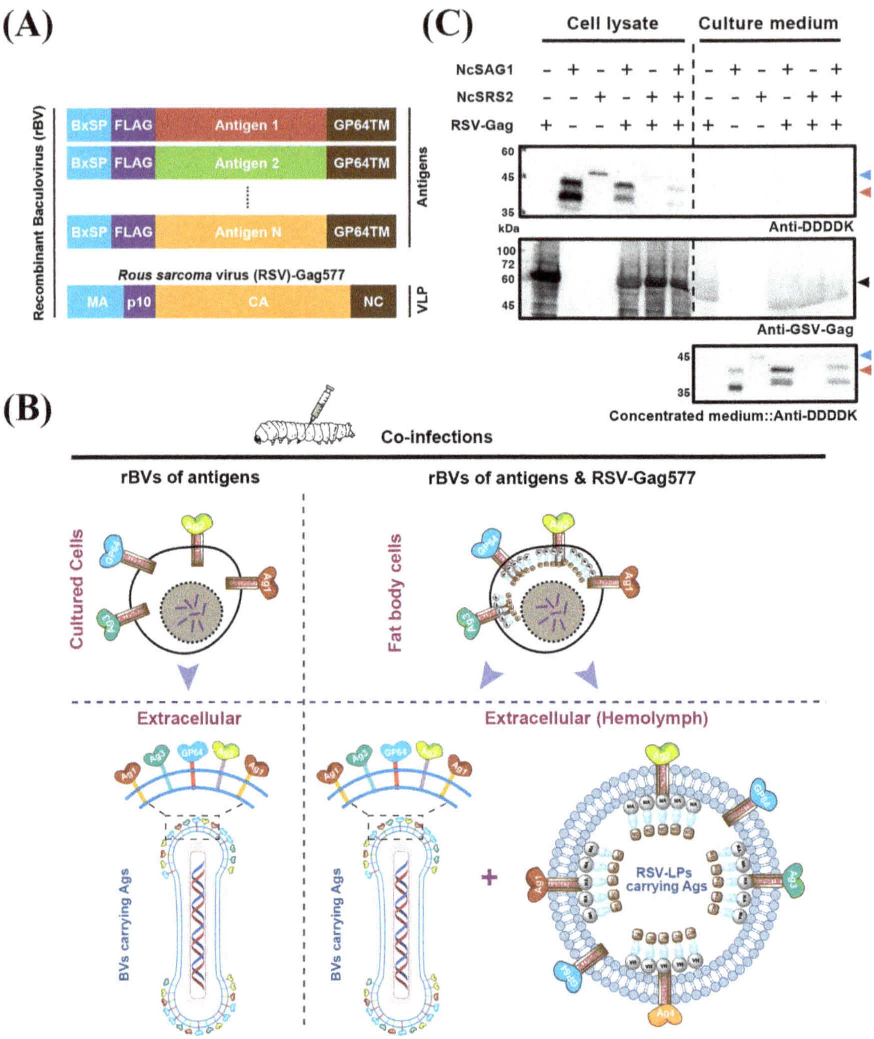

Fig. 7.2 Overview of the experimental setup for this study. (**a**) Design of constructs used for recombinant baculoviruses. BxSP, a signal peptide derived from silkworm bombyxin; FLAG, a DYKDDDDK tag used for affinity purification; GP64TM, the transmembrane region from the baculoviral GP64 protein; and MA/CA/NC, which represent the matrix, capsid, and nucleocapsid regions of the RSV gag protein (Group-specific antigen, amino acids 1–577). (**b**) Co-infected Bm cells with recombinant baculoviruses that encode either different antigens or RSV Gag proteins, the antigens containing a C-terminal transmembrane domain, expressed as membrane-bound proteins and displayed on the cell surface. Meanwhile, the Gag proteins self-assemble within the cytoplasm to form VLPs, which are subsequently released into the culture medium of Bm cells or into silkworm larval hemolymph. (**c**) Confirmation of expression levels and localization for various co-infection combinations. Protein samples from infected silkworm cells or culture medium (with or without ultracentrifugation in 20% sucrose) were analyzed using different antibodies: anti-DDDDK for detecting antigens (NcSAG1 and NcSRS2) and anti-RSV-gag for identifying VLPs. Specific protein bands for each protein are marked with arrows. (Adapted with permission (Xu et al. 2019a). Copyright 2019 Elsevier)

antigens in baculovirus systems (Makela and Oker Blom 2006) and other eVLPs (Schneider et al. 2018). As shown in Fig. 7.2c, protein expression following infection and co-infection (in equal ratios, either 1:1 or 1:1:1) with recombinant BmNPVs in cultured Bm5 cells was confirmed using Western blot analysis with antibodies targeting FLAG-tag and RSV-gag. Consistent with previous findings (Tsuji et al. 2011), both NcSAG1 and NcSRS2 were detected inside cells during double and triple co-infections, indicating that recombinant BmNPV co-infections were successfully established in silkworms. Although extracellular levels of these antigens were low (Fig. 7.2c, right panel), concentrating the collected medium using a 20% sucrose gradient and ultracentrifugation revealed distinct bands for NcSAG1 and NcSRS2 (Fig. 7.2c, lower panel), supporting the potential for antigen display or co-display on the surfaces of BmNPV or RSV-LPs.

The displayed antigens were identified by FLAG-affinity purification of pooled protein samples from fractions 5 through 7 of the ultracentrifuges, confirming successful binding (Xu et al. 2019a). An effective display is indicated by the co-precipitation of FLAG-tagged antigens with RSV-LP, as shown in Fig. 7.3a. The presence of each target protein before purification (input samples) was verified using Western blot analysis (Fig. 7.3b). Notably, when compared to RSV-LPs displaying only NcSRS2, the amount of NcSRS2 displayed was clearly reduced in the double-display VLPs (Fig. 7.3b, left panel). After the purification step, distinct RSV-gag bands were detected in the eluted (pulled down) protein samples, further confirming that antigens were successfully exhibited on the surface of the RSV-LPs.

The Neospora-sensitive Mongolian gerbil (*M. unguiculatus*) model (Ramamoorthy et al. 2005), known for acute infection, was used to investigate whether RSV-LPs displaying both NcSAG1 and NcSRS2 would improve antigenicity and provide better protection against *N. caninum* infection. Gerbils were immunized three times (Fig. 7.4a) in four groups: PBS, VLP alone, VLP-NcSRS2, and VLP-NcSAG1/NcSRS2. No noticeable differences in body weight were observed among the groups during vaccination. However, after exposure to 1×10^5 *Neospora* tachyzoites (Fig. 7.4b), there was a marked drop in body weight in the PBS and VLP groups. Notably, gerbils immunized with RSV-LPs displaying antigens had lower levels of *N. caninum* in their brain tissue than controls following infection, suggesting that these vaccines help prevent cerebral infection (Fig. 7.4c). To assess antibody production against NcSAG1 and NcSRS2, blood samples collected before the challenge were analyzed by ELISA using silkworm-expressed FLAG-NcSAG1 or FLAG-NcSRS2 as antigens. As shown in Fig. 7.4d, e, serum from gerbils immunized with antigen-displaying RSV-LPs exhibited specific antibody binding to purified antigens when compared with controls.

7.2.2.2 Multivalent Antigen-Displaying eVLPs and Their Applications

The influenza vaccines are definitely required in situations involving pandemic or endemic threats. Here, hemagglutinin subtypes 1 (HA1) and 5 (HA5) were displayed together on the surface of RSV-LPs to function as a dual-component

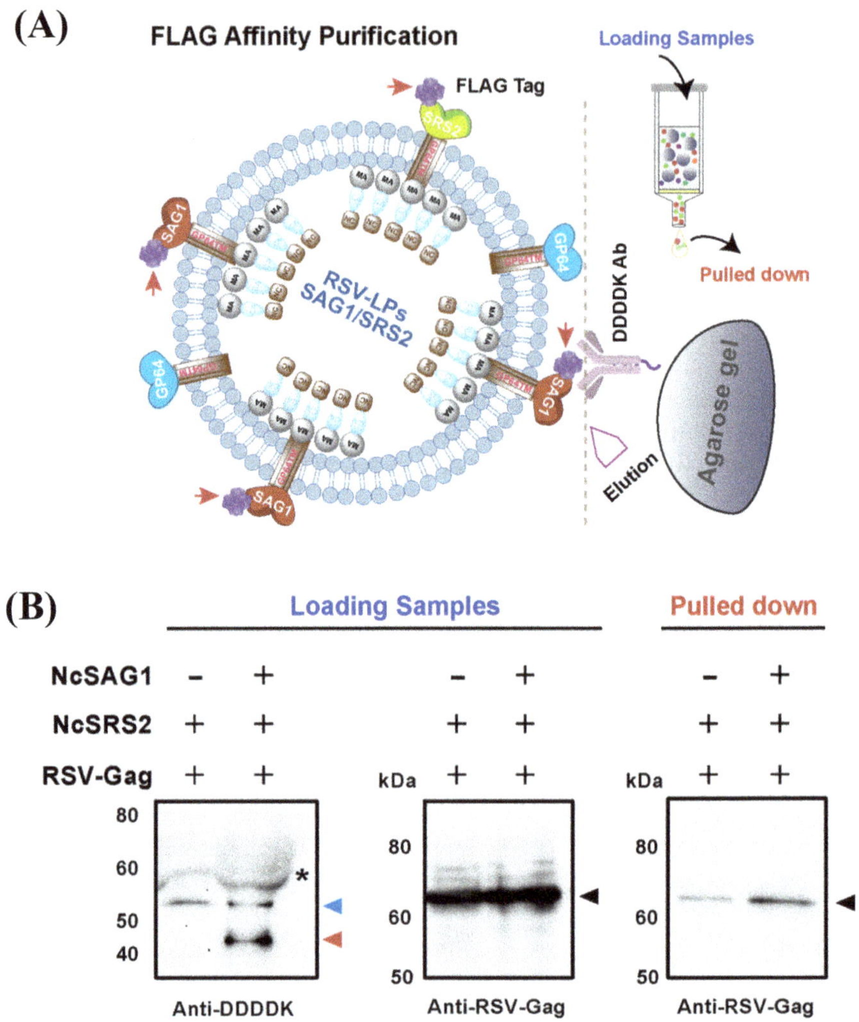

Fig. 7.3 Assessment of RSV-LP displaying NcSAG1/NcSRS2 through Flag tag-based affinity purification. If surface display occurs, RSV-LP should co-purify with Flag-tagged NcSAG1/NcSRS2 during Flag-tag affinity chromatography (**a**). Western blotting was used to detect proteins purified from silkworm hemolymph (concentrated with 20% sucrose) after co-infection with either a single antigen (NcSRS2) or both antigens (NcSAG1 and NcSRS2) along with RSV-Gag (**b**). Blue, red, and black arrows mark the successful detection of NcSAG1, NcSRS2, and RSV-gag, respectively, while an asterisk (*) designates a nonspecific band. (Adapted with permission (Xu et al. 2019a). Copyright 2019 Elsevier)

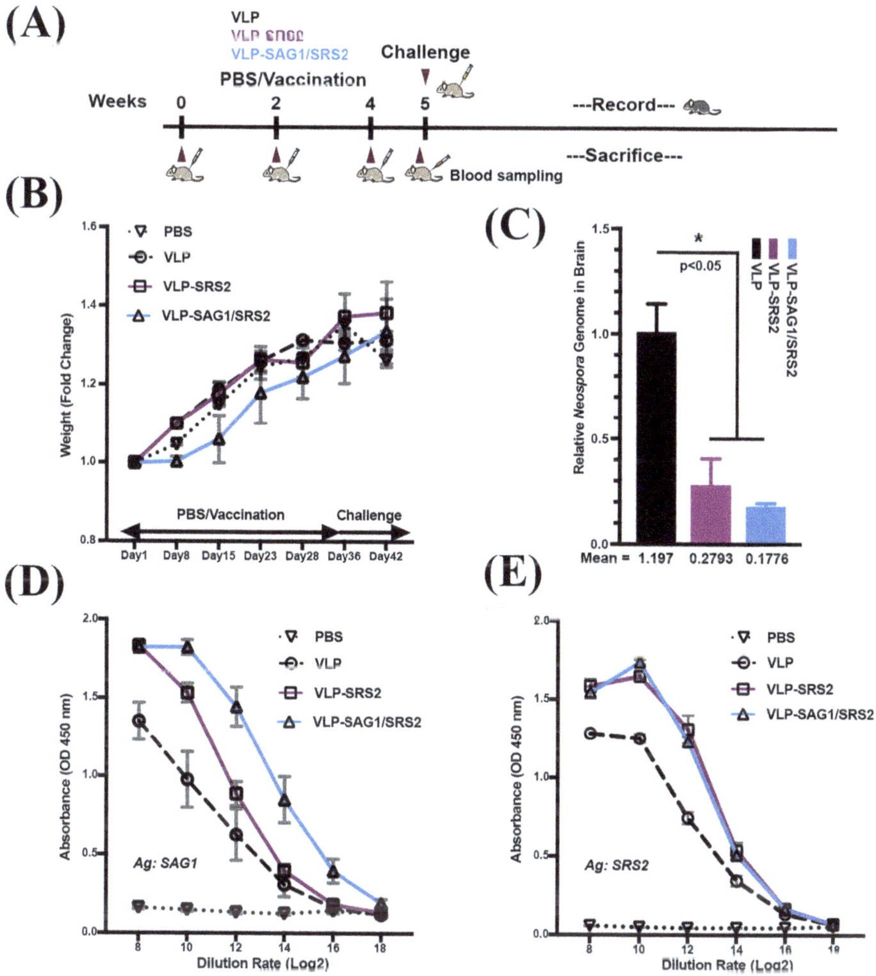

Fig. 7.4 Evaluation of RSV-LPs displaying antigens for vaccination against Neospora infection in a Mongolian gerbil model. (**a**) The experimental timeline included gerbil immunization, blood collection, Neospora challenge, and euthanasia. Four groups were tested: PBS control, VLP alone, VLP-NcSRS2, and VLP-NcSAG1/NcSRS2. Animals received a prime dose followed by booster immunizations at weeks 2 and 4. In week 5, gerbils were challenged intramuscularly with Neospora at a dose of 1×10^5 in the fifth week. Body weight changes (**b**) and brain parasite burden were assessed by quantitative PCR (**c**). Blood samples collected before challenge were analyzed for antibody titers against NcSAG1 (**d**) and NcSRS2 (**e**) using ELISA. Silkworm-expressed Flag-tagged NcSAG1 and NcSRS2 served as antigens. Serum dilutions were expressed in Log2, and absorbance was measured at 450 nm after color development. (Adapted with permission (Xu et al. 2019a). Copyright 2019 Elsevier)

vaccine (Goffar et al. 2022). HA1- and HA5-displaying bivalent eVLPs (eVLPs/BivHA) were able to cause hemagglutination in erythrocytes from chicken, rabbits, and sheep. In contrast, VLPs displaying only HA1 did not affect sheep erythrocytes, and those with only HA5 had no effect on rabbit erythrocytes. These results highlight the dual-action capability of VLPs/BivHA as a bivalent vaccine.

Interleukin-12 (IL-12) was used to stimulate natural killer (NK) cells. In addition to IL-12, multiple antigens were displayed on the surface of VLPs to boost the vaccine's immune response (Nerome et al. 2023). The IL-12 gene was linked to the stalk of neuraminidase (NA), which extended outward from the VLP surface (Fig. 7.5a). To scaffold the antigen-displaying VLPs, the M2 protein from influenza A H3N2 (IFA/H3N2) was used. The N- and C-terminal regions of the NA stalk were fused to the M2 and IL-12 genes, respectively (CC-AcNPV: IL12-N M2-AcNPV, Fig. 7.5b). Three types of recombinant baculoviruses were engineered: one with

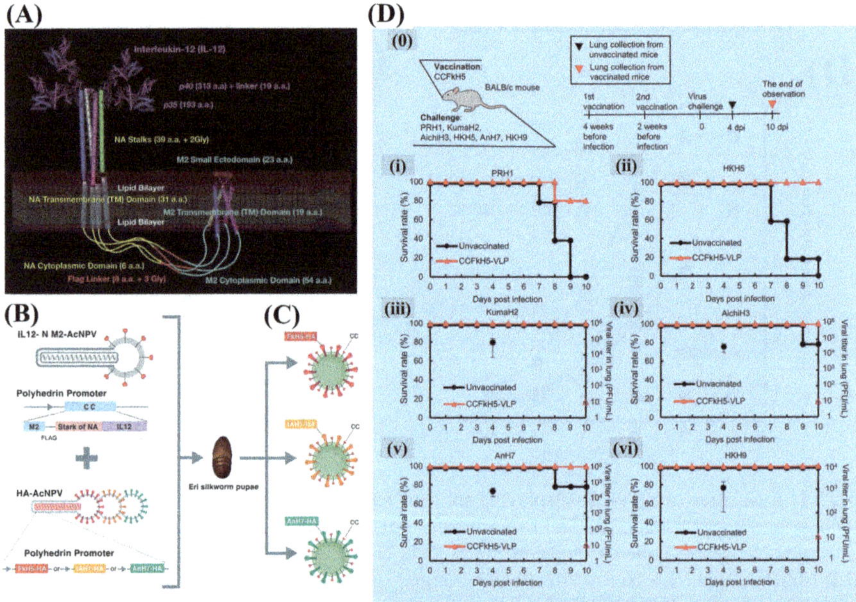

Fig. 7.5 A chimeric cytokine-displaying universal influenza VLP vaccine (Nerome et al. 2023). (**a**) Diagram depicting the design of the CC molecule, which is composed of IL-12, a NA stalk lacking its head (specifically, the cytoplasmic domain of NA), and the M2 from the influenza virus. (**b**) Coexpression of CC-AcMNPVs alongside either IAH1-HA-AcNPV, FkH5-HA-AcNPV, or AnH7-HA-AcNPV in Eri silkworms. (**c**) Representation showing VLPs displaying various hemagglutinins (HAs). (**d**) Mouse survival following vaccination with CCFkH5-HA-VLP against different influenza A virus subtypes in BALB/c mice (0). Subtypes used in challenge experiments include PRH1 (i), HKH5 (ii), KumaH2 (iii), AichiH3 (iv), AnH7 (v), and HKH9 (vi) viruses. After two doses of CCFkH5-VLP were administered to BALB/c mice at 2-week intervals, the animals were exposed to the viruses and monitored for 10 d. Lung samples were collected at 4 and 10 dpi from both unvaccinated (black circles) and vaccinated (red triangles) groups

Iowa H1N1 (IAH1-HA), another with Fukushima H5N1 (FkH5 HA), and a third with Anhui H7N9 (AnH7-HA) (HA-AcNPV, Fig. 7.5b). These baculoviruses were co-infected into Eri silkworm pupae along with CC-AcNPV (Fig. 7.5c). Transmission electron microscopy (TEM) showed that the purified VLPs measured about 150–200 nm in diameter; the spike structure of HA on CC-AcNPV particles was approximately 11 nm, increasing to about 14 nm when co-infected with FkH5-AcNPV (Nerome et al. 2023). Mice immunized with IL-12-displaying VLPs exhibited notably higher numbers of interferon (IFN)-γ–producing cells per million splenocytes compared to unvaccinated controls. When FkH5-HA–displaying VLPs were tested as vaccines against six different virus subtypes (Fig. 7.5d(0)), mice challenged with highly virulent strains (PRH1 and HKH5) showed strong protection: all unvaccinated mice died within 10 days postinfection (dpi), while those vaccinated had survival rates of 80% and 100% after PRH1 and HKH5 challenges, respectively (Fig. 7.5d(i, ii)). For moderate-virulence strains, all unvaccinated mice survived the KumaH2 and HKH9 challenges (Fig. 7.5d(iii, vi)). Additionally, 80% of unvaccinated mice survived the AichiH3 and AnH7 challenges (Fig. 7.5d(iv, v)). At 4 dpi, lung viral titers in unvaccinated groups ranged from 10^3 to 10^4 PFU/lung for all tested viruses (Nerome et al. 2023). At the same time, vaccinated mice showed no detectable viral replication in their lungs at 10 dpi (Nerome et al. 2023).

7.2.3 eVLPs as Carriers for Targeted Delivery

VLPs also show strong potential as carriers for drug delivery. They demonstrate remarkable physical and chemical stability, maintaining their integrity under different pH levels and temperatures, and are both biocompatible and biodegradable (Ding et al. 2018). Furthermore, VLPs can be intentionally modified or engineered to load, activate, and release therapeutic substances (Ikwuagwu and Tullman-Ercek 2022). These particles are well-suited for encapsulating small molecules and transporting them effectively (Goldmann et al. 2000). In addition to carrying drugs, VLPs constructed from the Influenza virus M1 protein can serve as scaffolds that stimulate immune responses while simultaneously delivering molecules (Fig. 7.1d) (Deo et al. 2015). To enhance delivery to specific cells, VLPs have been engineered for targeted delivery. For example, a single-chain variable fragment (scFv), hCC49—known to bind specifically to tumor-associated glycoprotein-72 (TAG-72)—was anchored onto RSV-derived VLPs by linking it with either glycosylphosphatidylinositol (Deo et al. 2016; Deo et al. 2014; Goldmann et al. 2000; Ikwuagwu and Tullman-Ercek 2022; Tsuji et al. 2011) or the transmembrane region of influenza virus hemagglutinin (Kato et al. 2015) (Fig. 7.1e). Using this approach, model compounds such as Sulforhodamin B have been efficiently packaged and delivered to their target cells (Deo et al. 2014).

As a more sophisticated approach, IL2 was displayed alongside hCC49 using the HA-transmembrane domain, displaying it on the surface of RSV-LPs (Fig. 7.6a(i–iii)). This modification enabled the activation of T-cells or NK cells. RSV-LPs

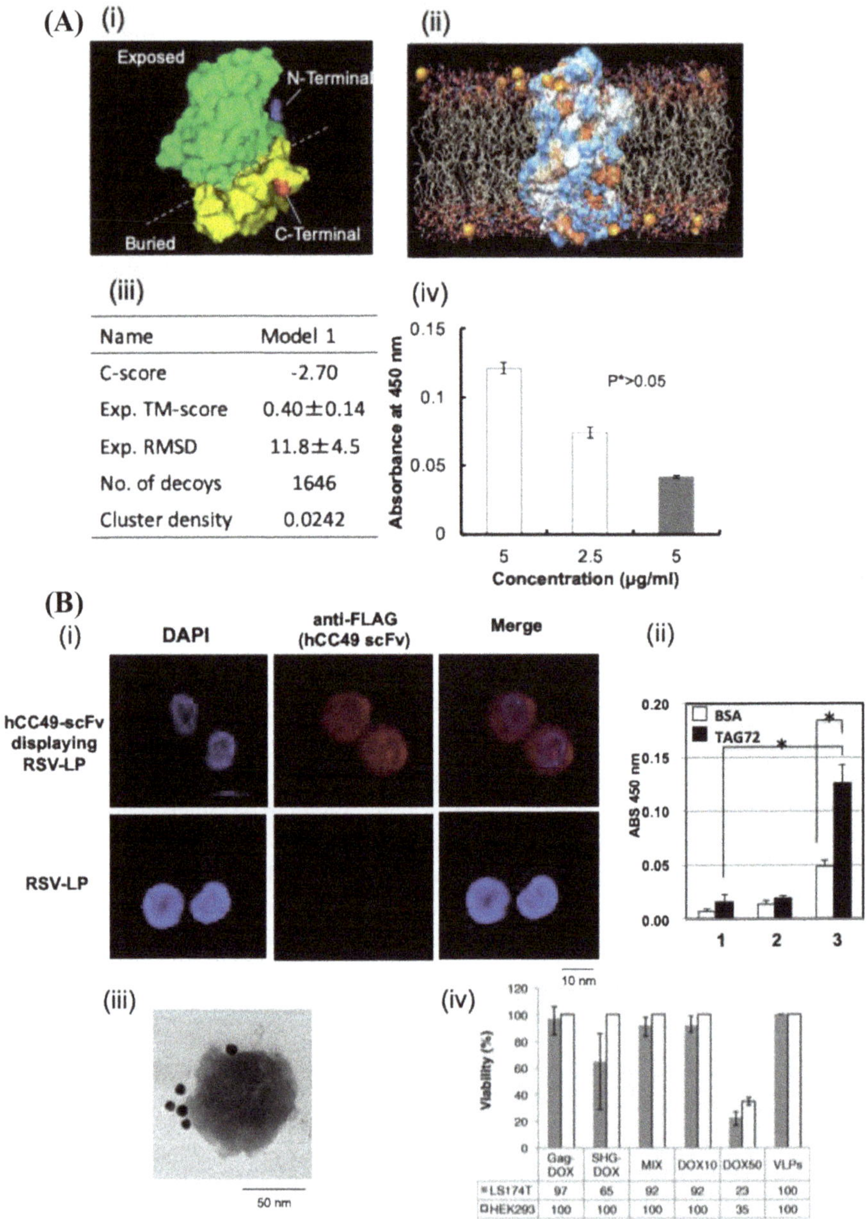

Fig. 7.6 Computational analysis of IL-2 structurally anchored to HAs (**a**) and examination of hCC49 scFv-displaying RSV-LPs (b). (**a**-i) The IL-2 molecule (shown in green) is localized within the transmembrane domain of HA2 (yellow), as visualized with the PyMOL software. The blue and red colors highlight the N- and C-termini of IL-2, respectively, while the white dashed line marks the segment buried inside the lipid membrane. (**a**-ii) IL-2's integration into the lipid bilayer is depicted using PyMOL and the protein structure. (**a**-iii) I-TASSER generated various scoring metrics for model-1. (**a**-iv) The specificity of IL-2 for soluble IL-Rα is shown, with white bars

engineered to display both molecules showed binding to colon cancer cells and successfully attracted macrophages in chemotaxis chamber experiments (Fig. 7.6a(iv)) (Deo et al. 2016). The presence of hCC49 on RSV-LPs was confirmed (Fig. 7.6b(i–iii)), and these particles effectively targeted TSG-72 found on the surface of cancer cells (Fig. 7.6b(ii)). Furthermore, doxorubicin (DOX) was encapsulated within hCC49-displaying RSV-LPs. When these DOX-loaded hCC49-RSV-LPs were administered to LS174T cancer cells, cell viability decreased by 35%, while there was no such effect on 100% of HEK293 cells (Fig. 7.6b(iv)) (Kato et al. 2015). This provides evidence that hCC49 binds to TAG-72 on cancer cells and efficiently delivers DOX.

Recent advances have positioned VLPs as a promising platform for gene therapy with diverse potential applications. For example, thymidine kinase-loaded VLPs incorporating HIV CD4-specific receptors have been shown to selectively target and eradicate HIV-infected cells when administered alongside ganciclovir (Peretti et al. 2006). Additionally, engineered HIV-derived VLPs have successfully engaged resting CD4 T cells, enabling gene modulation through silencing strategies (Geng et al. 2014). Extensive research efforts are currently focused on exploiting VLPs for drug delivery and vaccine development, as highlighted in several comprehensive reviews (Comas-Garcia et al. 2020; Mellid-Carballal et al. 2023).

7.3 Nonenveloped VLPs

The neVLPs lack an external envelope, which makes it difficult to display antigens by simply coexpression of them with structural proteins. To display antigens on the exterior of neVLPs, a binding interaction must be established between antigens and structural proteins. There are two main strategies to achieve this: covalent bonding and noncovalent bonding. With covalent bonding, a "Catcher" and a "Tag" are attached to either the N-terminus of the structural protein or the C-terminus of the antigen. The two modified bacmids—one with the Catcher on the structural protein

Fig. 7.6 (continued) representing scFv and IL-2 s on VLPs, and gray bars indicating those VLPs that do not display. (**b**-i) Representative immunofluorescence images of LS174T cells incubated with RSV-LPs either displaying the hCC49 antibody or lacking the display. Nuclei were counterstained with DAPI to visualize cellular architecture. (**b**-ii) Quantitative ELISA demonstrating the binding affinity between hCC49-displaying RSV-LPs and TAG-72. Samples 1, 2, and 3 correspond to mock hemolymph, hemolymph containing hCC49-displaying RSV-LPs, and purified hCC49-displaying RSV-LPs, respectively. Statistical significance was determined ($p < 0.01$). (**b**-iii) IEM visualization of RSV-LPs displaying hCC49. (**b**-iv) Assessment of cell survival for LS174 cells (gray) and HEK293 cells (white) following exposure to Gag-DOX (RSV-LPs containing DOX), SHG-DOX (RSV-LPs carrying both DOX and hCC49), a combination of Gag-DOX and SHG-DOX (MIX), DOX at 10 μg/mL (DOX10), DOX at 50 μg/mL (DOX50), and unmodified VLPs. (Adapted with permission (Deo et al. 2016) for (**a**) and (Kato et al. 2015) for (**b**), copyright 2016 Elsevier and copyright 2015 Springer Nature, respectively)

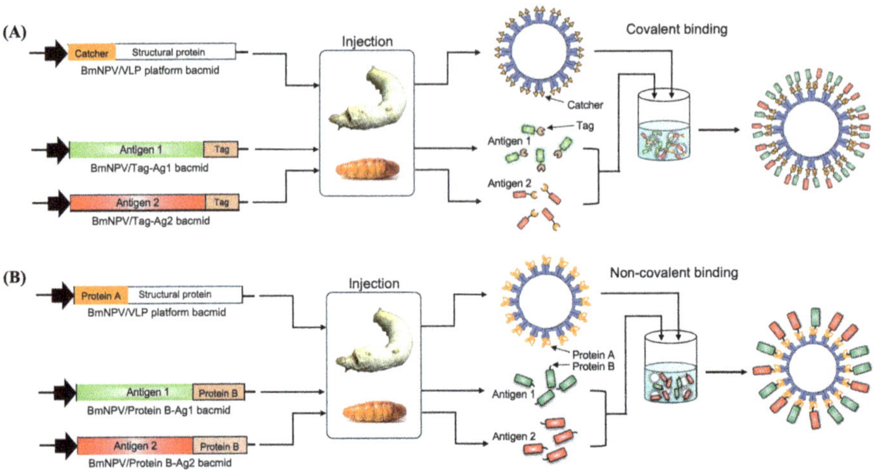

Fig. 7.7 Antigen-display strategies on neVLP surfaces. (**a**) Scaffold VLPs fused with Catcher domains and antigens with specific tags were separately expressed and purified and then combined in vitro, allowing tag-catcher covalent binding to display various antigens on the neVLP surface. (**b**) Scaffold VLPs linked to protein A and antigens linked to protein B were individually expressed and purified and then mixed in vitro, utilizing the protein A–protein B interaction to co-display multiple antigens on the neVLP surface. (Adopted with permission (Park 2025). Copyright 2025 Elsevier)

and the other with the Tag on the antigen—are expressed separately in silkworm larvae or pupae, purified, and then chemically bound together using the covalent Catcher-Tag interaction (Fig. 7.7a). This covalent bond is stable and allows antigens to be attached modularly.

On the other hand, noncovalent bonding usually relies on protein-protein interactions or affinity-based binding. In this approach, "Protein A" is fused to the N-terminus of the structural protein, and "Protein B" is fused to the C-terminus of the antigen (Fig. 7.7b). The proteins are expressed separately and then combined in vitro to display the antigens on the neVLP surface. Because of the nature of protein-protein interactions, these VLPs often end up being larger than those created by the covalent method.

7.3.1 Covalent Conjugation

Proteins maintain their structure through various stabilizing interactions, such as hydrophobic forces, ionic bonds, and hydrogen bonding. Disulfide bonds, covalent links formed between cysteine residues, are a well-established mechanism for enhancing protein stability. In recent years, additional types of internal cross-links have been identified. Notably, researchers have discovered an isopeptide bond in the fibronectin-binding protein (FbaB) from the Gram-positive bacterium *Streptococcus*

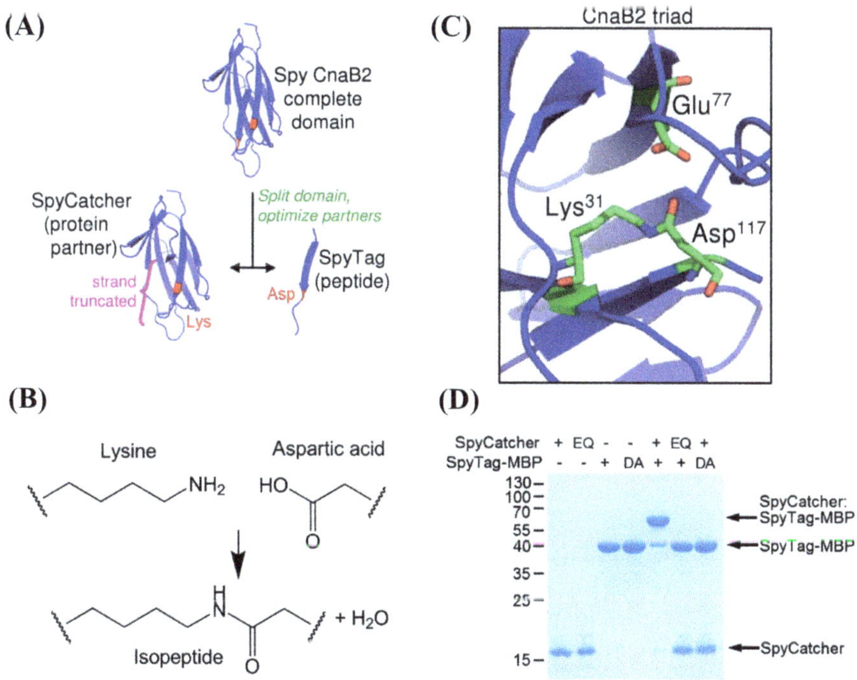

Fig. 7.8 Spontaneous formation of intermolecular amide bonds by SpT [155]. (**a**) Diagram showing the assembly of Spy CnaB2 domain. The CnaB2 domain from *S. pyogenes* was separated into two parts: an N-terminal segment (SpC) and a C-terminal segment (SpT). Lysine and aspartic acid residues are highlighted in red. (**b**) Schematic showing the amide linkage that forms between lysine and aspartic acid. (**c**) Important amino acids in CnaB2 are depicted as green sticks in the PDB structure 2X5P. (**d**) Demonstration of the covalent interaction between SpT and SpC. SpT bound to maltose binding protein (SpT-MBP) was combined with SpC and incubated for 3 h. SDS-PAGE analysis with Coomassie staining represents samples including a control, SpC E77Q (EQ) mutant, and D117A (DA) SpT-MBP. (Adapted with permission (Zakeria et al. 2012). Copyright 2012 National Academy Sciences)

pyogenes (Kang et al. 2007). The second immunoglobulin-like collagen adhesin domain (CnaB2) of FbaB is vital to a process resembling phagocytosis, enabling *S. pyogenes* to be internalized by endothelial cells (Amelung et al. 2011). CnaB2 can be divided into two segments: a 13-amino-acid peptide at the C-terminus referred to as SpyTag (SpT) and a 138-amino-acid N-terminal protein fragment (approximately 15 kDa), called SpyCatcher (SpC) (Zakeria et al. 2012) (Fig. 7.8a). The SpT peptide contains a reactive aspartic acid at position 117, which rapidly forms a covalent isopeptide bond with the lysine at position 31 in SpC, a protein derived from CnaB2 (Fig. 7.8b). In this reaction, glutamic acid at position 77 in SpC and aspartic acid at position 117 in SpT are both critical for creating the covalent isopeptide bond (Fig. 7.8c). Altering Glu^{77} (EQ in Fig. 7.8d) or Asp^{117} (DA in Fig. 7.8d) prevents SpT and SpC from forming the isopeptide bond (Fig. 7.8d). This bond formation occurs under a broad range of temperatures and pH values, is unaffected by various buffers and detergents, and is irreversible (Zakeria et al. 2012).

In *Streptococcus pneumoniae*, the D4 immunoglobulin-like domain found in the RrgA protein is responsible for forming an isopeptide bond between the amino acids lysine at position 742 (Lys^{742}) and asparagine at position 854 (Asn^{854}). This D4 domain is divided into two separate components: a short peptide sequence covering residues 734 to 745, referred to as SnoopTag (SnT), and a larger protein segment spanning residues 749 to 860, known as SnoopCatcher (SnC) (Veggiani et al. 2016). For the bond formation to occur, Lys^{742} within SnT and Asn^{854} within SnC are crucial; if either of these residues is altered, the isopeptide bond cannot be formed, effectively stopping the reaction.

7.3.2 *In Vivo and In Vitro Covalent Conjugation in Silkworm*

As shown in Fig. 7.9a, a covalent isopeptide bond is formed between lysine and aspartic acid residues, resulting in the assembly of the SpT and SpC complex. To explore whether proteins fused to SpT/SpC could be expressed effectively, various constructs were designed to generate recombinant baculovirus using the pFastBac vector (Fig. 7.9b). Genes encoding various reporter proteins—including EGFP, Venus, and mCherry—were introduced and fused to SpT or SpC as needed. The recombinant BmNPVs were subsequently generated and analyzed in both cultured Bm5 cells and in larval or pupal silkworms. To evaluate the presence of uncoupled and coupled SpT/SpC-reporter proteins, single- and co-infection approaches were used in vitro and in vivo, as follows (Fig. 7.9c).

Recombinant BmNPV bacmids generated from SpT/SpC designs—including SpC-Venus SpT-mCherry and Venus-SpT—were injected into silkworms at the third day of the instar larval stage (Fig. 7.10a) and into pupae (Fig. 7.10b). At 6 dpi, fluorescent signals produced by Venus or mCherry proteins were detected using UV illumination. In addition to whole silkworm larvae, lysates from fat body tissues infected with baculovirus (at 5 dpi) were also analyzed for fluorescence (Fig. 7.10c). The results confirmed that all engineered SpT/SpC constructs were successfully expressed, as evidenced by their fluorescent signals. When recombinant BmNPVs expressing SpC-Venus and SpT-mCherry were co-injected, both green and red fluorescence were visible in larvae and pupae under UV light. Furthermore, expression could be readily identified by observing green or pink coloration in epidermal or fat body tissue lysates at 5 or 6 dpi. These findings demonstrate that the designed constructs function effectively inside of the silkworm body (in vivo).

To analyze the SpT/SpC protein pair, their conjugation efficiency was evaluated in vitro by using purified SpC-EGFP and SpT-mCherry proteins, as well as in vivo by directly assessing the protein products obtained from co-infected silkworm larvae, as illustrated in Fig. 7.11a. His_8-Strep-tag II (HS)-tagged SpC and Flag-Strep-tag II (FS)-tagged SpT enable the isolation of SpT/SpC-tagged proteins using specific tag-affinity purification. When SpC-EGFP and SpT-mCherry were

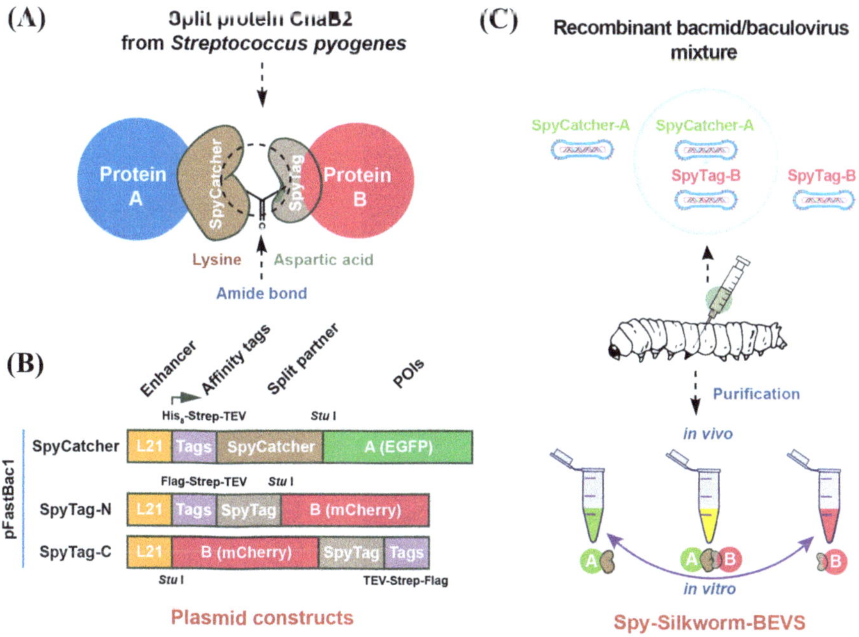

Fig. 7.9 Overview of the SpT/SpC-Bacmid/Baculovirus expression vector system (SpyBEVS) (Xu et al. 2019b). (**a**) Formation of a lysine-aspartic acid amide bond between SpT/SpC-fused pairs. (**b**) SpC and SpT fused constructions for expression using the pFastBac1 vector, either at the N- or C-terminus. The L21 sequence, a 21-base-pair leader from lobster tropomyosin cDNA, serves as an enhancer to boost protein production in the BEVS system (Kato et al. 2016). Additional elements include an 8× histidine tag (His_8), a Strep-tag, a TEV protease recognition site, and a FLAG tag (DYKDDDDK). GGGS linkers are incorporated between tags to enhance protein flexibility. Enhanced green fluorescent protein (EGFP), Venus, and mCherry are utilized as model reporter proteins in ligation assays for each construct. (**c**) The experimental process for the use of silkworm instar larvae. Recombinant baculoviruses containing SpT/SpC-fused proteins are generated using the Bac-to-Bac system. These are then injected into 5th instar silkworm larvae, either individually or in combination, to express SpT/SpC-fused proteins for both in vitro (following protein purification) and in vivo (via co-infection in silkworms) ligation experiments

incubated together in PBS at room temperature for 1 h, an adduct was detected after separation by both SDS-PAGE (Fig. 7.11b) and Western blot (Fig. 7.11c) with high specificity, indicating successful conjugation between the SpT/SpC protein pair. Additionally, increasing the amount of SpT-mCherry resulted in a greater amount of adduct, as shown in Fig. 7.11b, c. To confirm in vivo bioconjugation, fat body tissue lysates underwent Strep-based purification, ensuring that all Strep-tagged products were co-eluted. As anticipated, the SDS-PAGE results from Fig. 7.11b, d confirmed that similar patterns of in vitro ligation products were also observed in silkworms. Furthermore, the ratio of viruses used for co-infection with SpC-EGFP and SpT mCherry (1:2) corresponded well with the ratio of proteins expressed. Notably, the overexpression of SpT-mCherry saturated the lower levels of SpC-EGFP in the silkworms, resulting in sufficient assembly of the SpC-EGFP:SpT mCherry complex.

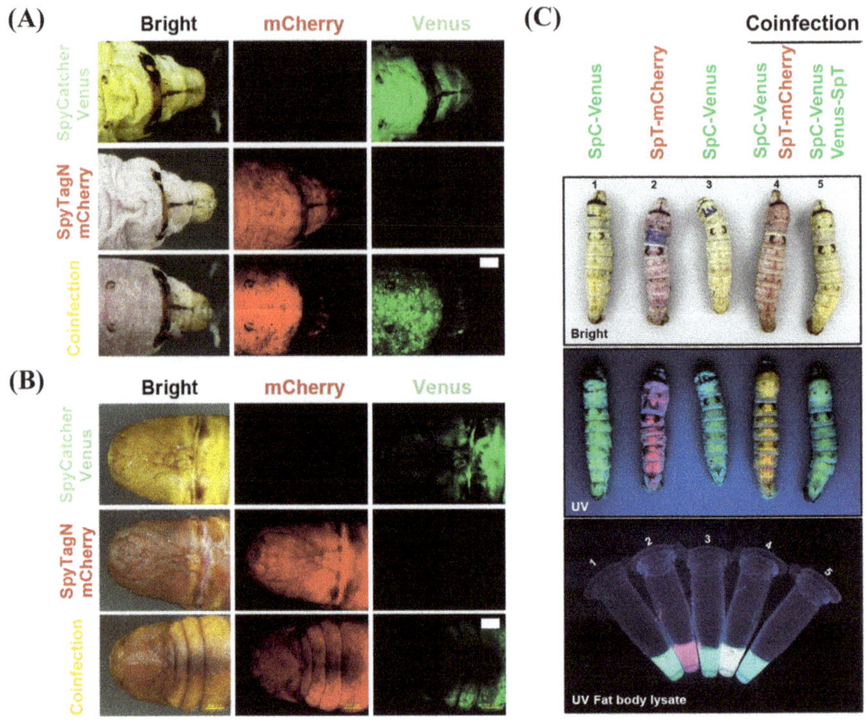

Fig. 7.10 In vivo expression of SpT- and SpC-tagged proteins in silkworm (Xu et al. 2019b). Recombinant bacmids created from SpT/SpC constructs—specifically SpC-Venus and SpT-N-mCherry—were injected into silkworms at third day fifth instar larvae (**a**) and pupae (**b**). The presence of fluorescent proteins (EGFP, Venus, or mCherry) was confirmed by7observing their fluorescence under ultraviolet (UV) light. Additionally, both whole larvae and fat body tissue lysates were examined five days after infection with the recombinant baculoviruses to detect fluorescence (**c**). The top panel shows larvae under normal light; the middle panel shows larvae under UV light; and the bottom panel shows fat body lysate under UV light

Interestingly, as depicted, several intermediate/partial products appeared in both in vivo and in vitro conjugation, which may be attributed to protein degradation or conformational changes following the ligation of SpC-EGFP and SpT-mCherry.

To further investigate the capabilities of SpT/SpC in protein conjugation the co-infection strategy to assess its flexibility and efficiency when one or more recombinant BmNPVs expressing SpT-tagged proteins were conjugated with the SpC-EGFP partner as shown in Fig. 7.12a, b. As a fusion partner, the domain III region (EDIII) of the E protein from all four dengue virus serotypes was selected, as this domain is known for its unique immunogenic properties. The amino acid sequence alignment of EDIIIs is shown in Fig. 7.12c. Each EDIII was genetically fused to the SpT tag at the C-terminus, resulting in constructs named 1EDIII-SpT, 2EDIII-SpT, 3EDIII-SpT, and 4EDIII-SpT. Western blot analysis of all EDIII-SpT variants expressed in soluble form in fat body lysates from silkworm larvae infected with recombinant

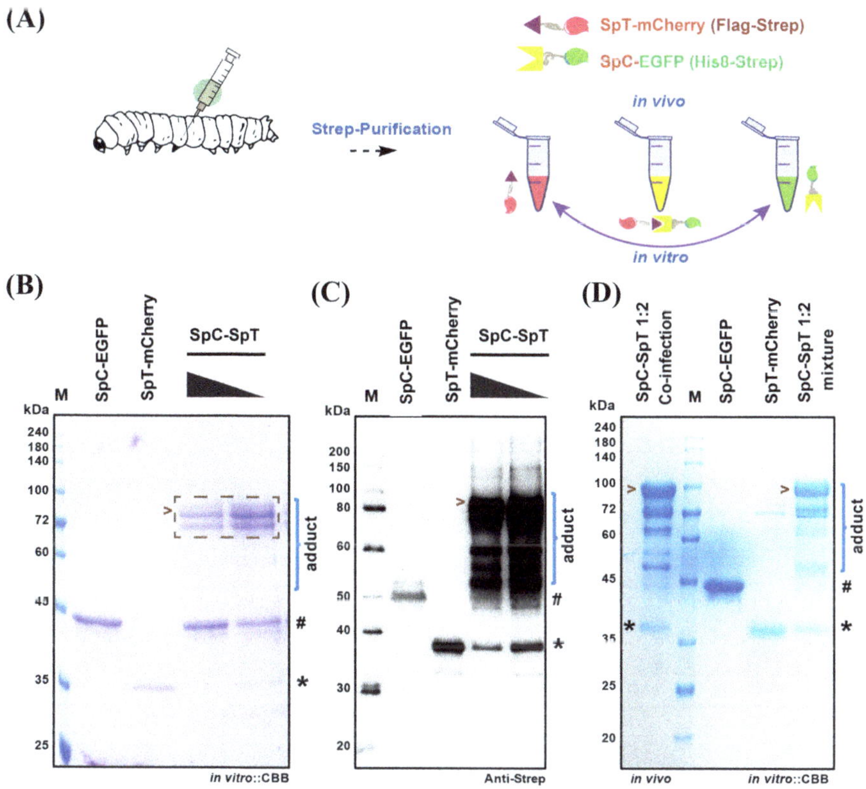

Fig. 7.11 In vitro and in vivo protein conjugation of SpT/SpC pairs using the silkworm (Xu et al. 2019b). (**a**) Illustration showing the expression and conjugation of SpT-mCherry and SpC-EGFP proteins in the silkworm. Each protein was purified from the fat bodies of infected silkworms and incubated in decreasing proportions of SpC-EGFP to SpT-mCherry (SpC/SpT) to examine in vitro coupling efficiency, as depicted Coomassie Brilliant Blue staining (**b**) and Western blot analysis using anti-Strep tag antibody (**c**). For in vivo conjugation, silkworm larvae were co-infected with two types of recombinant baculoviruses expressing SpC-EGFP and SpT-mCherry (in a 1:2 ratio, SpC/SpT). The resulting proteins were purified using a Strep affinity column and analyzed by SDS-PAGE (**d**). The symbols # and * denote SpC-EGFP and SpT-mCherry, respectively. The ligated products are indicated by a brace, while the fully formed conjugate is marked with arrows

BmNPVs and confirmed all EDIII expression in silkworms (Fig. 7.12d). Ligated adducts were analyzed by CBB staining (Fig. 7.12e) and by western blot using anti-Strep tag (Fig. 7.12f). Notably, a co-infection strategy utilizing five recombinant BmNPVs—comprising SpC-EGFP and four different EDIII-SpTs (nEDIII-STs)—was implemented to assess the capacity for in vivo ligation involving multiple protein conjugation (Fig. 7.12g). The products generated by this approach were confirmed by western blot analysis, which included controls for SpC-EGFP alone and for SpC-EGFP combined with nEDIIIs. These results demonstrate that the SpT/SpC platform facilitates efficient, robust, and highly specific protein-protein conjugation in silkworms.

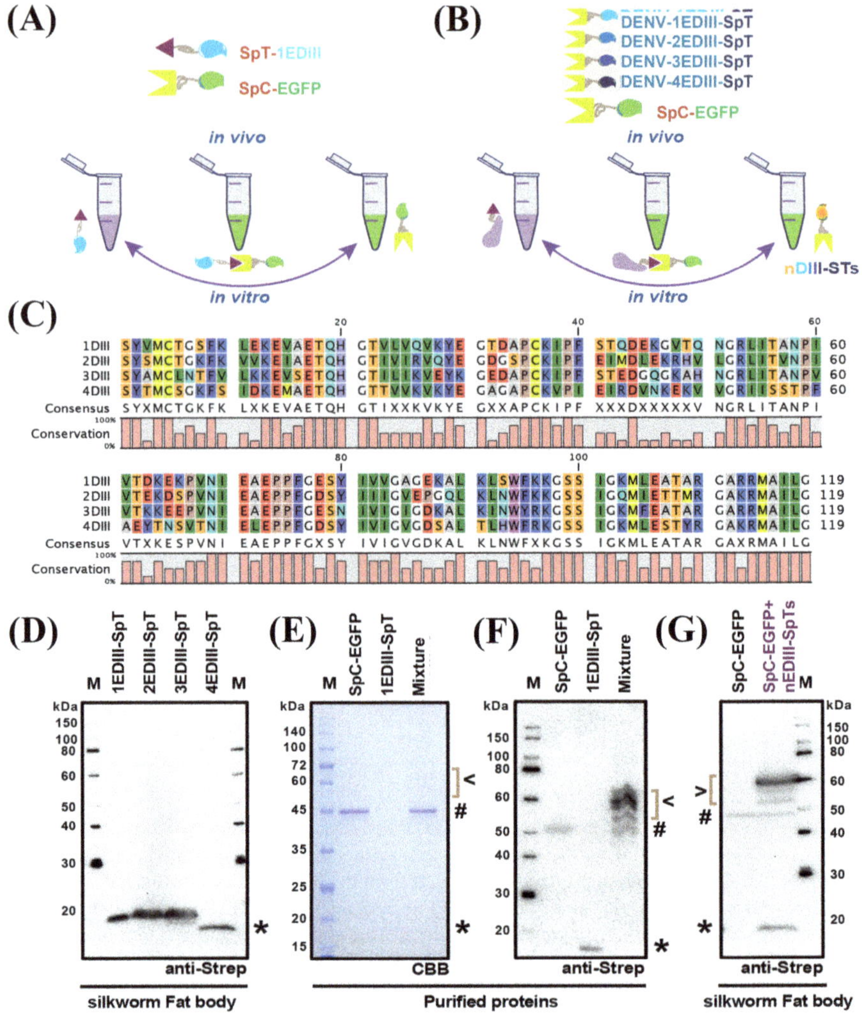

Fig. 7.12 Protein binding and labeling in silkworms (Xu et al. 2019b). Diagrams illustrating both in vitro and in vivo protein conjugation are depicted in (**a**) (SpT-1EDIII) and (**b**) (SpT-nEDIIIs) with SpC-EGFP. (**c**) Amino acid sequence comparison of dengue virus E protein domain III (EDIII) across all four serotypes. (**d**) Western blot analysis of each EDIII-SpT expressed in the fat body of silkworm larvae. For in vitro ligation, purified SpC-EGFP was conjugated with 1EDIII-SpT, and the resulting conjugate was validated using CBB staining (**e**) or anti-Strep western blot (**f**). (**g**) Co-infection with five recombinant BmNPVs—one with SpC-EGFP and four with EDIII-SpTs (nEDIII-SpTs). These products were detected through western blot analysis alongside a control group with only SpC-EGFP infection. The symbols # and * denote SpC-EGFP and nEDIIIs, respectively; newly ligated products are indicated by braces, and the fully formed conjugates are marked with arrows

7.3.3 Engineering neVLPs for Multiple Antigen Display

A SpT/SpC conjugation for protein binding was demonstrated to be effective both with purified proteins in vitro and with proteins coexpressed in silkworm larvae in vivo through viral co-infection (Xu et al. 2019b). To further explore a versatile platform for VLPs, VP2 capsid protein from canine parvovirus (CPV)—a nonenveloped, $T = 1$ icosahedral particle comprising 60 subunits and measuring 26 nm in diameter—as a model was selected. CPV-LP has a history of successful expression in insect cells or silkworm pupae, and it is known for its strong immunogenic properties (Chang et al. 2020; Elia et al. 2012; Feng et al. 2014; Hurtado et al. 1996; Jin et al. 2016; Yuan and Parrish 2001). Additionally, the three-dimensional structure of CPV is available (PDB: 4QYK), and four key loop regions (Loop 1–Loop 4) have been characterized for their roles in VLP formation (Hurtado et al. 1996; Tsao et al. 1991). Notably, Loop 2 (R216-G235) is not required for VLP assembly, making it an ideal candidate for further modification to display antigens or epitopes (Brown et al. 1994; Feng et al. 2011; Hurtado et al. 1996; Xu et al. 2014). Here, several chimeric VP2 proteins were constructed by inserting SpT at various positions, and it was found that VLPs derived from the VP2-SpT-Loop2 variant exhibited a marked increase in the surface display of a SpC-tagged GFP as a model protein. To enable orthogonal and multi-antigen display, two loop domains were further modified by incorporating either SpT or SnC. The complete VLP was engineered using a VP2 variant tagged with both SpT and SnT, allowing for orthogonal ligation with binding partners in vivo and in vitro. This work introduces a versatile VLP platform capable of displaying multiple antigens as needed, laying the foundation for future studies on VLP-based vaccines in diverse animal–pathogen models.

7.3.3.1 Rational Design of Insertion of SpT into Loop Regions of VP2

The VLPs showed intact structures and were distributed within the 20–30 nm range by transmission electron microscopy (TEM), which aligns with earlier findings (Choi et al. 2000; Gilbert et al. 2006). These results directly demonstrate that the VP2-derived SpT–SpC complex can be effectively produced in silkworms and utilized as a VLP-based display platform. Nevertheless, successful protein display on SpT–VP2 VLPs was limited. The absence of gold particles on the VLPs observed in immuno-TEM using an anti-His6 antibody suggests that the positioning of SpT requires further refinement to enhance surface protein display.

Based on the VP2 structure (PDB ID: 4QYK) and previous studies (Brown et al. 1994; Feng et al. 2011; Hurtado et al. 1996; Xu et al. 2014), two loops—Loop 2 (S226_G227) and Loop X (T391_G392)—have been identified as promising sites for SpT insertion (Rueda et al. 1999; Xie and Chapman 1996). In theory, these loop regions should be present at a 1:1 ratio on both the protein and the assembled VLPs, with 60 copies per VLP, as illustrated in Fig. 7.13a (monomer, 1×) and Fig. 7.13b (pentamer, 5×). Given the lengths of SpC (SpyCatcher002, 113 amino acids) and

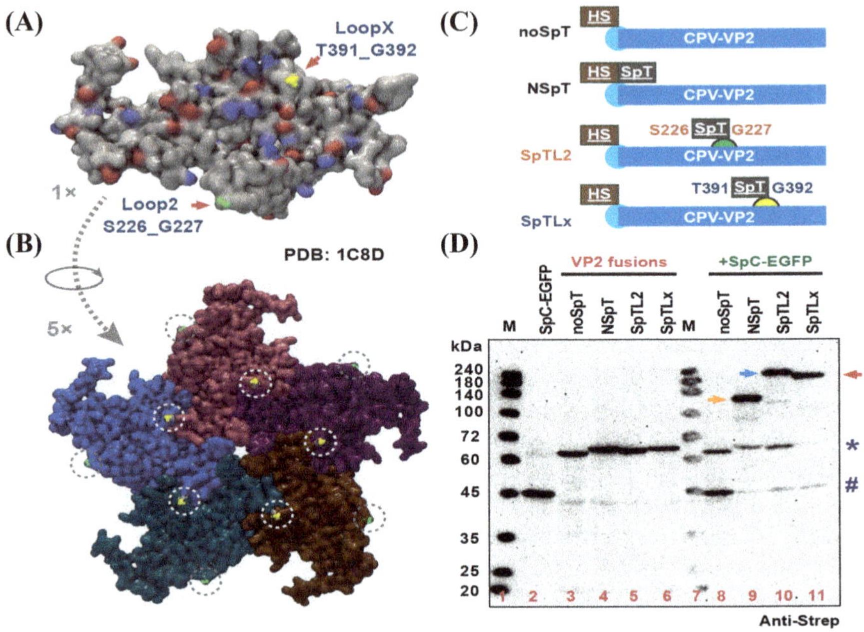

Fig. 7.13 Strategic design of the SpT exposure site on VLPs derived from the VP2 structural protein (Xu et al. 2023). The configuration of a VP2 protein monomer (**a**) and a partially assembled form (**b**) is illustrated (PDB: 4QYK). Two predicted loop segments, S226_G227 and T391_G392, are marked as Loop 2 (green) and Loop X (yellow), respectively, with arrows, and are expected to be displayed on the VLP surface, with 60 copies per VLP. The SpT sequence was engineered to be inserted at various positions within VP2. The study included several VP2 variants, as shown in (**c**): noSpT (without SpT insertion), NSpT (attached at the N-terminus), SpTL2 (inserted at Loop 2), and SpTLx (inserted at Loop X). Using Western blot analysis (**d**, anti-Strep antibody), the expression and interaction between all VP2 SpT-fusion constructs and SpC-EGFP in cultured Bm5 cells were confirmed. The symbols # and * denote SpC-EGFP and the different VP2 variants, respectively, while arrows highlight possible conjugated products formed during co-infection of SpC–SpT pairs

SpT (SpyTag002, 14 amino acids) (Brune et al. 2016; Keeble et al. 2017), the shorter SpT peptide appears more appropriate for insertion into the VP2 loop without disrupting the VLP assembly. To evaluate this approach, several VP2 chimeric variants were designed: one without SpT (noSpT, 68.1 kD), and others with SpT inserted at different positions—N-terminus (NSpT, 70.8 kD), Loop 2 (SpTL2, 70.5 kD), and Loop X (SpTLx, 70.5 kD) (Fig. 7.13c). The resulting conjugates from various co-infection combinations (recombinant BmNPV ratio 1:1) between each VP2 variant and SpC-EGFP were also confirmed in Fig. 7.13d. As expected, each VP2 variant demonstrated effective binding to SpC-EGFP, signifying successful ligation in Bm5 cell cultures. The conjugation efficiency correlated with relative expression levels, as evidenced by a marked reduction in free SpT-VP2 or SpC-EGFP in co-infection samples (Lanes 9–11) compared to single infection samples (Lanes 2, 4–6).

The VP2 variants were expressed in silkworm larvae and purified to assess their in vitro binding and display capabilities. As shown in Fig. 7.14a, all variants were purified in significant quantities (on the milligram scale) from 10 silkworms. Notably, two loop variants, SpTL2 and SpTLx, yielded slightly less protein (~2.3 mg and ~1.1 mg, respectively), likely due to increased loss during Strep-column purification. To determine whether these purified VP2 proteins could efficiently assemble into VLPs, VLPs were purified using a standard 20% sucrose cushion (Huhti et al. 2010; Kato et al. 2023). Following ultracentrifugation at 122,000 × *g*, both the supernatant (containing monomers and non-VLPs) and the pellet (containing VLPs) were analyzed by SDS-PAGE. Fig. 7.14b indicates that NSpT-VP2 and SpTL2-VP2 predominantly formed VLPs. However, SpTLx-VP2 did not show any visible VLP bands, suggesting that insertion of SpT into the LoopX region interfered with correct VLP assembly, as further supported by the lack of intact VLPs observed in TEM images (Hurtado et al. 1996).

Next, whether the SpT-tagged VLPs could serve as platforms for displaying SpC-fused proteins (SpC-EGFP) by directly mixing them in a PBS-buffered solution was investigated. As shown in Fig. 7.14c, specific ligation occurred between the SpT proteins (NSpT, SpTL2, and SpTLx) and SpC-EGFP, with no shifted bands present in the noSpT-VP2 control. Since Fig. 7.14b showed that SpTLx-VP2 does not reliably form VLPs, the resulting adducts from SpTLx-VP2 incubated with SpC-EGFP are likely derived from monomeric SpTLx-VP2 rather than assembled VLPs. After purifying the SpC-EGFP and NSpT or SpTL2 mixtures using a 20% sucrose cushion, a marked increase in SpC-EGFP display on VLPs was observed in the SpTL2 group, as indicated by the relative band intensity between unmodified and EGFP-decorated VLPs (EGFP-VLP) (Fig. 7.14d). Display efficiency, expressed as a percentage and as the number of proteins per VLP, was estimated from three independent binding experiments (Figs. 7.14e, f). Figure 7.14g illustrates that protein display abundance increased from 5.43% in NSpT-VP2 to 80.3% in SpTL2-VP2, corresponding to 3 to 48 proteins per VLP. This demonstrates that inserting SpT into the VP2 loop region significantly enhances the surface display of proteins on VLPs. TEM validation of VLP formation for SpTL2-VP2 and the NSpT-VP2 control is shown in Fig. 7.14h.

7.3.3.2 SpT-SnT Double-Conjugated VLPs

To advance the chemically orthogonal display of proteins on SpT-VP2-based VLPs, SnoopTag-SnoopCatcher (SnT-SnC) chemistry was applied to SpTL2-VP2, as depicted in Fig. 7.15a, to enable dual plug-and-display (Brune et al. 2016). Following this approach, various DNA plasmids encoding His6-Strep (HS)-tagged VP2 proteins with SnT_{L2}-SpT_{Lx} (❸, 72.7 kD), SpT_{L2}-SnT_{Lx} (❹, 72.7 kD), a version of VP2 lacking the HS-tag (ΔStrepTag) designated ΔStrepTag-$STSnT_{L2}$-SpT_{Lx} (❺, 69.1 kD), as well as SpC-EGFP (❶, 43.8 kD) and SnC-mCherry (❷, 43.5 kD) were constructed. Expression of each construct, except for ❺, was confirmed in cultured Bm5 cells by individual or co-infection (Fig. 7.15b). When all three (❶,❷,❸:

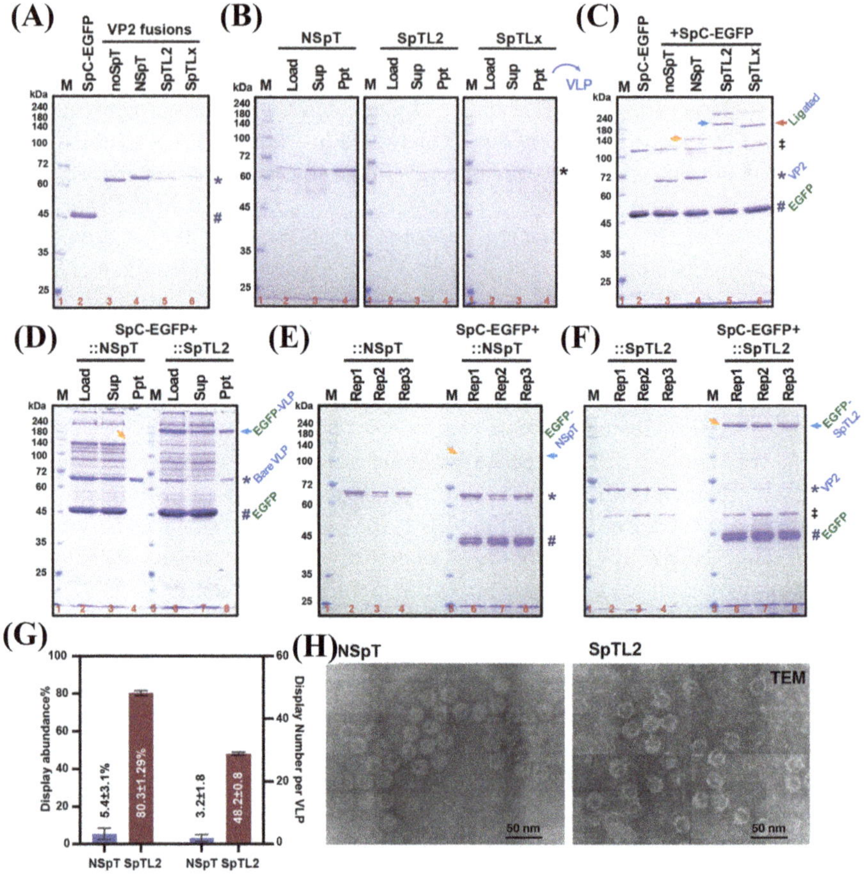

Fig. 7.14 Assessment of improved display in SpT-inserted VP2-based VLPs (Xu et al. 2023). All SpC-EGFP and VP2 forms—without SpT (noSpT), with SpT at the N-terminus (NSpT), in Loop2 (SpTL2), and in LoopX (SpTLx)—were produced and isolated from the fat bodies of recombinant baculovirus-infected silkworms (**a**). The VLPs were purified by ultracentrifugation using a 20% sucrose cushion (**b**). The efficiency of in vitro covalent binding between SpC (#) and SpT (*) variants was performed using saturated SpC-EGFP (#) proteins, as in (**c**). Ligated complexes are indicated with arrows, while the double dagger (‡) marks a purification impurity from SpC-EGFP. The reaction mixtures of SpC-EGFP and the VP2 variants NSpT and SpTL2 were further purified by ultracentrifugation with 20% sucrose. The resulting VLPs, including both unconjugated (*) and EGFP-labeled (EGFP-VLP, arrow) particles, were visualized in CBB (**d**). The binding experiments were conducted in triplicate (**e, f**), and the data on EGFP-VLP precipitation were statistically analyzed and presented in (**g**). The results showed a significant rise in EGFP-displayed VLPs, from roughly 5.4% (3 per VLP) to 80% (48 per VLP). The symbol (‡) also denotes a purification impurity for SpTL2. The formation of VLPs for the NSpT and SpTL2 variants was confirmed by TEM (**h**)

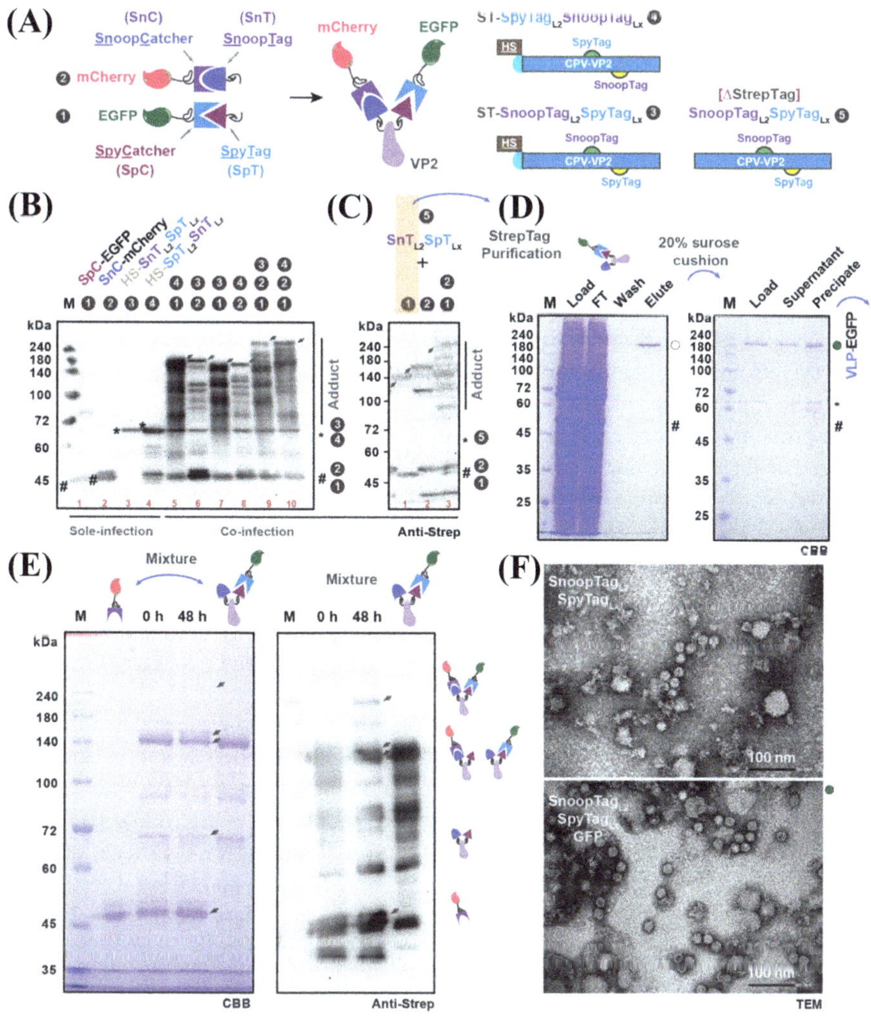

Fig. 7.15 Engineering and evaluation of simultaneous protein display on VP2-derived VLPs incorporating SpT and SnT inserts (Xu et al. 2023). (**a**) A diagram illustrates the specific covalent interactions between SpT and SpC, as well as SnT and SnC. Five different constructs were generated to enable multiprotein display: SpC-EGFP (❶), SnC-mCherry (❷), VP2-SnT$_{L2}$SpT$_{Lx}$ (❸), VP2-SpT$_{L2}$SnT$_{Lx}$ (❹), and VP2-SnT$_{L2}$SpT$_{Lx}$ (ΔStrepTag, ❺). (**b**) Confirmation of conjugation of these constructs individually, in pairs, or in triplets in Bm5 cell cultures. When cells were co-infected, higher-molecular-weight adducts appeared, as indicated by arrows. (**c**) Western blot analysis with an anti-Strep antibody further verified the expression and binding among VP2-SnT$_{L2}$SpT$_{Lx}$ (ΔStrepTag, ❺), SpC-EGFP (❶), and SnC-mCherry in Bm5 cells. (**d**) Purification of co-infected silkworm larvae expressing VP2-SnT$_{L2}$SpT$_{Lx}$ (ΔStrepTag, ❺) and SpC-EGFP (❶) by ultracentrifugation through a 20% sucrose cushion. The resulting EGFP-displayed VLPs (VLP-EGFP) were presented in the pellet. In vitro, the binding of purified SnC-mCherry to [SnT$_{L2}$][SpT$_{Lx}$-SpC-EGFP] (VLP-EGFP) (**d**) was assessed over 48 h, with results shown by SDS-PAGE (**e**, left) and Western blot using anti-Strep (**e**, right). (**f**) Transmission electron microscopy images of VLPs (top: SnT$_{L2}$SpT$_{Lx}$; bottom: SnT$_{L2}$SpT$_{Lx}$::GFP). Scale bar: 100 nm

SpC-EGFP, SnC-mCherry, and SnT_{L2}-SpT_{Lx}; 160 kD) or (❶,❷,❹: SpC-EGFP, SnC-mCherry, and SpT_{L2}-SnT_{Lx}; 160 kD) were co-infected, the resulting adducts (marked by arrows) exhibited higher molecular weights than those observed in double infections (❶,❹: 116.5 kD; #2,3: 116.2 kD; ❶,❸: 116.5 kD; ❷,❹: 116.2 kD). To streamline the purification of antigen-displaying VLPs, fusion tags were removed from SnT_{L2}-SpT_{Lx}-VP2, particularly focusing on tags from SpC/SnC-tagged antigens. Utilizing the HS-tag-free VP2 (❺, 69.1 kD), Western blot analysis of co-infection samples—❶,❺ (112.9 kD), ❷,❺ (112.6 kD), and ❶,❷,❺ (156.4 kD)—demonstrated specific VP2 adducts covalently bound with SpC-EGFP (❶), SnC-mCherry (❷), or both (❶ and ❷), as indicated by arrows (Fig. 7.15c). These findings is similar to those in Fig. 7.15b. The results from the SpT/SnT dual chemistry system indicate that ligation among SpT, SpC, SnT, and SnC is both specific and orthogonal in the silkworm cells.

To improve the yield of each protein partner, SpC-EGFP::SnT_{L2}-SpT_{Lx} (❶,❺) was purified from the fat body extracts of ten silkworm larvae. The larvae were co-infected with recombinant BmNPVs at a ratio of approximately 1:2 (SpC-EGFP:SnT_{L2}-SpTLx). This optimized condition was expected to enhance the purification of Strep-tagged SpC-EGFP, ensuring that the majority of SpC-EGFP::SnTL2-SpTLx-VP2 was present in both the lysates and the final eluate. As expected, the results shown in Fig. 7.15d (left panel) confirmed the successful purification of the ligated SpC-EGFP::SnT_{L2}-SpT_{Lx}-VP2 adduct (VP2-EGFP), with only a minor proportion of SpC-EGFP detected in the final product. The assembly of VP2-EGFP-SnT_{Lx} VLPs (VLP-EGFP displaying free SnT) was validated by their presence in the pellet following ultracentrifugation through a 20% sucrose cushion (Fig. 7.15d, right panel), which was unexpected since chimeric VP2 with SpT-L2 insertion previously failed to self-assemble into VLPs (Xu et al. 2023). To further assess the in vitro binding capability of SnT exposed on the VLP surface, purified VLP-EGFP-SnT was incubated with SnC-mCherry for 48 h at 4 °C, and results are shown in Fig. 7.15e. The appearance of high-molecular-weight bands in both CBB staining and Western blot indicated the successful formation of VLPs displaying both SnC-mCherry and SpC-EGFP. Additionally, the integrity and efficient assembly of purified SnT_{L2}-SpT_{Lx}-VP2 (bare VLP carrier) and SpC-EGFP::SnT_{L2}-SpT_{Lx}-VP2 (SpC-EGFP-displaying VLP) were confirmed by transmission electron microscopy (Fig. 7.15f). Combined with VLP-EGFP purified via 20% sucrose cushion (Fig. 7.15d, f), the successful generation of a VP2-VLP platform capable of displaying multiple proteins using orthogonal SnT-SnC and SpT-SpC chemistries.

7.3.4 *Improvement of Display Efficiency in SpyTac-Tagged VLPs*

Covalent conjugation is advantageous for its stability compared to noncovalent methods and makes it easier to analyze the results (Li et al. 2020; Veggiani et al. 2016; Zakeria et al. 2012). The Tag/Catcher system (such as SpyTag/SpyCatcher) is attractive for making modular nanoparticle vaccines because the process is quick and straightforward. This system works by forming a strong isopeptide bond between two proteins, making the connection tough against chemical and physical stress. Also, the reaction is flexible and can happen under different conditions like changes in buffer, pH, or temperature (Reddington and Howarth 2015; Zakeria et al. 2012). However, with regard to antigen display, the number of antigen molecules displayed on the surface of VLPs—that is, the antigen display efficiency—is a crucial issue. In other words, it is better that as many antigen molecules as possible be displayed on the surface of the VLPs. Norovirus-like particles (NoV-LPs) were used to investigate this. While many icosahedral VLPs have been used to display other proteins, NoV-LPs are not often used this way. Most studies focus on NoV-LPs as vaccine candidates against norovirus. When the main capsid protein VP1 is expressed, NoV-LPs naturally form particles around 30–40 nm in diameter. These particles are highly stable across a wide range of pH, temperature, and salt concentrations (Samandoulgou et al. 2015). They can be stored for months if chilled or for a week at room temperature (Huhti et al. 2010). Their ability to self-assemble into icosahedral shapes makes them highly immunogenic and capable of displaying multiple copies of antigens.

In this section, NoV-LPs from norovirus GII.4 were engineered by adding a SpT to the end of the capsid protein, allowing them to bind other proteins. The effectiveness of this connection was tested with and without a spacer between VP1 and SpT. Protein display on NoV-LP/SpT was demonstrated using fluorescent proteins fused to SpC. Protein linkers are typically categorized into three groups: flexible (such as GGGGS and $(Gly)_n$), rigid (such as EAAAK and PAPAP), and cleavable linkers (for example, PLGLWA and AGNRVRRSVG). Of all synthetic protein linkers, the flexible "$(GGGGS)_n$" and the rigid "$(EAAAK)_n$" linkers are the ones most widely utilized in research (Chen et al. 2013). The 'EAAAK' linker forms an alpha helix, providing rigidity and restricting rotational movement between protein domains compared to flexible linkers (Wriggers et al. 2005). Here, the rigid "EAAAK" linker was chosen as a spacer to ensure consistent separation and orientation between the C-terminus of VP1 and the SpT peptide, theoretically maintaining the SpT's outward orientation.

The 13-amino acid SpT peptide was fused to the 3′ end of the norovirus VP1 gene, either directly or via the rigid, alpha helix-forming "EAAAK" linker (Fig. 7.16a). The variant without the linker is referred to as "VP1-(c-SpT)," while those containing two and three linker repeats are named "VP1-L_2(c-SpT)" and "VP1-L_3(c-SpT)," respectively. In the "VP1-L_2(c-SpT)" and "VP1-L_3(c-SpT)" constructs, the "EAAAK" linker is bordered by GS-linkers. All constructs were expressed in silkworm larvae and subsequently purified. The theoretical molecular weights were 61.1 kDa for VP1-(c-SpT), 62.2 kDa for VP1-L_2(c-SpT), and 62.7 kDa for VP1-L_3(c-SpT). Most of the expressed proteins were soluble (Fig. 7.16b) and were purified using both sucrose gradient centrifugation and His-tag affinity chromatography (Fig. 7.16c). Interestingly, even though these constructs lacked a polyhistidine tag, the assembled proteins still bound strongly to nickel-based ligands, likely due to the many histidine residues distributed across the NoV-LP surface

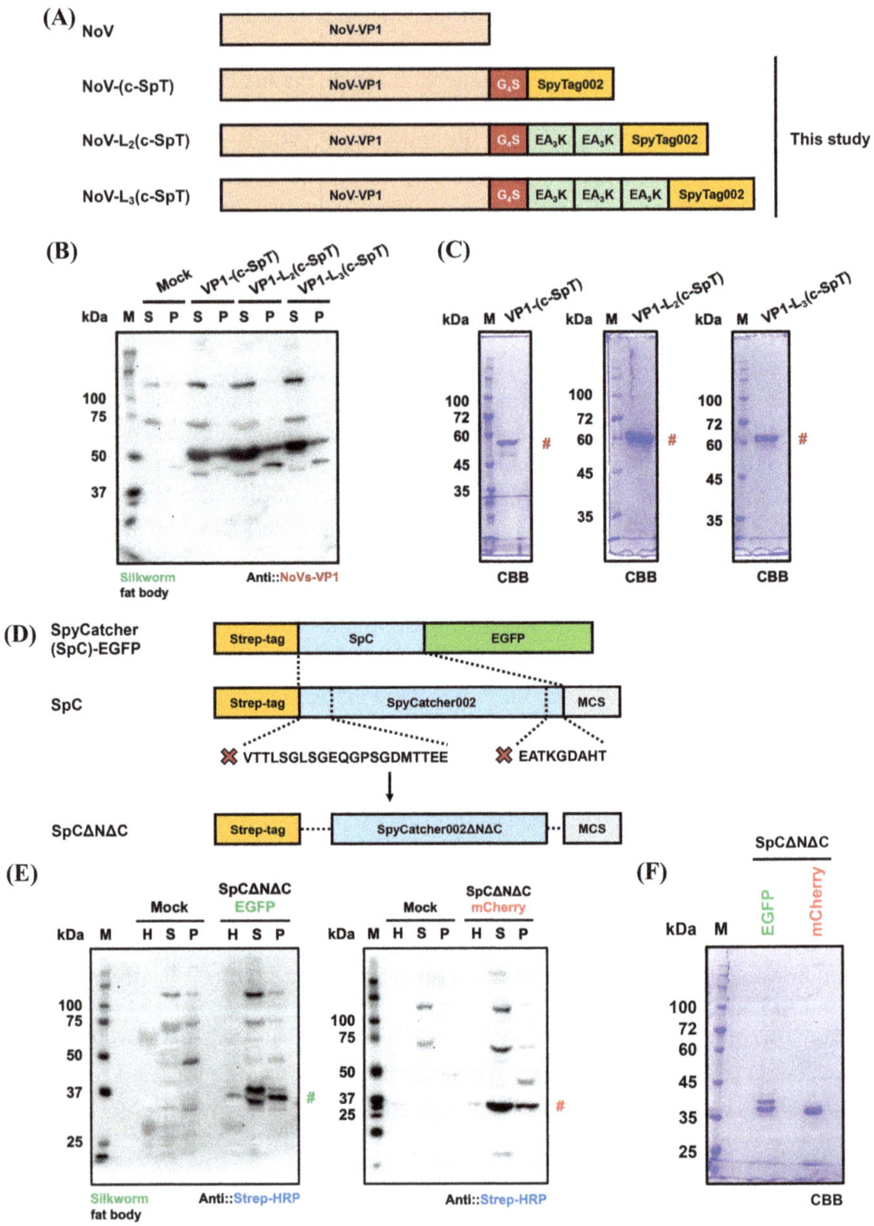

Fig. 7.16 Design and expression of VP1-SpT variants in silkworms. (**a**) Norovirus VP1-SpT constructs. The SpT sequence was fused to the C-terminus of the norovirus VP1 gene using linkers of different lengths. (**b**) Western blot analysis of the VP1-SpT variants expressed in silkworm larvae, detected with anti-VP1 (NS14) antibodies. (**c**) SDS-PAGE with CBB staining of purified VP1-SpT variants. Proteins were purified from silkworm fat bodies using sucrose gradient centrifugation and then dialyzed overnight in PBS buffer (pH 7.4). (**d**) Construction of SpC and its truncated version (SpCΔNΔC) fusion proteins, where both N- and C-terminal portions were deleted to create

(Boonyakida et al. 2022; Chen and Huang 2020). SDS-PAGE analysis showed that the purified proteins appeared as a single major band at the expected ~60 kDa size, with purity above 85% (Fig. 7.16c) (Boonyakida et al. 2023).

SpC was fused to the model proteins EGFP and mCherry so that the protein of interest (POI) could be displayed on the surface of NoV-LPs/VP1-SpT variants. Since removing both the N- and C-terminal regions of SpC does not affect the formation of isopeptide bonds with SpyTag (Li et al. 2014), a truncated version of SpC (SpCΔNΔC) (Fig. 7.16d) was constructed, operating under the assumption that eliminating unnecessary structural components would improve conjugation efficiency. The SpCΔNΔC construct was then fused to the 5′ end of either the EGFP or mCherry gene, subsequently expressed in silkworm larvae, and purified using Strep-tag affinity chromatography. The calculated molecular weight of SpCΔNΔC-EGFP and SpCΔNΔC-mCherry was 37.6 kDa. Western blot results showed that these recombinant proteins were predominantly soluble and localized exclusively in the fat body tissue (Fig. 7.16e). When SpCΔNΔC-mCherry was purified, it displayed a single band at the expected size; SpCΔNΔC-EGFP showed two closely sized bands, consistent with the observations from the Western blot analysis (Fig. 7.16f).

As for control, full-length SpC-fused EGFP (SpC-EGFP, 45 kDa) was displayed to the surface of NoV-(c-SpT). To improve the display efficiency, conjugation conditions were optimized, but only about 15–20% efficiency was achieved, meaning 27–36 SpC-EGFP molecules displayed per NoV-(c-SpT) particle. Since NoV-LP forms a $T = 3$ structure, it was believed that the limited exposure of the SpT on the surface was causing the low binding efficiency (Boonyakida et al. 2023). To improve efficiency to at least 50%, two or three "EAAAK" linkers were added to the ends of VP1 and SpT to make the tag more accessible for binding. After confirming the formation of VLPs, they were incubated with an excess amount of SpC-EGFP and kept overnight at 4 °C (Fig. 7.17a). This resulted in 17% efficiency for NoV-(c-SpT)/SpC-EGFP, but NoV-L_2(c-SpT)/SpC-EGFP and NoV-L_3(c-SpT)/SpC-EGFP achieved 47.4% and 51.4% efficiency, respectively (Boonyakida et al. 2023) (Fig. 7.17b). Removing extra β-strands from SpC did not affect its ability to form an isopeptide bond with SpT and might even improve efficiency. By deleting 21 and 9 amino acids from the N- and C-termini of SpC (SpCΔNΔC), better results were shown when modifying VLPs with SpCΔNΔC-EGFP. The display efficiency for SpCΔNΔC-EGFP on NoV-(c-SpT) increased to around 35% (NoV::EGFP). After overnight incubation at 4 °C, NoV-L_2(c-SpT)/SpCΔNΔC-EGFP (NoV-L_2::EGFP) reached 60% efficiency, and NoV-L_3(c-SpT)/SpCΔNΔC-EGFP (NoV-L_3::EGFP)

Fig. 7.16 (continued) SpCΔNΔC. MCS stands for multiple cloning site; red X indicates sites of amino acid deletions. (**e**) Western blot results for SpCΔNΔC fused to EGFP and mCherry, expressed in silkworms. (**f**) Purified SpCΔNΔC EGFP and SpCΔNΔC-mCherry proteins. M: molecular weight marker; H: clarified hemolymph from silkworms; S: supernatant (soluble fat body lysate); P: pellet (insoluble fraction); CBB: Coomassie Brilliant Blue staining. The '#' symbol marks the target recombinant proteins. (Adapted with permission (Boonyakida et al. 2023). Copyright 2023 Americal Chemical Society)

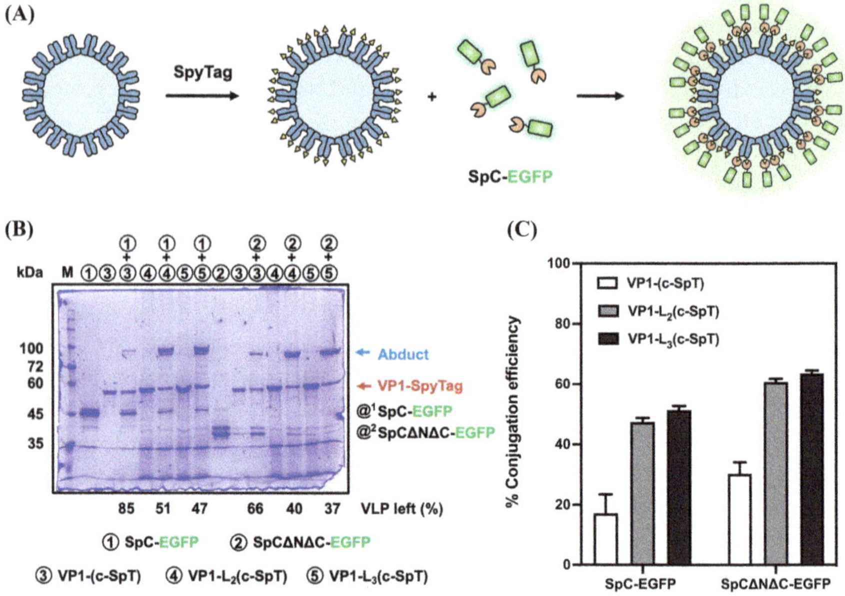

Fig. 7.17 Bioconjugation of NoV-LP using the SpT/SpC. (**a**) Illustration showing Bioconjugation of NoV-LP SpT variants with the desired protein (e.g., EGFP or mCherry). (**b**) SDS-PAGE results after the bioconjugation. (**c**) Conjugation yields, determined by analyzing the intensity of protein bands on the gel. (Adopted with permission (Boonyakida et al. 2023). Copyright 2023 American Chemical Society)

reached 63% efficiency (Fig. 7.17c). However, NoV-L_2(c-SpT)/SpC-EGFP and NoV-L_3(c-SpT)/SpC-EGFP formed large aggregates after several days, as evidenced by strong green fluorescence, which made further analysis difficult. In contrast, VLPs modified with SpCΔNΔC-EGFP did not form aggregates. As a result, NoV-L_2::EGFP and NoV-L_3::EGFP were selected for future display because they exhibited greater than 50% efficiency and did not aggregate.

The bioconjugation process was monitored to determine the optimum reaction time. At each interval, samples were heated at 95 °C for 5 min to stop the reaction and then subjected to SDS-PAGE analysis. Densitometric analysis showed that the bioconjugation of NoV-L_2(c-SpT) (Fig. 7.18a, b) or NoV-L_3(c-SpT) (Fig. 7.18c, d) to SpCΔNΔC-EGFP reached a steady state after 4–5 h incubation.

Transmission electron microscopy (TEM) and dynamic light scattering (DLS) were used to analyze the structural integrity and shape of NoV-LPs before and after VLP modification. According to TEM observations (Fig. 7.19a), NoV-(c-SpT), NoV-L_2(c-SpT), and NoV-L_3(c-SpT) appeared unchanged compared to both wild-type NoV-LP and modified VLPs. In earlier work, silkworm-derived NoV-LPs were reported to have a mean hydrodynamic diameter of 34.5 nm as measured by DLS (Boonyakida et al. 2022). The modified NoV-(c-SpT) had an average diameter of 35 nm, closely matching prior findings. The mean diameters of NoV-L_2(c-SpT) and

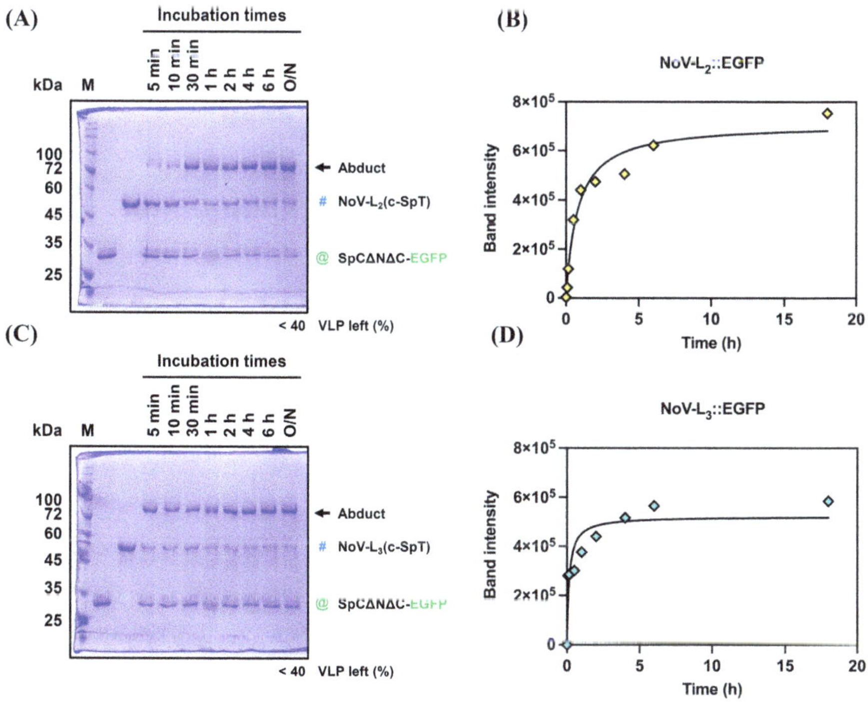

Fig. 7.18 SDS-PAGE and time course of band intensity of abduct. (**a**) SDS-PAGE assay of VP1-L_2(c-SpT) and SpCΔNΔC-EGFP. (**b**) Time course of EGFP-displayed VP1-L_2(c-SpT) band intensity. (**c**) SDS-PAGE assay of VP1-L_3(c-SpT) and SpCΔNΔC-EGFP. (**d**) Time course of EGFP-displayed VP1-L_3(c-SpT) band intensity. (Adopted with permission (Boonyakida et al. 2023). Copyright 2023 American Chemical Society)

NoV-L_3(c-SpT) increased to 39.5 nm (a 12.6% rise) and 40.3 nm (a 14.8% rise), respectively, indicating that the presence of a linker likely improves SpT peptide display on the particle surface for conjugation (Fig. 7.19b). The unmodified NoV-(c-SpT) (35.1 nm) and its EGFP-conjugated counterpart (NoV::EGFP, 37.2 nm) differed by 6% in average diameter (Fig. 7.19c). Upon conjugating SpCΔNΔC-EGFP to NoV-L_2(c-SpT) and NoV-L_3(c-SpT), DLS detected further size increases. Specifically, the diameter of NoV-L_2(c-SpT) (now NoV-L_2::EGFP) increased by 10% (from 39.5 nm to 43.4 nm) (Fig. 7.19d), and NoV-L_3(c-SpT) conjugated with SpCΔNΔC-EGFP (NoV-L_3::EGFP) grew by 15.5% (from 40.3 nm to 46.5 nm) (Fig. 7.19e). Overall, DLS data revealed a stepwise increase in size after modification.

To better understanding how display efficiency is improved, the structures of VP1-(c-SpT), VP1-L_2(c-SpT), and VP1-L_3(c-SpT) were predicted using an AlphaFold2-based method, with the GII.4 Minerva strain structure (PDB 6OUU) serving as the template. This software was used to demonstrate how adding an "EAAAK" linker between VP1 and SpT can impact the tag's exposure on the protein surface. According to the models (Fig. 7.20a–c), the S and P domains folded

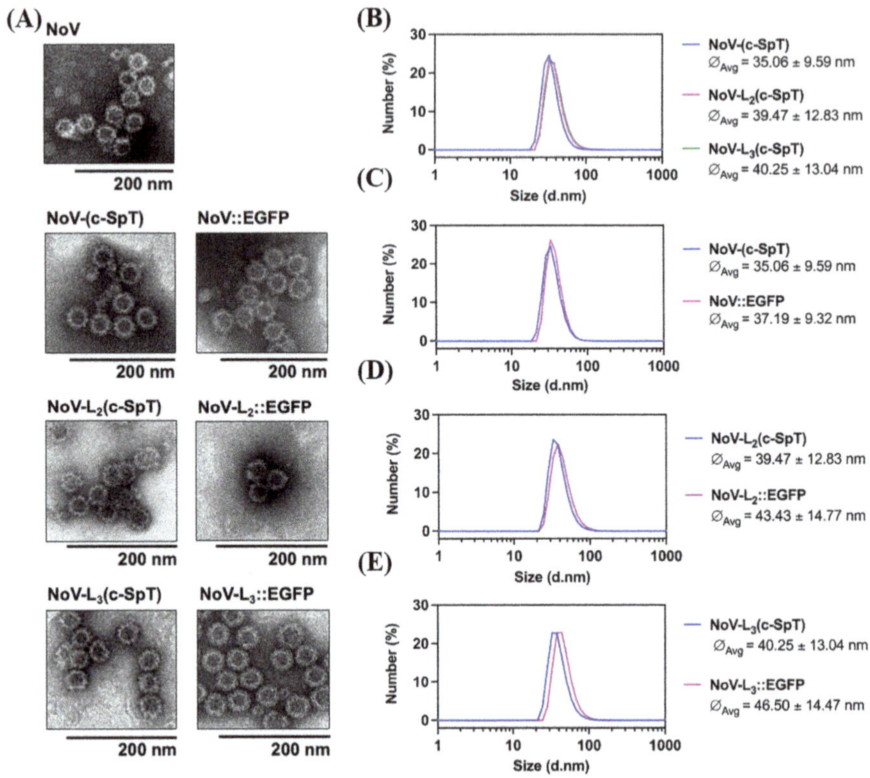

Fig. 7.19 Transmission electron microscopy (TEM) and dynamic light scattering (DLS) analysis of NoV-LPs. (**a**) TEM micrograph images of NoV-SpT VLP variants, both before and following surface modification. (**b**) DLS profiling of the unmodified variants. DLS profiling for NoV-(c-SpT) (**c**), NoV-L_2(c-SpT) (**d**), and NoV-L_3(c-SpT) (**e**) before and after EGFP bioconjugation. (Adapted with permission (Boonyakida et al. 2023). Copyright 2023 American Chemical Society)

similarly in all variants due to the shared template. However, when the models were aligned, noticeable differences appeared at the C-terminal region (Fig. 7.20d). In the VP1-(c-SpT) model, the SpT peptide was found close to the shell domain, which might restrict complex formation because the VP1 C-terminus sits in the grooves formed by the three- and five-fold axes of $T = 3$ norovirus particles (Devant and Hansman 2021; Hansman et al. 2012). For SpC to bind to SpT, it must access these grooves. By introducing the “EAAAK” linker, the SpT in VP1-L_2(c-SpT) and VP1-L_3(c-SpT) is positioned farther from the C-terminus, potentially increasing tag exposure and facilitating complex formation.

From this computational model, spatial modulation played a role in two main aspects: (1) how efficiently proteins were conjugated and displayed on the VLPs, and (2) the spatial distance between conjugation sites on the VLPs. Using a longer linker created more space, allowing for easier conjugation between SpT and SpC. Conversely, a shorter linker resulted in a denser VLP surface layer, forming a compact corona.

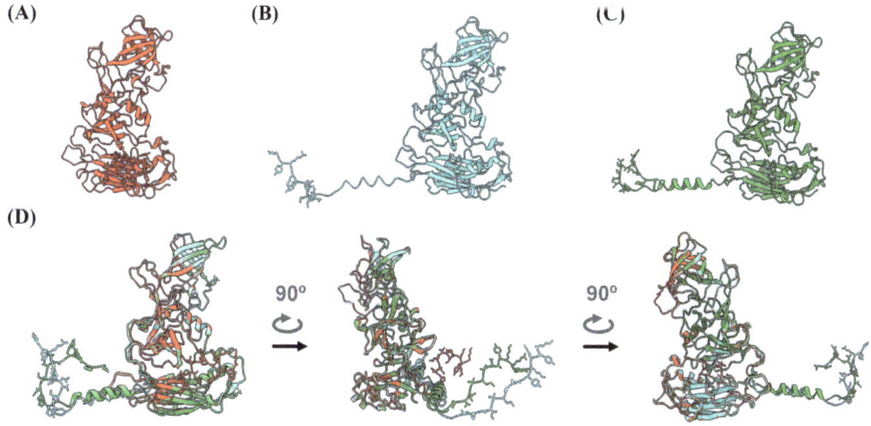

Fig. 7.20 Transmission electron microscopy (TEM) and dynamic light scattering (DLS) analysis of NoV-LPs. (**a**) TEM micrograph images of NoV-SpT VLP variants, both before and following surface modification. (**b**) DLS profiling of the unmodified variants. DLS profiling for NoV-(c-SpT) (**c**), NoV-L_2(c-SpT) (**d**), and NoV-L_3(c-SpT) (**e**) before and after EGFP bioconjugation. (Adapted with permission (Boonyakida et al. 2023). Copyright 2023 American Chemical Society)

7.3.5 *Bivalent VLPs Displaying* Plasmodium yoelii *Merozoite Surface Protein and Circumsporozoite Protein*

To block multiple infections of pathogens or viruses, multiple antigens must be prepared. Consequently, displaying diverse antigens on a single VLP surface is remarkably effective. The chemical conjugation method using SpT/SpC offers a straightforward approach for creating multivalent VLPs. In this section, the display of multiple antigens on the surface of VLPs are discussed, including those associated with malaria and dengue fever.

Malaria, a serious disease transmitted by mosquitoes, is caused by *Plasmodium* parasites. Annually, this illness puts hundreds of millions of people at risk. In 2022, the World Health Organization reported approximately 249 million malaria cases and about 608,000 deaths. Although progress has been made in controlling the disease, new challenges have emerged, such as *Plasmodium* strains that resist medications and mosquitoes that are resistant to insecticides (Haldar et al. 2018; Kleinschmidt et al. 2018; Wicht et al. 2020). To address this, the Malaria Eradication Research Agenda (malERA) recommended a comprehensive global strategy at its 2016 summit, aiming to reduce and eventually eliminate malaria, especially in regions where it is highly prevalent. Among the most promising approaches is developing an effective vaccine that can block the parasite at various points in its life cycle (Rabinovich et al. 2017).

Merozoite surface protein-1 (MSP1) is a key target for developing vaccines aimed at the blood stage of *Plasmodium* parasites (Duffy and Patrick Gorres 2020). As the most abundant protein on the merozoite surface, MSP1 is crucial for the parasite's invasion of red blood cells (Das et al. 2015; Moss et al. 2012). Initially produced as a large polypeptide in schizonts, it is cleaved into four fragments—p83, p42, p38, and p30—which stay together on the parasite's surface (Child et al. 2010). During red blood cell infection, the p42 fragment is further processed into p33 and p19. The p19 fragment remains attached to the merozoite via a glycosylphosphatidylinositol (GPI) anchor (Dijkman et al. 2021). Antibodies against $MSP1_{19}$ can prevent merozoite binding to red blood cells and inhibit parasite growth, making it a valuable antigen in vaccine research (Hui and Hashimoto 2007; Perera et al. 1998; Wilson et al. 2011).

Circumsporozoite protein (CSP) is the main surface protein of the *Plasmodium* sporozoite and is essential for the parasite's entry into liver cells. CSP has a central repeat domain, which varies among species. For example, *P. falciparum*'s CSP has 35–38 NANP repeats and several NVDP motifs (Oyen et al. 2018); *P. vivax* features strain-specific nine-amino-acid repeats; and *P. yoelii* has QGPGAP and QQPP repeats (Lal et al. 1987). CSP has been the main focus for pre-erythrocytic malaria vaccine development.

A new vaccine approach aims to enhance protection by targeting multiple stages of the *Plasmodium* parasite's life cycle. In this section, a dual-antigen vaccine that addresses both the liver (pre-erythrocytic) and blood (erythrocytic) stages was designed using *P. yoelii* CSP (PyCSP) and *P. yoelii* $MSP1_{19}$ ($PyMSP1_{19}$) as antigens. These antigens were displayed on the surface of a modified NoV-LP that incorporated a SpyTag peptide to enhance antigen display.

The SpyTagged-NoV-LP[NoV-L_2(c-SpT)] was produced following previously reported methods but used anion exchange chromatography (AEX) for the secondary purification step (Boonyakida et al. 2023). This strategy yielded nearly 3 mg of NoV-LP from 10 silkworm larvae. The purified protein was a single dominant band at approximately 60 kDa with a purity exceeding 90% (Fig. 7.21a–d).

For the display of CSP (PyCSP) and $MSP1_{19}$ from the rodent parasite *P. yoelii* ($PyMSP1_{19}$) on NoV-LP surfaces, a N- and C-terminal truncated SpC was fused to the N-terminus of each target protein (SpCΔNΔC-PyCSP and SpCΔNΔC-$PyMSP1_{19}$). These constructs were produced using the silkworm expression system (Boonyakida et al. 2023). Immunoblotting revealed that both proteins were predominantly soluble and localized to the silkworm fat body (Fig. 7.21e). All antigenic proteins were purified through Strep-tag affinity chromatography and examined by SDS-PAGE. The total amounts of SpCΔNΔC-PyCSP and

Fig. 7.21 (continued) and $PyMSP1_{19}$ expressed in silkworm fat body. (**e**) Western blot analysis of SpyCatcher (SpCΔNΔC)-fused PyCSP and $PyMSP1_{19}$ expressed in silkworm larvae using StrepTactin-HRP conjugate. S refers to the soluble fraction of the fat body, and P to the insoluble (precipitate) fraction. CBB-stained SDS-PAGE of SpCΔNΔC-PyCSP (**f**) and SpCΔNΔC-$PyMSP1_{19}$ (**g**) purified from silkworm fat body lysate using a Strep-Tactin affinity column. The purified target proteins are indicated by a red hashtag. M—protein marker; L—loading sample; FT—flowthrough; W—washing; CBB—Coomassie brilliant blue staining. (Adapted with permission (Boonyakida et al. 2024). Copyright 2024 American Chemical Society)

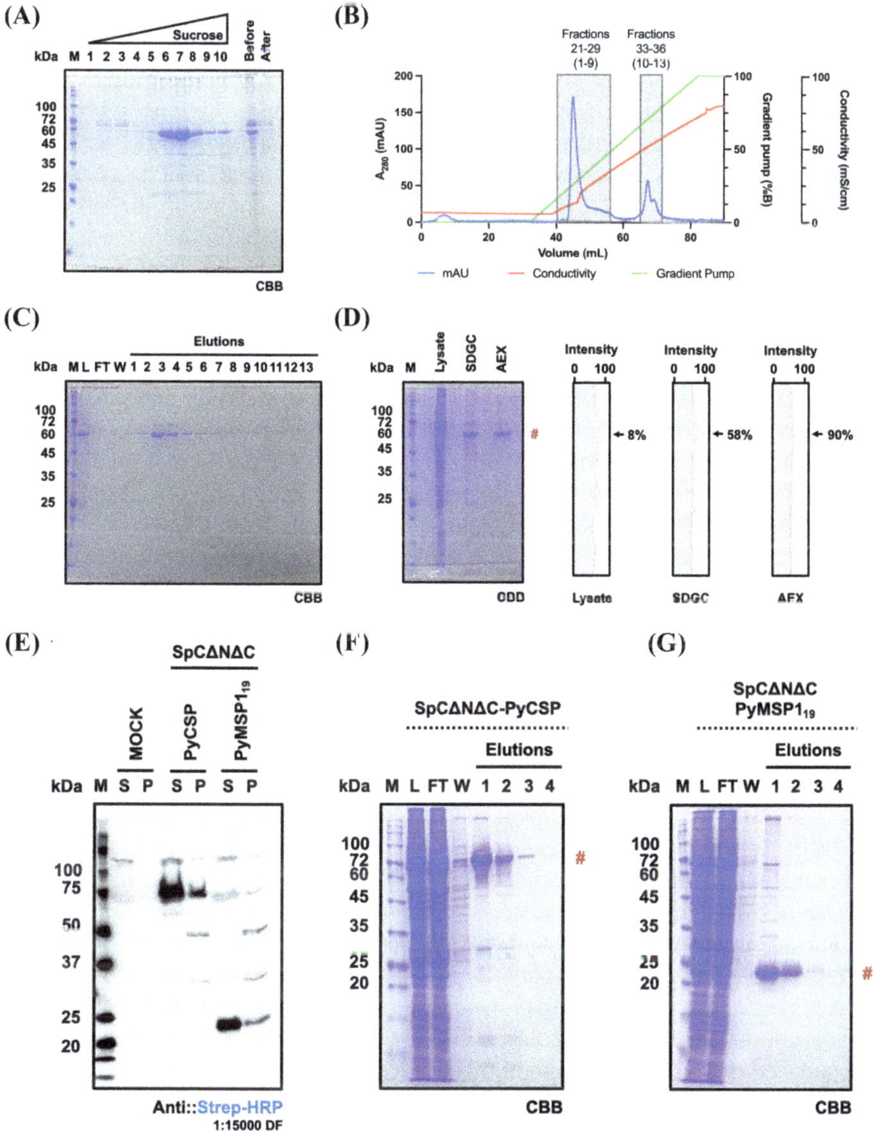

Fig. 7.21 (**a–d**) Purification of NoV-L$_2$(c-SpT) from the lysate of silkworm fat body. The NoV-LPs were purified from silkworm fat body lysate using a two-stage process involving sucrose gradient centrifugation followed by anion exchange chromatography. (**a**) SDS-PAGE of the samples of sucrose gradient centrifugation. (**b**) The chromatogram of an anion exchange (AEX) column. (**c**) CBB-stained SDS-PAGE of the loading (L) sample, the flowthrough (FT), the wash (W), and the eluted fractions. (**d**) CBB-stained SDS-PAGE analysis of NoV-LPs before and after each purification step. Protein purity post-purification was assessed by densitometry, with gray representing the background for each lane. The peak of NoV-L$_2$(c-SpT) in each lane is marked by a black arrow. M—protein marker; SDGC—sucrose density gradient centrifugation; AEX—anion exchange chromatography; CBB—Coomassie brilliant blue staining. (**e–g**) Purification of PyCSP

SpCNΔC-PyMSP1_{19} were roughly 1 mg and 1.2 mg, respectively, from six silkworm larvae (Fig. 7.21f, g).

CSPs was reported to be efficiently produced in *E. coli.* Still, those derived from *E. coli* are usually found in inclusion bodies, necessitating additional purification steps with chaotropic agents to extract and solubilize the proteins. This adds complexity to the process, requiring on-column refolding and removal of the chaotropic agent (Kolodny et al. 2001; Plassmeyer et al. 2009). In contrast, in the silkworm, full-length PyCSP was predominantly expressed as a soluble form in the fat body lysate, indicating proper folding and robust expression. Using Strep-Tag affinity purification, approximately 150 μg of PyCSP per silkworm larva were obtained with about 85% purity in a single purification step. The soluble nature of PyCSP simplified the purification compared to *E. coli*-derived CSPs. Interestingly, although SpCΔNΔC-PyCSP was expected to be about 50 kDa, SDS-PAGE analysis showed an apparent molecular weight of 75 kDa (Fig. 7.21f). This phenomenon is consistent with previous reports that CSP migrates at a higher apparent molecular weight in SDS-PAGE, likely due to the high content of polar amino acids (such as asparagine and glutamine) in the CSP sequence (Amelia et al. 2019; Chawla et al. 2019; Janitzek et al. 2016). Up to now, recombinant MSP$1_{19,}$ has been produced in several systems, such as *E. coli*, yeast, and mammalian cells (Cunha et al. 2001). However, the yield from these platforms is seldom discussed. Approximately 200 μg of purified PyMSP1_{19} was obtained per silkworm larva through a straightforward affinity purification, achieving over 80% purity for subsequent conjugation studies (Boonyakida et al. 2024). Furthermore, the observed molecular weight of SpC-fused PyMSP1_{19} closely matched the predicted value (~21.3 kDa) (Fig. 7.21g).

These findings demonstrate notably high yields of purified proteins, with about one milligram of target protein obtained from only 5–6 silkworm larvae. Importantly, this is the first to report the expression of CSP and MSP1_{19} in silkworm larvae, with efficient one-step purification using the Strep-Tactin affinity method, achieving over 80% purity from the fat body lysate.

The conjugation of SpC-PyAntigens on the surface of NoV-L_2(c-SpT) was optimized in vitro at different molar ratios. The input stochiometric ratio of SpC-PyAntigens to the NoV-L_2(c-SpT) was varied to investigate the maximum display number of the antigens. The molar ratio of SpyTagged-NoV-LP was fixed at 3 μM (or molar equivalent of 1) throughout all conjugation experiments, while SpCΔNΔC-PyCSP or -PyMSP1_{19} was varied from 1.5 to 12 μM (or molar equivalents from 0.5 to 3). At the molar ratio of 3:1.5 (or 1:0.5 of NoV-L2(c-SpT) to SpCΔNΔC-PyMSP119), the conjugation efficiency was about 50%. This corresponds to an estimated 90 molecules of PyMSP119, as determined by densitometric analysis. NoV-LP exhibited T = 3 assembly (180 subunits per particle). When the ratio increased to 3:3 (or 1:1), efficiency rose to approximately 70%, or 126 molecules of PyMSP119 per particle. The highest conjugation efficiency obtained was about 80% (144 molecules of PyMSP1_{19} per particle) and no noticeable difference in the efficiency was observed even at the ratio of 3:12 (or 1:4) (Boonyakida et al. 2024).

However, in the case of the NoV-L_2(c-SpT) to SpCΔNΔC-PyCSP reactive pair, despite the increment of PyCSP molarity, the conjugation efficiency was approximately 45% (81 molecules of PyCSP per VLP, denoted as NoV::PyCSP), and no aggregation was observed. Although the percent conjugation increased from about 42 to 49% of the efficiency when the molar concentration of PyCSP increased from 1.5 to 6 μM (or from 1:0.5 to 1:2), this change was not significant (Boonyakida et al. 2024). These results suggest that full occupancy could not be achieved due to the steric hindrance on the surface of NoV-LPs. Thus, not all VLP subunits were conjugated to the target antigens.

To prepare a bivalently displayed on the surface of NoV-LPs, the conjugation partner CSP and $MSP1_{19}$ was reacted in varying molar proportions. The ratio of PyCSP-to-Py$MSP1_{19}$ was 1:1, 1:2, 1:4, and 2:1; then, NoV-L_2(c-SpT) was added to the mixture, incubating at 4 °C (Fig. 7.22a). After incubating overnight, CSP conjugation efficiencies on the VLP surface ranged from 10% to 17% at 2:1 and 1:1 PyCSP: Py$MSP1_{19}$ ratios, while Py$MSP1_{19}$ conjugation efficiencies were between 53% and 67%, depending on the antigen combination (Fig. 7.22b). The highest CSP conjugation efficiency was 17%, corresponding to roughly 30 CSP molecules per NoV-LP at a 2:1 ratio (3 μM PyCSP and 1.5 μM Py$MSP1_{19}$). In comparison, Py$MSP1_{19}$ showed a 53% efficiency, corresponding to about 95 molecules per particle. With a 1:1 ratio (1.5:1.5, 3:3, or 6:6 of PyCSP to Py$MSP1_{19}$), CSP displayed on the surface of NoV-LPs at a relatively low level. When both antigens were at 1.5 μM, CSP occupied about 15% of available sites, and Py$MSP1_{19}$ occupied 55%. At 3 μM, CSP conjugation dropped to 10%, while Py$MSP1_{19}$ rose to 67%. Increasing both proteins to 6 μM further decreased CSP conjugation to 6%, whereas Py$MSP1_{19}$ reached 80%. As the proportion of Py$MSP1_{19}$ was increased to 1:2 and 1:4, CSP's conjugation efficiency fell to approximately 8% and 1%, respectively.

To increase the number of PyCSP molecules displayed, the conjugation reaction was performed in steps. The CSP was first incubated with the VLP on ice for either 5 or 15 min before the addition of Py$MSP1_{19}$. The reaction then continued overnight at 4 °C (Fig. 7.22a, highlighted in a yellow box). For this procedure, the concentrations were 3 μM SpyTagged-NoV-LP, 1.5 μM PyCSP, and 1.5 μM Py$MSP1_{19}$. Using this sequential approach, CSP conjugation efficiency increased to approximately 25% (45 CSP molecules per NoV-L_2(c-SpT)), nearly twice that observed with the simultaneous antigen conjugation method. The Py$MSP1_{19}$ conjugation efficiency was about 50% (90 $MSP1_{19}$ molecules per VLP) (Fig. 7.22c).

Densitometric analysis indicated that the fusion moiety on the 9-kDa SpC had a significant effect on the extent of protein attachment to the modified VLPs. The size and structure of each fusion protein—CSP (40 kDa) and $MSP1_{19}$ (10 kDa)—influenced binding dynamics, leading to competition between SpC-PySCP and SpC-Py$MSP1_{19}$ for binding to NoV-L_2(cSPT). Stepwise conjugation method revealed that pre-incubating the NoV-LP with PyCSP alone for 5 or 15 min before adding Py$MSP1_{19}$ increased the number of CSP molecules bound to the VLP surface. Nonetheless, after overnight incubation, roughly 25% of the VLPs remained unconjugated, likely due to steric hindrance from closely spaced antigens on the surface.

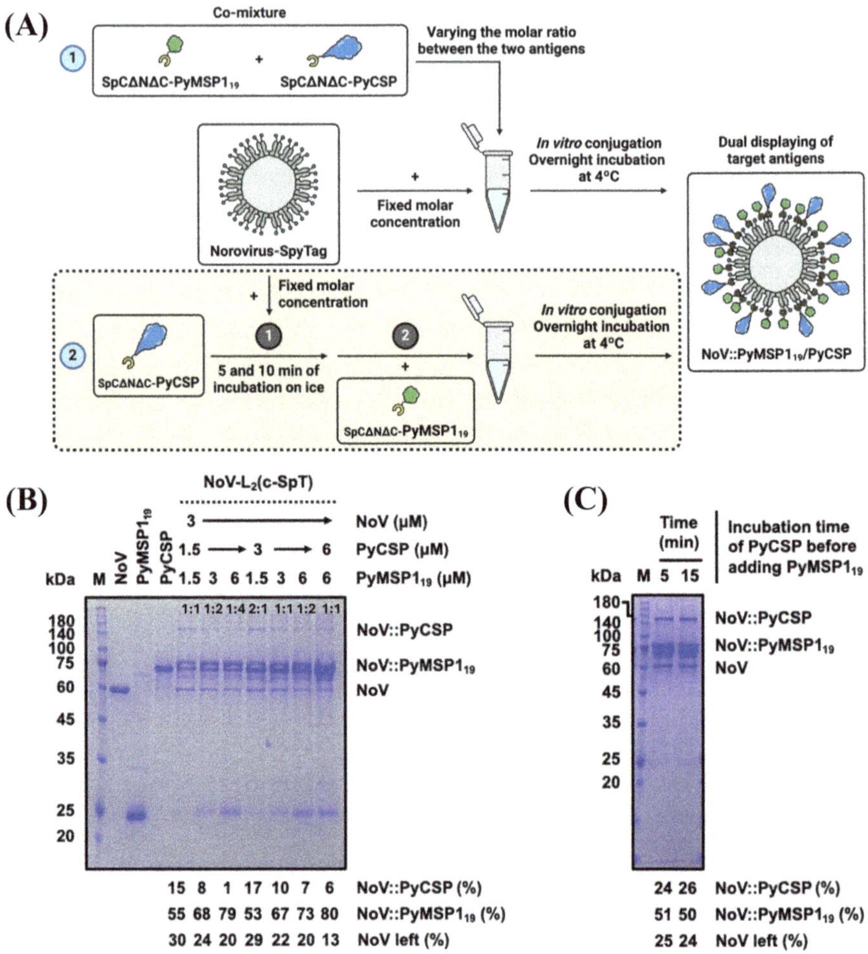

Fig. 7.22 Bivalent NoV-LPs displaying PyMSP1$_{19}$ and PyCSP (referred to as NoV::PyMSP1$_{19}$/PyCSP). (**a**) Illustration of the step-by-step process for preparing a dual-antigen display in vitro. In this approach, NoV-LPs displaying both antigens were generated by conjugating the VLPs with varying molar amounts of PyMSP1$_{19}$ and PyCSP, while keeping the VLP concentration fixed. To enhance PyCSP display on the VLP surface, PyCSP (1.5 μM) was first introduced to NoV-L$_2$(c-SpT) (3 μM), and the mixture was incubated on ice for either 5 or 15 min. After this, PyMSP1$_{19}$ (1.5 μM) was added, and the reaction was allowed to proceed overnight at 4 °C. The method for preparing dual-antigen-displaying VLPs is highlighted in the yellow box in the schematic. SDS-PAGE with CBB staining shows the display efficiency of PyMSP1$_{19}$ and PyCSP on the VLP surface when the VLP is simultaneously conjugated with both antigens (**b**), as well as the increased PyCSP display achieved with the step-by-step conjugation method, as shown in the yellow-highlighted schematic (**c**). (Adapted with permission (Boonyakida et al. 2024). Copyright 2024 American Chemical Society)

7.3.6 Tetravalent VLPs Displaying Four Dengue Serotypes' Envelope Domain IIIs

Dengue continues to pose a major public health threat worldwide, with cases and deaths rising in regions beyond its traditional tropical and subtropical regions. In 2024, the number of reported cases reached 14.1 million, resulting in 9508 deaths (Haider et al. 2024). The disease is caused by the dengue virus (DENV), which is spread mainly by female *Aedes aegypti* and *A. albopictus* mosquitoes. The World Health Organization (WHO) identifies dengue as the most widespread vector-borne viral infection. By 2025, the World Health Organization (WHO) has approved two live-attenuated dengue vaccines for commercial distribution: Dengvaxia produced by Sanofi Pasteur and Qdenga developed by Takeda (Lee et al. 2025). In addition to these live-attenuated options, researchers are exploring other dengue vaccine types, including inactivated, DNA-based, subunit, peptide-based, mRNA, viral vector, and nanoparticle-based vaccines (Lee et al. 2025; Park et al. 2022). Subunit vaccines are unique in that they do not include any live virus components; instead, they are composed solely of specific antigens to trigger an immune response. This characteristic makes them generally safer for individuals with weakened immune systems. Nevertheless, one major limitation of subunit vaccines is their low immunogenicity, which often necessitates the use of adjuvants or specialized delivery systems to enhance their effectiveness (Fernandez et al. 2015).

The DENV genome encodes both structural and nonstructural proteins, with the capsid (C), pre-membrane (prM), and envelope (E) proteins forming the core structure of the virus. DENV exists in four closely related but antigenically unique serotypes: DENV-1, DENV-2, DENV-3, and DENV-4. Severe dengue fever is often associated with a secondary infection by a different serotype, primarily due to a process called antibody-dependent enhancement (ADE) (Gromowski et al. 2008; Modis et al. 2004; Phan et al. 2022). Vaccine development efforts have primarily targeted these structural components, exploring various combinations such as CprME and prME (Utomo et al. 2019; Utomo et al. 2022, Utomo et al. 2020). Additionally, modifications to the E protein, its specific domains, or coexpression with prM have been employed to create recombinant tetravalent vaccine candidates (Liu et al. 2014; Thoresen et al. 2024; Urakami et al. 2017). Among the E protein's three domains, envelope domain III (EDIII) is frequently selected for subunit vaccine strategies because it facilitates receptor binding and is recognized by antibodies that neutralize specific DENV serotypes. This protein consists of four distinct regions: envelope domain I (EDI), envelope domain II (EDII), envelope domain III (EDIII), and a transmembrane anchor. Notably, EDIII contains specific sites targeted by highly effective neutralizing antibodies. Furthermore, antibodies that bind to DENV's EDIII region are less likely to cause antibody-dependent enhancement ADE, making EDIII an attractive target for dengue vaccine design (Nilchan et al. 2024; Seesen et al. 2022; Shukla et al. 2017). Given the simultaneous circulation of all four DENV serotypes, an effective dengue vaccine should be tetravalent and centered on EDIII to reduce the likelihood of severe disease caused by ADE

(Halstead et al. 1970), such as dengue hemorrhagic fever and dengue shock syndrome.

Protein nanoparticles that self-assemble—such as ferritins, nanocages, and VLPs—can display multiple copies of antigens or epitopes because of their repeating structural units (Lopez-Sagaseta et al. 2016; Zabel et al. 2013). This repetitive arrangement on their surface promotes effective cross-linking of B cells to antigens, which, in turn, enhances both B-cell and T-cell immune responses (Jegerlehner et al. 2002; Win et al. 2011). Of these nanoparticles, VLPs stand out as a valuable platform for displaying antigens and are known for their self-adjuvanting properties (Liu et al. 2022; Sharma et al. 2020; Vuitika et al. 2024). Therefore, integrating VLPs with dengue antigenic proteins could significantly advance the development of effective dengue vaccines.

7.3.6.1 Scaffolds for Displaying Four Types of EDIIIs of Four Dengue Serotypes

Two types of VLPs composed of Canine parvovirus (CPV) structural protein VP2 and DENV capsid protein C2 were used as a scaffold for tetravalent EDIII of all four DENV serotypes. DENV capsid is vital, forming a homodimer with each unit containing four α-helices (α1–α4) (Jones et al. 2003). Assembling the DENV capsid has been demonstrated both in vitro—using oligonucleotides or particular buffers—and in vivo (Muthuraman et al. 2024; Neves-Martins et al. 2023). DENV capsid proteins fused with the EDIII domain have also been investigated as tetravalent vaccine candidates (Valdes et al. 2017). Within the Parvoviridae family, CPV has been used as a platform for displaying foreign epitopes (Feng et al. 2011). This virus is nonenveloped, icosahedral, and measures roughly 25 nm in diameter. CPV contains three structural proteins—VP1, VP2, and VP3—as well as two nonstructural proteins, NS1 and NS2. The VP2 protein is the predominant component of the viral capsid and can form CPV-like particles (CPV-LPs) that closely resemble the wild-type parvovirus in both size and structure (Feng et al. 2014; Tsao et al. 1991). These CPV-LPs serve as a scaffold for the display of foreign proteins (Xu et al. 2023). To enhance the effectiveness of epitope display and stimulate a stronger immune response, epitopes are engineered into specific loop regions of the CPV-LPs (Rueda et al. 1999; Xu et al. 2023). The study used two structural proteins: CPV-VP2 and DENV-C2, to express self-assembling CPV-like particles (CPV-LPs) and DENV capsid-like particles (DENV C-LPs) using silkworms. These VLPs were then engineered to display the EDIIIs from all four dengue serotypes (1EDIII–4EDIII, tetEDIII) in vitro using SpT/SpC linkage. This research marks the first creation of heterologous tetravalent CPV-LPs (tetCPV-LPs) and homologous tetravalent DENV C-LPs (tetDENV C-LPs), all displaying tetEDIII. The investigation compared the morphology, immune response, and antigen display efficiency among the three types of VLPs.

SpC-fused 1EDIII–4EDIII (SpC-EDIIIs) constructs were expressed to serve as bioconjugation counterparts for linking SpT-CPV-LP and SpT-DENV C-LP, enabling tetEDIII display. Both the SpT-CPV-LP (Fig. 7.23a) and SpT-DENV C-LP (Fig. 7.23b) designs, along with four SpC-EDIIIs constructs (Fig. 7.23c), were produced in the fat bodies of silkworm larvae. Protein expression was verified via Western blot analysis, employing anti-Strep-tag II antibodies for detecting SpT-CPV-VP2 and SpC-EDIIIs (Fig. 7.23d), and anti-capsid antibodies for detecting SpT-DENV-C2 (Fig. 7.23e). The observed molecular weights for SpT-CPV-VP2 (~70 kDa), SpC-EDIIIs (~17 kDa), and SpT-DENV-C2 (~15 kDa) matched their calculated theoretical values. SpT-CPV-VP2 and SpC-EDIIIs were purified using the Strep-Tactin XT 4Flow high-capacity affinity chromatography column, while DENV-C2 was purified using the FLAG affinity chromatography column. The purified SpC-EDIIIs (~17 kDa) and SpT-CPV-VP2 (~70 kDa) (Fig. 7.23f), as well as SpT-DENV-C2 (~15 kDa) (Fig. 7.23g), were confirmed by SDS-PAGE, followed by Coomassie brilliant blue staining (CBB).

For the bioconjugation reaction, overall conjugation efficiency (Y_{VLP}), defined as the proportion of VLPs successfully displaying the antigen, is defined as follows:

$$Y_{\mathrm{VLP}} = \frac{\Delta \mathrm{VLP}}{\mathrm{VLP}_0} = \left(1 - \frac{\mathrm{VLP}_f}{\mathrm{VLP}_0}\right) \times 100$$

$$\Delta \mathrm{VLP} = \mathrm{VLP}_0 - \mathrm{VLP}_f$$

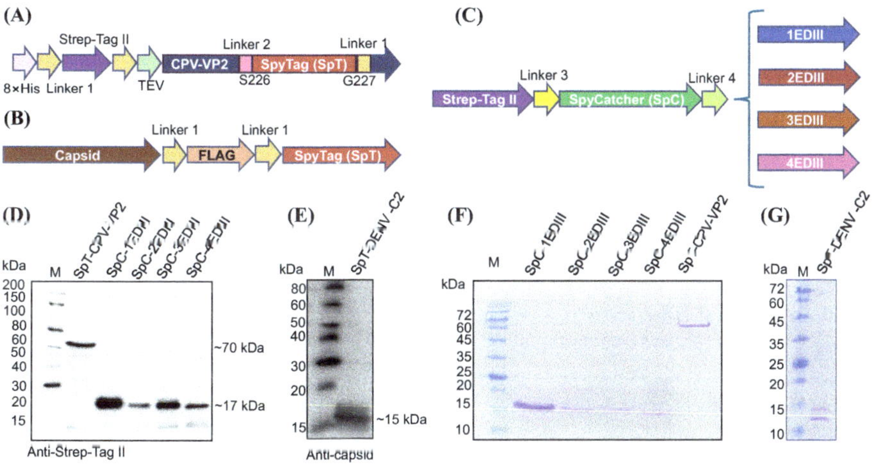

Fig. 7.23 Construction, expression, and purification of SpT-CPV-VP2, SpT-DENV-C2, and SpC-EDIIIs. For the CPV construct, SpT was inserted between amino acids S226 and G227 within loop 2 at the C-terminus of the Strep-tag II fused CPV-VP2 (**a**); for DENV-C2, SpT was added to the C-terminus of the FLAG tag-fused protein (**b**); and EDIIIs were added to the Strep-tag II-SpC fused protein (**c**). Recombinant protein expression was confirmed by Western blot using anti-Strep-tag II antibody for SpT-CPV-VP2 and SpC-EDIIIs detection (**d**), and an anti-capsid antibody for identifying DENV-C2 (**e**). SDS-PAGE with CBB staining showed purified SpC-EDIIIs and SpT CPV-VP2 (**f**), as well as concentrated SpT-DENV-C2 (**g**). (Adapted with permission (Muthuraman et al. 2025b). Copyright 2025 American Chemical Society)

where VLP_0 and VLP_f denote the initial and final amount of VLPs after the reaction, respectively. ΔVLP indicates the utilized amount of VLPs for the bioconjugation, determined using densitometry analysis of molecular band intensities. For SpT-CPV-VP2 conjugated with SpC-2EDIII, the display efficiency increased with higher molar ratios, reaching 94.7% at a 1:5 ratio, as confirmed by Western blot results. Similarly, with SpT-DENV-C2 revealed a display efficiency of 71.4% at a 1:1 ratio and nearly complete efficiency at ratios of 1:2 and above. The optimal display for SpT-DENV-C2 and SpC-2EDIII was achieved at a 1:3 ratio, as determined by band intensity analysis (Muthuraman et al. 2025b). These findings suggest that CPV-VP2 is approximately twice as effective as DENV-C2 in displaying antigens. When testing serotypes 1, 3, and 4EDIII, the highest levels of displayed EDIII for CPV-VP2 were observed at a 1:5 ratio, while DENV-C2 showed optimal ratios between 1:3 and 1:5 for different serotypes. In summary, the most effective molar ratios for antigen display were 1:5 for SpT-CPV-VP2 to SpC-EDIIIs and 1:3 for SpT-DENV-C2 to SpC-EDIIIs (Muthuraman et al. 2025b).

To display all four serotype tetEDIII proteins on CPV-LP and DENV-C2, the molar ratios used were 1:5 for SpT-CPV-VP2 to SpC-tetEDIII, and 1:3 for SpT-DENV-C2 to SpC-tetEDIII. The conjugation yielded similar results for tetCPV-VP2 and tetDENV-C2, with successful conjugation of SpC-tetEDIII, as confirmed by Western blot analysis employing serotype-specific EDIII antibodies. On the tetCPV-VP2 surface, each displayed EDIII had an approximate molecular mass of 17 kDa (Fig. 7.24a(i–iv)). The overall molecular weight of tetCPV-VP2 was about 90 kDa,

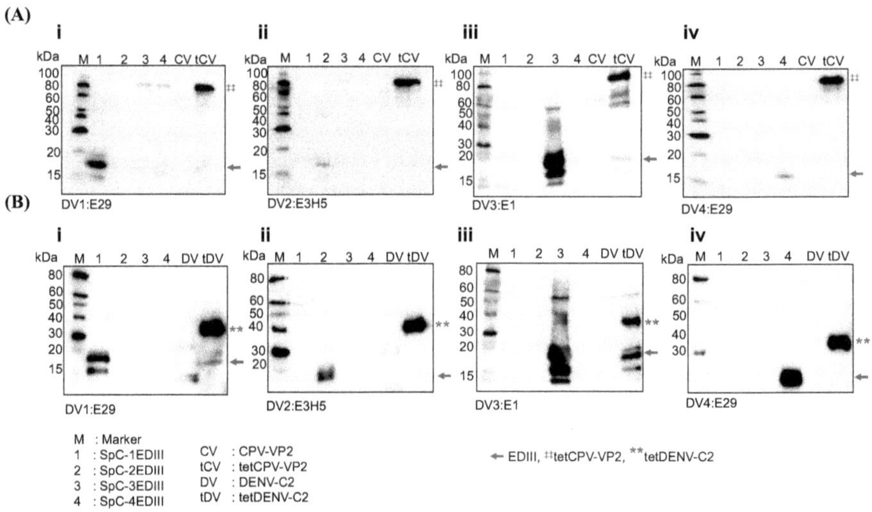

Fig. 7.24 Western blot analysis of tetEDIII-conjugated CPV-VP2 (**a**) and DENV-C2 (**b**). The EDIII proteins displayed on the VLPs were identified using serotype-specific antibodies: E29 clone for serotype-1 (i), E3H5 clone for serotype-2 (ii), E1 clone for serotype-3 (iii), and E29 clone for serotype-4 (iv). (Adapted with permission (Muthuraman et al. 2025b). Copyright 2025 American Chemical Society)

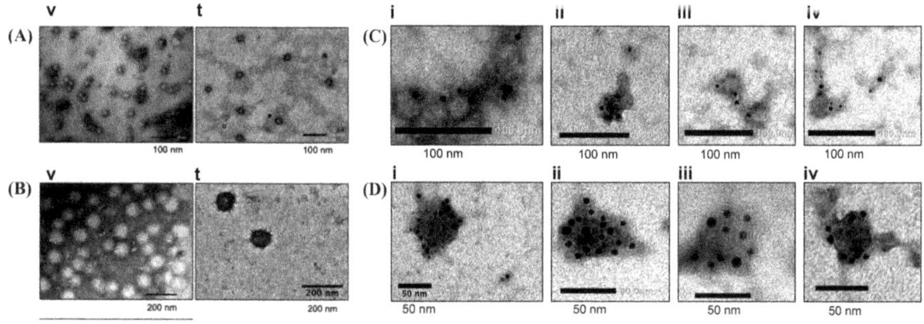

Fig. 7.25 TEM and Immunoelectron microscopy (IEM) images of VLPs and tetEDIII-displaying VLPs. Negative stain TEM images of CPV-LPs (**a**-v) and tetCPV-LPs (**a**-t), DENV C-LPs (**b**-v) and tetDENV C-LPs (**b**-t). IEM images of tetCPV-LPs (**c**) and tetDENV C-LPs (**d**). Detection of each EDIII using serotype-specific primary antibodies: E29 clone for serotype-1 (i), E3H5 clone for serotype-2 (ii), E1 clone for serotype-3 (iii), and E29 clone for serotype-4 (iv), followed by labeling with goat anti-mouse IgG(H + L)-conjugated gold nanoparticles (5 nm). (Adapted with permission (Muthuraman et al. 2025b). Copyright 2025 American Chemical Society)

attributed to the addition of bioconjugated-EDIII. Likewise, the EDIII displayed on tetDENV-C2 was approximately 17 kDa, and the conjugated tetDENV-C2 band shifted to roughly 35 kDa (Fig. 7.24b(i–iv)). Although the corresponding EDIII was detected by serotype-specific antibodies, multiple bands appeared for 1EDIII and 3EDIII antibodies, indicating some protein degradation but minimal cross-reactivity among serotypes.

CPV-VP2 and DENV-C2 were assembled into CPV-like particles (CPV-LPs) (Fig. 7.25a(v)) and DENV-C2-like particles (DENV C-LPs) (Fig. 7.25b(v)). The CPV-LPs and DENV C-LPs had average sizes of 20.4 ± 2.7 nm (n = 19) and 66.7 ± 7.4 nm (n = 19), respectively. Additionally, tetEDIII-displayed CPV-LPs (tetCPV-LPs) (Fig. 7.25a(t)) and tetEDIII-displayed DENV C-LPs (tetDENV C-LPs) (Fig. 7.25b(t)) showed similar shapes. According to dynamic light scattering (DLS) analysis, the hydrodynamic diameters of CPV-LPs and tetCPV-LPs were 21.7 nm and 25.2 nm, respectively, with polydispersity indices (PDI) of 0.289 and 0.405, respectively. For DENV C-LPs and tetDENV C-LP, the sizes were 76.8 nm and 78.9 nm, and PDIs were 0.490 and 0.796, respectively (Muthuraman et al. 2025b). In both cases, the VLPs displaying EDIIIs showed an increase in size: CPV-LPs grew by 3.5 nm and DENV C-LPs by 2.1 nm. The PDI for CPV-LP indicates a fairly consistent particle size, as visualized in the TEM image distributions. The difference in PDI after tetEDIII attachment was 0.116 for CPV-LPs and 0.306 for DENV C-LPs. In contrast, the PDI of DENV C-LPs shows a broader particle size distribution, likely due to size heterogeneity, as indicated by the standard deviation (± 7.4 nm) of the measurements. This could also result from aggregation of the DENV C LPs, as previously observed (Muthuraman et al. 2024). Displaying the EDIII antigen increased the polydispersity of both VLPs in a proportional manner. CPV-LPs and tetCPV-LPs showed zeta potential values of 35.8 ± 2.25 mV and 46.35 ± 2.62 mV, respectively. For DENV C-LPs and tetDENV C-LPs, the zeta potentials were 42.83 ± 4.92 mV and 44.8 ± 3.82 mV, respectively. After bioconjugation, both VLPs displayed higher zeta potentials, indicating successful antigen display. The increase in zeta potential was 10.6 mV for CPV-LPs and 2.0 mV for DENV C-LPs, suggesting that CPV-LPs displayed more EDIIIs on their surfaces compared to DENV C-LPs.

The presentation of tetEDIII on the surfaces of CPV-LPs (Fig. 7.25c(i–iv)) and DENV C-LPs (Fig. 7.25d(i–iv)) was observed through Immunoelectron microscopy (IEM). The appearance of dark spots, representing 5 nm gold nanoparticles, marks the location of EDIIIs on both the tetCPV-LP (Fig. 7.25a(t)) and tetDENV C-LP (Fig. 7.25b(t)). The use of serotype-specific primary antibodies verified the presence of all four serotype EDIIIs on the surface of the tetVLPs (Fig. 7.25c(i–iv)) and DENV C-LPs (Fig. 7.25d(i–iv)).

BALB/c mice were immunized at 2-week intervals with prime and booster doses (Fig. 7.26a) to assess the immunogenicity of tetCPV-LPs and tetDENV C-LPs. EDIII-specific antibody titers in collected sera were quantified by ELISA. The ED50 values for anti-1EDIII, 3EDIII, and 4EDIII IgGs were significantly higher in sera from tetDENV C-LP–immunized mice compared to those from DENV C-LP (Fig. 7.26b(i, iii, iv)), whereas tetCPV-LP showed a nonsignificant increase relative to CPV-LP. For anti-1EDIII and 3EDIII IgGs, tetDENV C-LP elicited higher ED50 values than tetCPV-LP. No significant differences were observed among groups for anti-2EDIII IgG (Fig. 7.26b(ii)).

Neutralizing activity was assessed using the single-round infectious particle (SRIP) assay, a quick and efficient method to assess how well serum antibodies block dengue virus infection. Instead of using live viruses, it uses special SRIPs that can infect cells only once and carry a luciferase reporter gene (Izac et al. 2024; Matsuda et al. 2018; Yamanaka et al. 2021; Yamanaka et al. 2014). When serum containing these particles is added to cells, the amount of light produced indicates the extent of infection. If the serum neutralizes the virus well, the light signal is much lower. The IC75 value indicates the serum dilution that reduces infection by 75%, indicating strong neutralization. Both tetCPV-LP and tetDENV C-LP induced higher IC75 titers compared to CPV-LP and DENV C-LP, respectively (Fig. 7.26c(i–iv)). However, statistical significance in the DENV C-LP group was influenced by one outlier. For DENV serotype 1 SRIP (D1-SRIP), tetDENV C-LP exhibited a higher geometric mean IC75 titer than tetCPV-LP (Fig. 7.26c(i)), while similar titers were observed for serotype 3 SRIP (D3-SRIP) (Fig. 7.26c(iii)). Conversely, tetCPV-LP produced significantly greater IC75 values than tetDENV C-LP for serotypes 2 and 4 SRIP (D2-SRIP and D4-SRIP) (Fig. 7.26c(ii, iv)).

The optimal display of EDIIIs on the surface of VLPs is achieved at a molar ratio of 1:5 for SpT-CPV-LP and 1:3 for SpT-DENV C-LP, but SpT-CPV-LP displays nearly twice as much EDIII as SpT-DENV C-LP. The reduced binding efficiency in SpT-DENV C-LP may result from the internal orientation of certain loop regions that interact with viral RNA, which could limit the accessibility of SpT for conjugation with SpC-EDIII. Enhanced positioning of EDIII and improved exposure of antibody targets were observed in tetCPV-LP compared to tetDENV C-LP, likely due to engineered modifications in loop 2 of CPV. Structural models

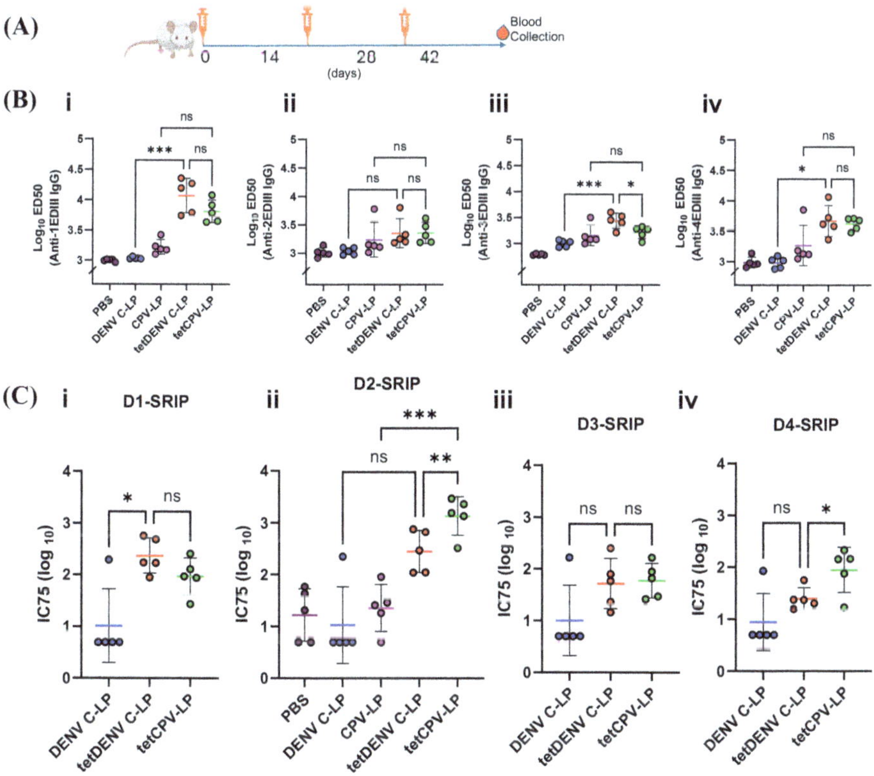

Fig. 7.26 Immunization protocol (**a**), antibody response (**b**), and SRIP-based neutralization assay (**c**). (**b**) Antibody generation in sera from immunized mice was assessed by ELISA against 1EDIII (i), 2EDIII (ii), 3EDIII (iii), and 4EDIII (iv). Specific IgG levels are represented as Log10 ED50 binding titers for each mouse. (**c**) Neutralizing activity of immunized sera was measured and represented as Log10 IC75 values for D1-SRIP (i), D2-SRIP (ii), D3-SRIP (iii), and D4-SRIP (iv). Data represent geometric mean ± geometric SD. One-way ANOVA with Tukey's multiple comparison test was used for statistical evaluation. A p-value <0.05 was considered statistically significant. Significance is indicated as $*p < 0.05$, $**p < 0.01$, and $***p < 0.001$. (Adapted with permission (Muthuraman et al. 2025b). Copyright 2025 American Chemical Society)

generated by Alphafold2 show that SpT is incorporated into the loop 2 region of SpT-CPV-VP2 (Fig. 7.27a), with linkers providing separation from other domains. However, further study is necessary to determine the required number of SpT-DENV-C2 (Fig. 7.27b) copies for capsid assembly. Alphafold2 also modeled SpC-EDIII fusion proteins, separated by linkers and tagged at the N-terminus (Fig. 7.27c–f). Visualization of SpT in the loop 2 region was performed using a CPV template (PDB ID 7M3M) (Goetschius et al. 2021) (Fig. 7.27g). Still, similar modeling for DENV C LP was not possible due to the absence of a suitable template. These structural differences may influence antibody binding and neutralization potency.

The tetCPV-LPs differ from the dengue virus in both the capsid and the displayed antigen, making them heterologous, whereas tetDENV C-LPs match the

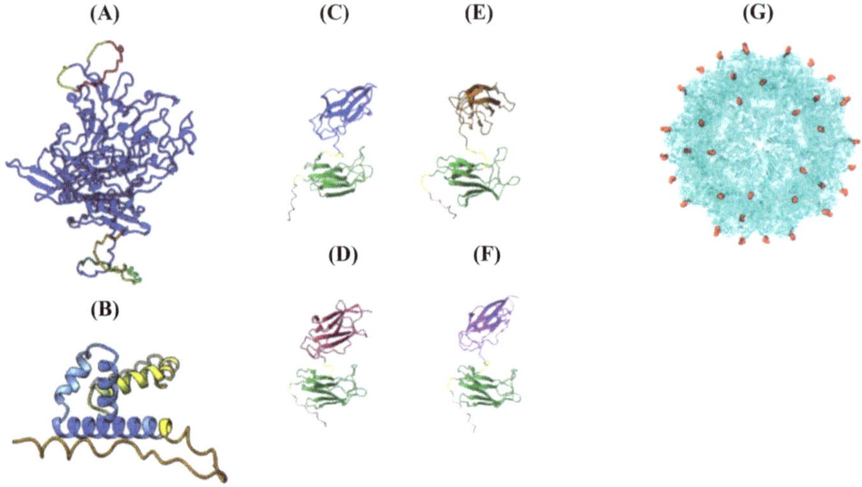

Fig. 7.27 Structural modeling of SpT-CPV-VP2, SpT-DENV C2, and EDIII constructs was performed using Alphafold. For SpT-CPV-VP2 (**a**), the SpT (red), flanked by yellow-colored linkers, was introduced into the loop2 region, preserving the overall structure. In SpT-DENV C2 (**b**), both the SpT and FLAG tags are separated by linkers and extend outward from the α-4 loop. The constructs SpC-1EDIII (**c**), SpC-2EDIII (**d**), SpC-3EDIII (**e**), and SpC-4EDIII (**f**) feature SpC (green) connected to EDIII and a Strep-Tag via yellow linkers. The assembled CPV template [PDB id: 7M3M] (**g**) displays SpT binding sites (marked in red), where the SpC-EDIIIs conjugate, allowing EDIII to be prominently displayed on the CPV-LP surface. (Adapted with permission (Muthuraman et al. 2025b). Copyright 2025 American Chemical Society)

native virus and are homologous. Homologous particles are expected to elicit a more targeted immune response, with improved recognition and memory. Both types of VLPs successfully displayed EDIII on their surfaces and elicited neutralizing antibodies against all four serotypes following intraperitoneal immunization. Results indicated that tetCPV-LP provided stronger neutralization against some serotypes, likely due to its superior display efficiency. Nevertheless, immune response and neutralization activity were similarly effective across several serotypes in both VLP types, making it difficult to determine which is better. Future research should explore enhancements in antigen display for tetDENV C-LP, as its homologous nature could offer advantages for immune memory. Both VLPs remain promising as vaccine candidates, but further investigation is needed to clarify the benefits of homologous versus heterologous approaches for dengue vaccines.

7.3.6.2 Tetravalent VLPs Displaying Distinct EDIII Structures from Four Dengue Virus Serotypes

When multiple antigens are displayed on the surface of VLPs, understanding how they are displayed will be a crucial factor in future vaccine design. For tetravalent VLPs, it is unclear how the arrangement of the four antigens on the VLP surface

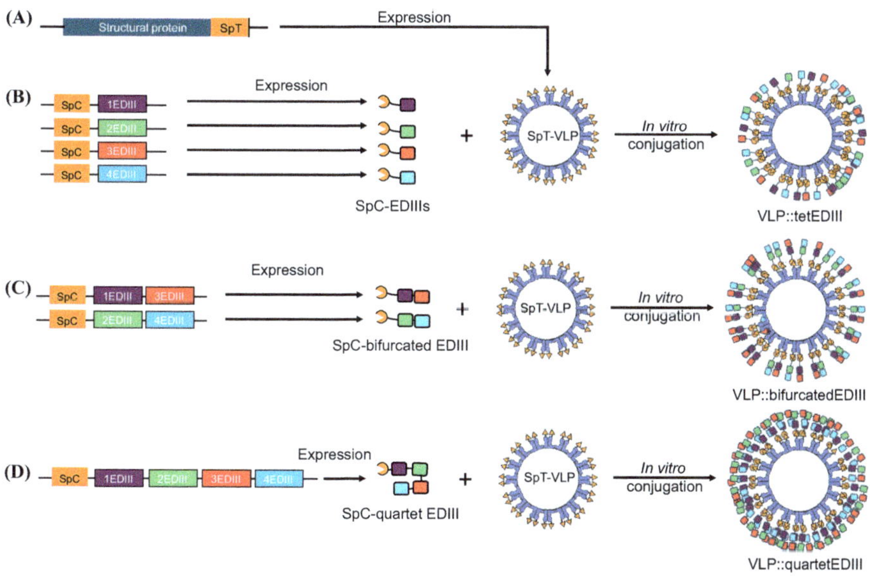

Fig. 7.28 Schematic diagram of various configurations of EDIII-displaying tetravalent. (**a**) SpT-fused structural protein is expressed, forming VLPs (SpT-VLP). (**b**) Each ScP-fused EDIII is expressed, forming four kinds of SpC-fused EDIIIs (SpC-EDIIIs). Upon incubation of SpC-EDIIIs and SpT-VLPs, isopeptide bonds are formed between SpT and SpC, yielding VLP::tetEDIII particles. (**c**) ScP-fused 1EDIII-3EDIII and ScP-fused 2EDIII-4EDIII are expressed, forming two kinds of distinct SpC-fused EDIII pairs (ScP-bifurcated EDIII). ScP-bifurcated EDIII and SpT-VLPs are conjugated, yielding VLP::bifurcated EDIII particles. (**d**) ScP-fused EDIIIs of DENV serotypes 1–4 as a polypeptide chain is expressed, forming a SpC-fused single polypeptide chain (SpC-quartetEDIII). SpC-quartetEDIII and SpT-VLPs bind via ScP/SpT chemistry, yielding VLP::SpC-quartetEDIII particles

affects the immune response. Dengue virus has four specific EDIII domains, each from a different serotype. When displaying EDIIIs on the surface of VLPs, consider the following three cases: First, each EDIII is displayed individually on the surface of VLPs (Fig. 7.28a), forming tetravalent-EDIII-displaying VLPs (Fig. 7.28b); second, two EDIII pairs are displayed in two distinct configurations, forming a bifurcated-EDIII-displaying VLPs (Fig. 7.28c); third, four types of EDIIIs linked into a single peptide chain are displayed, forming quartet-EDIII-displaying VLPs (Fig. 7.28d).

When these three structurally distinct EDIII are displayed on the VLP surface, the effects of VLP morphology, immune responses, and neutralization are of interest. In this section, tetravalent VLPs displaying tetravalent-EDIII or quartet-EDIII were compared. Additionally, tetravalent VLPs displaying bifurcated-EDIII or quartet-EDIII were designed and compared with respect to immune response.

7.3.6.3 Tetravalent VLPs Displaying Singlet tetEDIII and Quartet tetEDIII

This research explored the use of NoV-LPs as a platform to display EDIIIs from all four DENV serotypes, aiming to create a tetravalent VLP. The first involved binding each serotype's EDIII protein individually to the VLP (singlet tetEDIII) (Fig. 7.28b). The second approach fused all four EDIII proteins into a single polypeptide chain, which was then displayed on the VLP (quartet tetEDIII) (Fig. 7.28d). Both types of EDIII proteins were produced using silkworms and conjugated to NoV-LPs via the SpT/SpC bioconjugation system. The resulting tetravalent VLPs (NoV::tetEDIII for the singlet design and NoV::quartetEDIII for the quartet design) were subsequently tested for their ability to induce immune responses and generate neutralizing antibodies, with comparisons made to traditional subunit vaccines.

After conjugation reaction at 4 °C, displayed EDIIIs were confirmed by SDS-PAGE/CBB and verified by Western blot analysis that EDIIIs were bound to NoV-LPs, referred to as NoV::EDIII (Fig. 7.29a). Densitometric measurements from three separate conjugation experiments (the ratio of VLP and EDIII of 1:0.25, 1:0.5, and 1:1) revealed that as the total antigen concentration increased, the proportion of the NoV::tetEDIII adduct band (estimated molecular weight of ~75 kDa) also rose, indicating higher conjugation efficiency. Western blotting with antibodies specific to each EDIII serotype was examined. The NoV::tetEDIII produced using the VLP and EDIIIs ratio of 1:0.25q ratio (Fig. 7.29b) did not bind to the 2EDIII-specific antibody, but did show reactivity with antibodies for 1EDIII, 3EDIII, and 4EDIII, suggesting limited display of 2EDIII. When the ratio was increased to 1:1q, a thick band of all four anti-EDIII antibodies was observed, confirming the successful conjugation of all four serotype EDIIIs to the SpT-NoV-LPs (Fig. 7.29b). Analysis of NoV::tetEDIII and NoV bands showed that the highest conjugation efficiency, about 60%, was achieved with the 1:4 molar ratio reaction (NoV::tetEDIII$_{(1:1q)}$ band). Lower ratios, such as 1:10.25q and 1:0.5q, resulted in efficiencies of approximately 35% and 49%, respectively (Fig. 7.29c). To further validate the formation of the tetravalent NoV-LP (NoV::tetEDIII), a sandwich ELISA was performed for the NoV::tetEDIII$_{(1:1q)}$ sample, which confirmed proper antigen orientation (Fig. 7.29d–h). This strategy of fusing SpC or SpT peptides to EDIIIs and displaying them on nonenveloped VLPs or nanocarriers has previously been applied to EDIIIs from Zika virus, Powassan virus, and West Nile virus, all members of the *Flavivirus* genus. In one study, AP205 (a $T = 3$ icosahedral capsid scaffold) was used to present EDIII from West Nile virus on its surface, achieving a conjugation efficiency of approximately 50–51%, which equates to about 90 EDIII$_{WNV}$ molecules bound to AP205 (Stander et al. 2021). In addition to VLPs, lumazine synthase nanoparticles (LS-NP), which self-assemble, have also been used for antigen display (Georgiev et al. 2022). Structural studies show that LS forms icosahedral capsids with a triangulation number of 1 ($T = 1$), consisting of 60 identical protein subunits (Ladenstein and Morgunova 2020). Genetic engineering of LS-NP to display EDIIIs from Zika virus and Powassan virus demonstrated that this platform could reach full (100%)

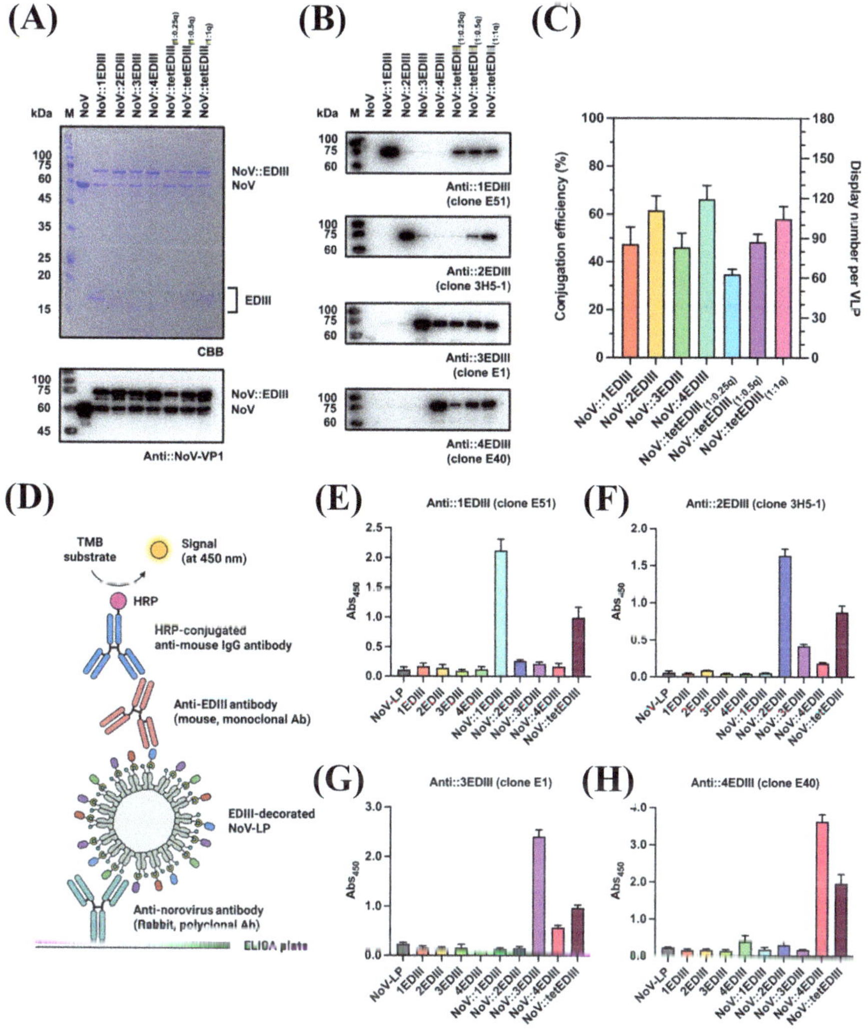

Fig. 7.29 Biochemical analyses of the tetEDIII-dispayed NoV-LP (NoV::tetEDIII). (**a**) SDS-PAGE/CBB staining of a norovirus VP1-specific antibody to detect EDIII-displayed NoV-LPs; (**b**) Western blotting of NoV::tetEDIII using serotype-specific anti-EDIII antibodies; (**c**) conjugation efficiency via densitometric analysis; (**d**) illustration of a sandwich ELISA setup for NoV::tetEDIII characterization; confirmation of the presence of EDIIIs from all four serotypes on the antigen-conjugated VLPs by ELISA utilizing clone E51 antibody (serotype 1-specific) (**e**), clone 3H5-1 antibody (serotype 2-specific) (**f**), clone E1 antibody (serotype 3-specific) (**g**), and clone E40 antibody (serotype 4-specific) (**h**). (Adapted with permission (Boonyakida et al. 2025). Copyright 2025 American Chemical Society)

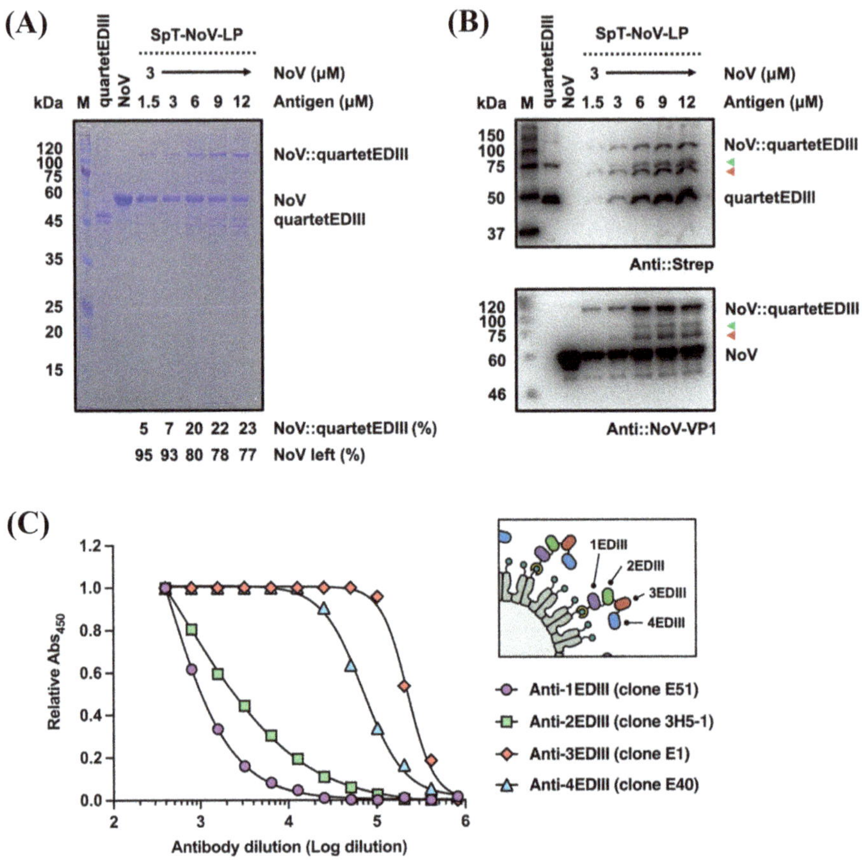

Fig. 7.30 Biochemical analyses of the quartetEDIII-displayed NoV-LP (NoV::quartetEDIII). SDS-PAGE/CBB staining (**a**) and Western blotting (**b**) analyses at various molar ratios of the quartetEDIII to NoV-LP; and (**c**) An indirect ELISA of NoV::quartetEDIII using serotype-specific anti-EDIII antibodies. (Adapted with permission (Boonyakida et al. 2025). Copyright 2025 American Chemical Society)

conjugation efficiency (Georgiev et al. 2022; Malonis et al. 2022), resulting in 60 EDIII molecules per LS-NP.

Subsequently, the SpC-quartetEDIII was conjugated with SpT-NoV-LP and incubated overnight. The conjugation was verified using SDS-PAGE/CBB staining and Western blot analysis (Fig. 7.30a, b). The appearance of an adduct band, labeled NoV::quartetEDIII, at approximately 110 kDa, confirmed successful conjugation. When using a 1:2 molar ratio (3 µM SpT-NoV-LP to 6 µM quartetEDIII), the efficiency of conjugation was 20%. Increasing the ratio to 1:4 resulted in a 23% conjugation rate (Fig. 7.30a). Western blotting with antibodies specific to EDIII serotypes further detected the presence of quartetEDIII on NoV::quartetEDIII (Boonyakida et al. 2025). Aside from the primary band at ~110 kDa, two additional protein bands were observed between quartetEDIII and NoV::quartetEDIII, indicated by red and green arrows (Fig. 7.30b). These bands are likely due to the breakdown of quartetEDIII during incubation, possibly caused by a silkworm-derived protease that was co-purified. From this conjugation, 20% conjugation efficiency was achieved, corresponding to 36 quartetEDIII molecules bound per VLP. Each quartetEDIII unit included one complete EDIII from each of the four serotypes (1–4EDIIIs), meaning that all four serotypes were uniformly presented. With 36 quartetEDIII units per VLP, a total of 144 EDIII proteins were displayed on the NoV-LP (NoV::quartetEDIII), arranged in a branched formation. In contrast, the NoV::tetEDIII construct only presented 108 EDIII proteins, and their distribution across the VLP surface may have been uneven. The quartetEDIII design, therefore, enhances the nano-assembly process, allowing for a greater and more balanced display of EDIII proteins from all four serotypes on each VLP.

ELISA binding assays using monoclonal antibodies against each DENV serotype's EDIII protein confirmed the antigenicity of NoV::quartetEDIII. All antibodies tested were able to bind NoV::quartetEDIII, demonstrating that the epitopes presented on the VLP were properly folded and accessible (Fig. 7.30c). The antibody reactivity among SpC-tagged EDIII, SpC-tagged quartetEDIII, and NoV::quartetEDIII was compared, revealing consistent binding profiles for EDIII-specific antibodies across all protein variants, indicating that the antigenic features were maintained and that the EDIII units within the quartetEDIII construct exhibited correct folding, comparable to that of individual EDIIIs (Boonyakida et al. 2025).

Transmission electron microscopy (TEM) showed a slight variation in the surface appearance between unmodified NoV-LPs (Fig. 7.31a) and those NoV-LPs conjugated with antigens (Fig. 7.31b1–b4, c1–c2), exhibiting additional densities along their hexahedral structures compared to the naked NoV-LPs. Although a branched structure was expected for NoV::quartetEDIII, this feature was not distinctly visible in the negative-staining TEM images (Fig. 7.31c2). Nevertheless, the external morphology of the NoV::quartetEDIII particles resembled that of the other EDIII-conjugated VLPs. The TEM observations indicated that both the modified and unmodified VLPs displayed a uniform size distribution, suggesting that antigen conjugation did not compromise the core capsid structure or VLP stability. The hydrodynamic diameters of the VLPs showed that all EDIII-conjugated VLPs had larger average diameters compared to the unmodified VLPs, indicating the addition of external structural elements (Fig. 7.31d). The polydispersity index (PDI) was also evaluated to assess particle distribution following conjugation. Most EDIII-conjugated VLPs had PDI values between 0.229 and 0.270, while the naked NoV LP had a PDI of 0.235, demonstrating that the particles remained predominantly monodisperse in PBS buffer (pH 7.6). However, NoV::3EDIII and NoV::4EDIII had higher PDI values of 0.429 and 0.441, respectively, suggesting greater polydispersity in these samples. Despite this, TEM images confirmed that the icosahedral cores of NoV::3EDIII and NoV::4EDIII remained intact and similar to those of the naked NoV-LPs (Fig. 7.31(a, b3, b4)). These findings imply that while the addition of 3EDIII and 4EDIII did not disrupt the structural integrity of the NoV-LPs, it did increase the particles' tendency to aggregate or self-associate. Overall, both TEM

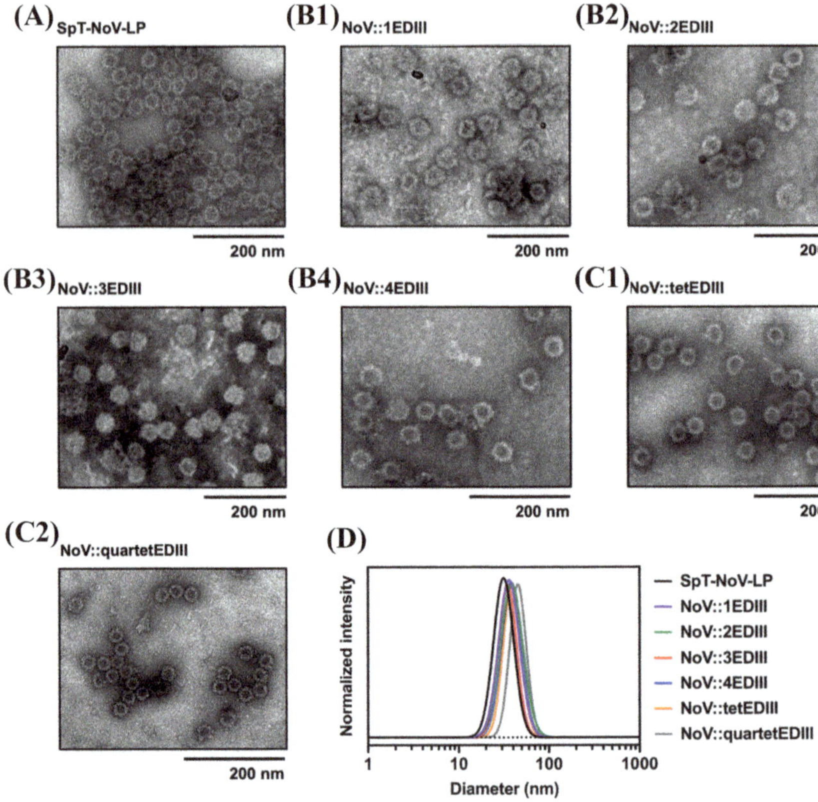

Fig. 7.31 Biophysical characterization of NoV-LPs. Shown are negative-staining TEM images of unmodified SpT-NoV-LPs (**a**), NoV::1EDIII (**b1**), NoV::2EDIII (**b2**), NoV::3EDIII (**b3**), NoV::4EDIII (**b4**), NoV::tetEDIII (**c1**), NoV::quartetEDIII (**c2**), along with results from dynamic light scattering (DLS) analysis (**d**). (Adapted with permission (Boonyakida et al. 2025). Copyright 2025 American Chemical Society)

and DLS analyses confirmed successful EDIIIs display on SpT-NoV-LP surfaces, with the conjugation approach preserving capsid structure and particle integrity.

The immunogenicity of NoV::tetEDIII and NoV::quartetEDIII was evaluated in comparison to EDIII and quartetEDIII to demonstrate the feasibility of displaying four EDIIIs on a single VLP and to assess the ability of these tetravalent VLPs to elicit strong antibody responses against DENV. Mice underwent a prime immunization followed by two booster shots spaced two weeks apart (Fig. 7.32a). Sera from mice were analyzed by indirect ELISA, using SpC-removed EDIIIs as the capture antigens. Across groups receiving subunit EDIII-based vaccines, the geometric mean titers (GMTs) of IgG antibodies were generally similar (Fig. 7.32b(i–iv)), except for the 2EDIII group, which showed a statistically significant difference compared to the quartetEDIII group (Fig. 7.32b(i)). Notably, both NoV::tetEDIII and NoV::quartetEDIII produced IgG responses that were on par with those induced

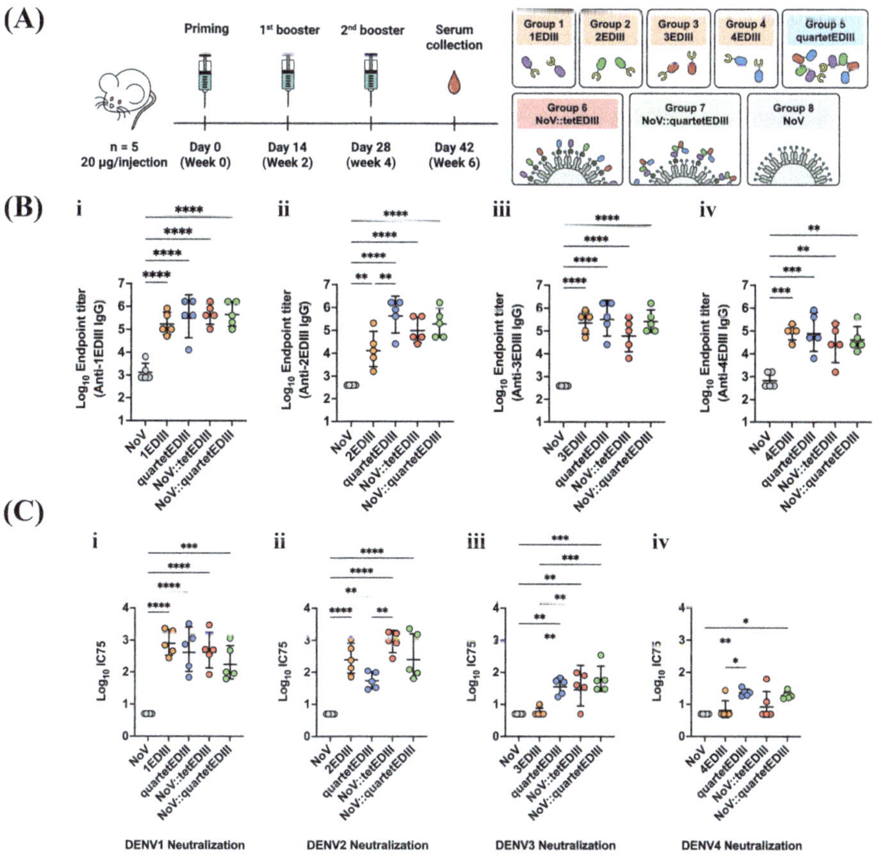

(D)

Vaccine Group	Neutralization titer (IC75)							
	DENV1		DENV2		DENV3		DENV4	
	GM-IC75 (range)	GMSD	GM-IC75 (range)	GMSD	GM IC75 (range)	GMSD	GM-IC75 (range)	GMSD
1EDIII	883.8 (279–2315)	2.49	-	-	-		-	-
2EDIII	-	-	267.4 (71–1071)	2.95	-	-	-	-
3EDIII	-	-	-	-	5.7 (5–10)	1.36	-	-
4EDIII	-	-	-	-	-	-	7.0 (5–27)	2.13
quartetEDIII	489.6 (68–3222)	4.72	55.8 (29–111)	1.85	36.5 (17–67)	1.82	23.1 (18–35)	1.30
NoV::tetEDIII	467.4 (83–2956)	3.57	913.1 (271–1799)	2.13	35.1 (5–98)	3.25	9.8 (5–62)	2.99
NoV::quartetEDIII	197.8 (61–1140)	3.51	304.3 (63–2296)	5.38	62.9 (30–350)	2.74	19.0 (15–30)	1.31
NoV	< 5	-	< 5	-	< 5	-	< 5	-

*GMT: Geometric Mean Titer of IC75

*GMSD: Geometric Mean SD Factor

Fig. 7.32 Mice immunization and SRIP neutralization assays. (**a**) Immunization schedule; (**b**) Endpoint titers of from immunization with total anti-1EDIII IgGs (i), total anti-2EDIII IgGs (ii), total anti-3EDIII IgGs (iii), and total anti-4EDIII IgGs (iv). One-way ANOVA with Tukey's multiple comparison test was used for statistical evaluation (*$p = 0.0332$; **$p = 0.0021$; ***$p = 0.0002$, ****$p < 0.0001$). (**c**) Neutralization activity of DENV serotype 1 (DENV1) (i), DENV serotype 2 (DENV2) (ii), DENV serotype 3 (DENV3) (iii), and DENV serotype 4 (DENV4) (iv). The data were represented as Log_{10} IC75 values. p-value <0.05, significant (*$p < 0.05$, **$p < 0.01$, and ***$p < 0.001$). (**d**) Neutralization titer of antisera. (Adapted with permission (Boonyakida et al. 2025). Copyright 2025 American Chemical Society)

by subunit EDIIIs and subunit quartet-EDIII. The NoV group, used as a negative control, confirmed that the NoV-LP scaffold itself induced little or no IgG response to DENV EDIII. Regarding the structure of quartetEDIII, it was expected that EDIIIs situated farther from the VLP surface would stimulate a stronger antibody response than those closer to the surface. Contrary to expectations, ELISA data revealed that 1EDIII, the domain nearest the VLP surface, produced a higher IgG response than 4EDIII (Fig. 7.32b(i, iv)), suggesting there is no direct correlation between the position of EDIII within quartetEDIII and the level of IgG produced.

The neutralizing capabilities of the generated antibodies were evaluated using the SRIP assay, and the results are reported as IC75 values. Antibodies produced in response to NoV::quartetEDIII demonstrated neutralizing effects against all DENV serotypes, while those from NoV::tetEDIII immunization were effective against serotypes 1, 2, and 3 (Fig. 7.32c). In the case of DENV serotype 4, only one mouse immunized with NoV::tetEDIII showed neutralizing activity. For the NoV::quartetEDIII group, antibodies exhibited particularly strong neutralization against DENV1 and DENV2, compared to DENV3 and DENV4 within the same group. A similar pattern was observed with NoV::tetEDIII, which appeared to induce even more robust neutralization of DENV2 Fig. 7.32c(i, ii), d). Both NoV::tetEDIII and NoV::quartetEDIII groups achieved comparable IC75 values for DENV3 neutralization, similar to results from the uncoupled quartetEDIII, whereas the 3EDIII group did not exhibit this activity (Fig. 7.32c(iii), d). The geometric mean IC75 (GM-IC75) values for NoV::quartetEDIII did not differ significantly from those of uncoupled quartetEDIII, indicating that both formulations generated similar immune responses. Furthermore, statistical comparison of GM-IC75s between NoV::quartetEDIII and NoV::tetEDIII showed no significant difference, even though NoV::tetEDIII resulted in IC75 values for DENV1 and DENV2 that were about 2-fold and 3-fold higher, respectively, than those from NoV::quartetEDIII. Notably, all mice immunized with NoV::quartetEDIII and uncoupled quartetEDIII displayed DENV4 neutralization activities with similar IC75 values, while the majority of mice in the 4EDIII and NoV::tetEDIII groups had little or no DENV4 neutralization activity (Fig. 7.32c(iv), d).

This research introduced an innovative strategy for dengue vaccine development by linking together EDIIIs from DENV serotypes 1 through 4 into a single polypeptide and binding it to a NoV-LP using the SpT/SpC bioconjugation method. All vaccine elements were successfully expressed in soluble form with proper folding using the silkworm system, as verified by immunoreactivity testing. These findings emphasize the suitability of silkworms as hosts for producing complex proteins efficiently and at scale. The newly designed construct, quartetEDIII displayed on the VLP (NoV::quartetEDIII), was assessed against NoV::tetEDIII for its ability to display antigens, provoke immune responses, and neutralize all four DENV serotypes. Transmission electron microscopy showed that binding the concatenated proteins did not compromise the integrity of the VLPs. NoV::quartetEDIII triggered the production of neutralizing antibodies effective against each DENV serotype, as demonstrated by the SRIP neutralization assay, underscoring its biomedical potential. Additionally, NoV::quartetEDIII streamlined the formulation by reducing the number of vaccine components from five to two, which offers benefits for large scale manufacturing and preparation. Nonetheless, further enhancements—such as refining the dosage and adjuvant, adding T-cell epitopes, and optimizing the structure—are necessary to strengthen the immune response, particularly for serotypes 3EDIII and 4EDIII.

7.3.6.4 Tetravalent VLPs Displaying Bifurcated tetEDIII and Quartet tetEDIII

Two distinct formats are considered. Two EDIII pairs are displayed in two distinct configurations, forming a bifurcated-EDIII, as two separate bivalently constructed constructs pairing 1EDIII with 3EDIII and 2EDIII with 4EDIII (bEDIII)-displaying VLPs (Fig. 7.28b). Third, four types of EDIIIs linked into a single peptide chain are displayed, forming quartet-EDIII (qEDIII)-displaying VLPs (Fig. 7.28c). For this research, the capsid protein from dengue virus serotype 2 (DENV-C2) was selected to generate in vivo self-assembled DENV capsid-like particles (DENV C-LPs), because it is a homologous VLP. This DENV C-LP serves as a scaffold for displaying the EDIII from all four DENV serotypes (tetEDIII) in vitro via SpT/SpC linkage. DENV C-LP displays their own four EDIIIs from all four serotypes. This section presents a comparative study of the structural characteristics, immunogenicity, and antigen-display efficiency of bifurcated-EDIII and quartet-EDIII displaying DENV C-LPs.

The SpT-fused C2 protein and two separate SpC-fused bifurcated EDIIIs (bE), which consist of two bivalently linked forms (1,3EDIII and 2,4EDIII) (Fig. 7.28c), and a quartet EDIII (qE), which is a single polypeptide containing all four EDIII domains from the DENV serotypes (Fig. 7.28d). These engineered proteins were produced in the fat body tissue of silkworm larvae and verified by Western blotting: the C2 capsid was detected using an anti-FLAG antibody (Fig. 7.33a), and the EDIIIs were identified using an anti-Strep-tag II antibody (Fig. 7.33b). The measured molecular weights for C2 (~15 kDa), 1,3E (~33.4 kDa), 2,4E (~33.1 kDa), and qE (~56 kDa) corresponded with theoretical expectations. Each purified proteins were confirmed by SDS-PAGE and CBB staining (Fig. 7.33c, d). The SpT-fused C2 served as the platform for the VLP, enabling the display of either SpC-fused qE or SpC-fused bE proteins. Western blot analysis showed distinct increases in molecular weight: C2 (approximately 15 kDa) linked to bE (containing 1,3E and 2,4E) increased a C2-bE band near 48 kDa, while C2 conjugated with qE (about 56 kDa) resulted in a C2-qE band at roughly 60 kDa. The display of all four serotypes on both C2-qE and C2-bE was further validated by Western blot using antibodies specific to each EDIII serotype (Fig. 7.33e(i–iv)). This approach also assessed cross-reactivity and conclusively determined the composition of the bE construct. In particular, DENV-1 and DENV-3 EDIII domains within the 1,3E were verified using DV1:E29 and DV3:E1 antibodies, respectively (indicated by blue arrows, Fig. 7.33e(i, iii)). Likewise, DENV-2 and DENV-4 EDIII within the 2,4E construct were identified with DV2:E3H5 and DV4:E29 antibodies (blue arrows,

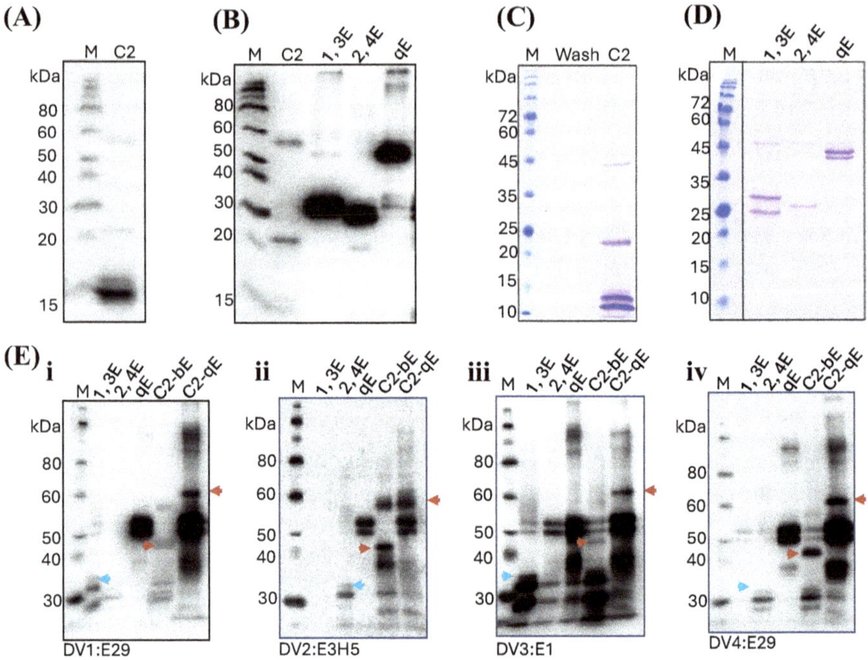

Fig. 7.33 Expression and purification of DENV C2 and EDIIIs, and their bioconjugation. Western blot images of EDIIIs and DENV C2 proteins (**a**, **b**) and SDS-PAGE/ CBB images of purified EDIIIs and DENV C2 proteins (**c**, **d**). Western blotting verified the expression of DENV C2 using an anti-FLAG antibody (**a**). The bifurcated and quartet EDIIIs were confirmed with an anti-Streptag II antibody (**b**). The purity of C2 (**c**), along with the bifurcated and quartet EDIII variants (**d**), was evaluated using CBB staining. Western blot images of the tetravalency and antigenic reactivity of EDIII proteins (blue arrows) and those displayed on the capsid (red arrows) (**e**). Antibodies specific for each serotype were used: E29 for serotype 1 (i), E3H5 for serotype 2 (ii), E1 for serotype 3 (iii), and E29 for serotype 4 (iv). Lanes M, C2, 1,3E, 2,4E, and qE denote molecular marker, DENV C2, 1EDIII-3EDIII, 2EDIII-4EDIII, and quartet EDIII, respectively. Lanes C2-bE and C2-qE in E indicate bifurcated EDIII-displayed DENV C2 and quartet EDIII-displayed DENV C2, respectively. (Adapted with permission (Muthuraman et al. 2025a). Copyright 2025 Elsevier)

Fig. 7.33e(ii, iv)). It is important to note that only the DV3:E1 antibody for serotype 3 produced nonspecific bands when analyzing the 2,4E construct (Fig. 7.33e(iii)), while all other results consistently verified the expected serotype-specific EDIII antigens.

The display of both quartet and bifurcated EDIIIs on DENV capsid-like particles (DENV C-LPs) was verified through IEM, while their physical characteristics were evaluated using dynamic light scattering (DLS). IEM imaging visualized the DENV C-LPs, which were confirmed by detection with anti-capsid (Fig. 7.34a) and anti-FLAG antibodies (Fig. 7.34b). The surface display of EDIII was further validated with anti-Strep-tag II antibodies for both bifurcated EDIII displayed on DENV C-LPs (C2-bE) (Fig. 7.34c), and quartet EDIII displayed on DENV C-LPs (C2-qE)

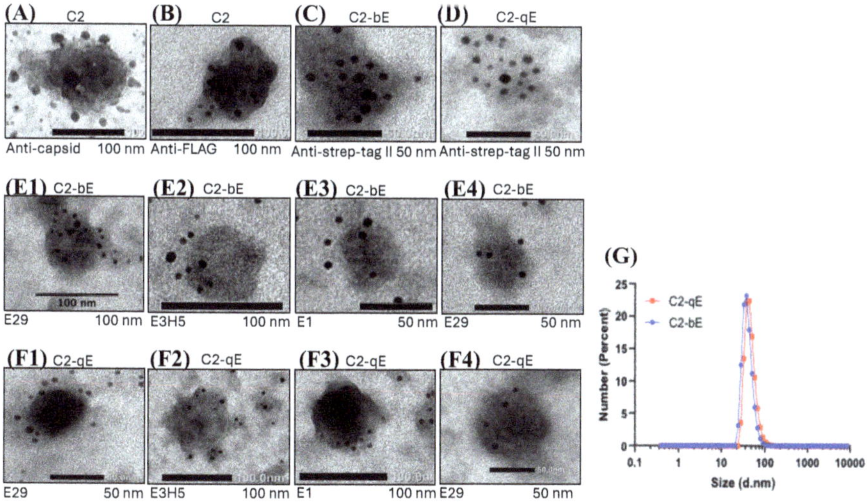

Fig. 7.34 Negatively stained IEM images of tetravalent EDIIIs. IEM was carried out using antibodies specific to each serotype. Gold-nanoparticle-conjugated secondary antibody (10 nm for anti-capsid and 5 nm for all EDIIIs) was used to visualize C2 and EDIIIs, respectively. The presence of DENV capsid-forming virus-like particles (VLPs) was confirmed with anti-capsid (**a**) and anti-FLAG (**b**) antibodies. Complete bifurcated EDIII display on C2-bE (**c**) and comprehensive quartet EDIII display on C2-qE (**d**) were visualized using an anti-strep tag II antibody. Analysis of C2-bE (**e1**–**e4**) and C2-qE (**f1**–**f4**) for tetravalent EDIII display was performed utilizing serotype-specific antibodies: E29 for serotype 1, E3H5 for serotype 2, E1 for serotype 3, and E29 for serotype 4. Finally, the hydrodynamic diameters of C2-bE and C2-qE were compared (**g**). (Adapted with permission (Muthuraman et al. 2025a). Copyright 2025 Elsevier)

(Fig. 7.34d). Additionally, the tetravalent display of EDIII on C2-bE (Fig. 7.34e1–e4)) and C2-qE (Fig. 7.34f1–f4)) was established by serotype-specific antibodies: E29 for DENV1 (Fig. 7.34e1, f1), E3H5 for DENV2 (Fig. 7.34e2, f2), E1 for DENV3 (Fig. 7.34e3, f3), and E29 for DENV4 (Fig. 7.34e4, f4). DLS measurements indicated hydrodynamic diameters of 40.46 ± 11.39 nm for C2-bE and 46.95 ± 14.52 nm for C2-qE (Fig. 7.34g). The respective polydispersity indices (PDIs) were 0.622 for C2-bE and 0.461 for C2-qE, indicating that C2-qE exhibits less size variation than C2-bE.

Serum samples obtained post-immunization (Fig. 7.35a) with bE, C2-bE, or C2-qE developed antibodies against all four serotypes (Fig. 7.35b(i–iv)). The titers of EDIII-specific antibodies were determined using ELISA, with assay plates coated with 1EDIII, 2EDIII, 3EDIII, or 4EDIII, excluding the SpC. Among the samples, unconjugated bE induced the highest antibody titers for every serotype. Notably, bE generated a significantly greater response than C2-qE for 2EDIII ($p = 0.0141$; Fig. 7.35b(iii)) and surpassed C2-bE for 4EDIII ($p = 0.0075$; Fig. 7.35b(iv)). When comparing the EDIII-displaying DENV C-LPs, there were no significant differences in overall seroconversion between C2-qE and C2-bE. Looking at geometric mean titers for individual serotypes, C2-bE had the strongest response to 1EDIII

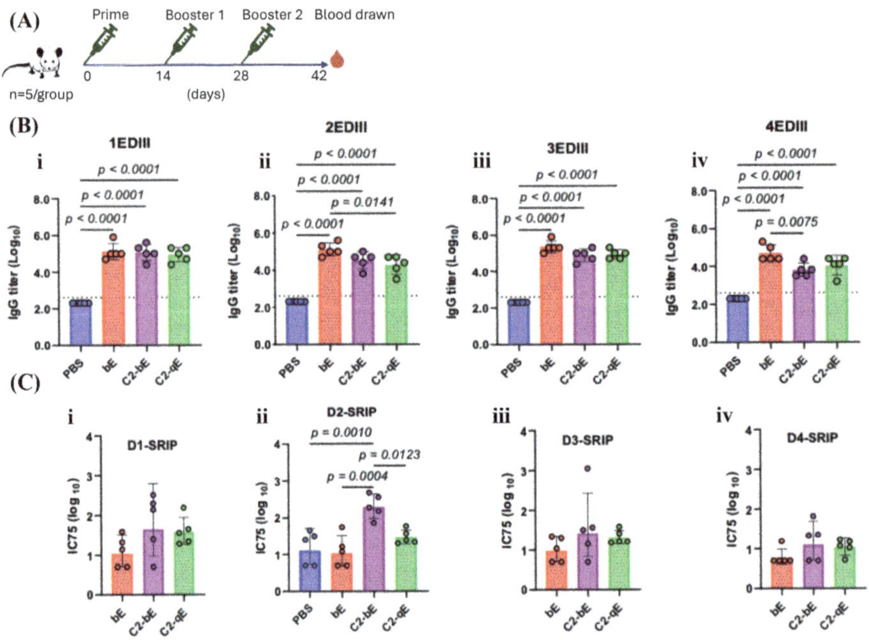

Fig. 7.35 Immunization and Assessment of Neutralizing Activity. The immunization protocol (**a**) administering 20 μg doses of the immunogens (C2-qE, C2-bE, or bE) on days 0, 14, and 28. Blood was drawn on day 42 for subsequent analysis. To evaluate antibody responses, serum samples underwent indirect ELISA using purified EDIIIs—excluding SpC—for each dengue serotype (**bi**–iv), and the resulting IgG titers were represented. The dotted line represents the assay's detection threshold. Serum samples were also subjected to the SRIP neutralization assay, with IC75 values plotted to illustrate neutralization across dengue serotypes (**ci**–iv). For both IgG titers and SRIP neutralization, results are represented as geometric means ± geometric standard deviations, and statistical significance was assessed using one-way ANOVA followed by Tukey's multiple comparison test. A p value below 0.05 was considered significant. (Adapted with permission (Muthuraman et al. 2025a). Copyright 2025 Elsevier)

(5.054) and 2EDIII (4.509), while C2-qE elicited higher titers for 3EDIII (4.945) and 4EDIII (4.020). Across all immunization groups, 4EDIII consistently triggered a weaker immune response. The highest antibody titer for unconjugated bE was observed against 3EDIII (5.364), whereas among the VLP constructs, peak responses were to 1EDIII for C2-bE (5.054) and to 3EDIII for C2-qE (4.945). All antigen groups produced significantly stronger immune responses ($p < 0.0001$) compared to the PBS negative control for each serotype.

The ability of the antibodies to neutralize was evaluated using the SRIP neutralization assay. Among all DENV serotypes and immunization groups, C2-bE consistently showed the highest neutralization potency (Fig. 7.35c(i–iv)). In particular, for D2-SRIP, C2-bE (2.294) demonstrated significantly greater neutralization compared to the unconjugated bE construct ($p = 0.0004$) and to C2-qE ($p = 0.0123$)

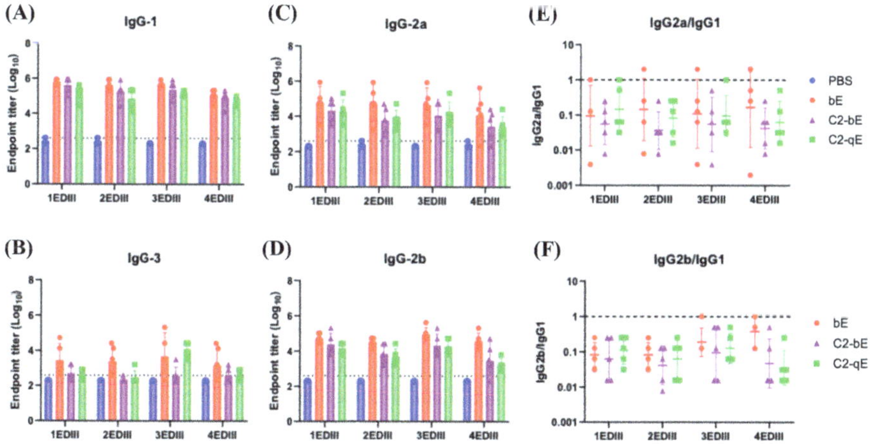

Fig. 7.36 Evaluation of IgG antibody subclasses. Endpoint titers for IgG1 (**a**), IgG3 (**b**), IgG2a (**c**), and IgG2b (**d**) targeting four EDIII serotypes were measured, and data are represented by the geometric means and geometric standard deviations. Additionally, the ratios of IgG2a to IgG1 (**e**) and IgG2b to IgG1 (**f**) were analyzed using ELISA. A dotted line in **a–d** marks the assay's detection threshold, while a dashed line in **e–f** represents the cut-off value for identifying the specific T helper (Th) cell immune response type. (Adapted with permission (Muthuraman et al. 2025a). Copyright 2025 Elsevier)

(Fig. 7.35c(ii)). For other groups, the highest geometric mean IC75 values were observed for bE (1.035) and C2-qE (1.579) against D1-SRIP. Similar to the antibody titer findings, no significant difference in neutralization was seen between C2-qE and C2-bE for the remaining serotypes. Importantly, although the unconjugated bifurcated EDIII (bE) produced higher antibody titers, it had the weakest neutralization activity across all four serotypes compared to both the C2-bE and C2-qE displayed EDIIIs.

The evaluation of IgG subclasses (IgG1, IgG2a, IgG2b, and IgG3) demonstrated unique antibody production profiles following immunization (Fig. 7.36a–d). Among all DENV serotypes and immunization groups (bE, C2-bE, and C2-qE), IgG1 displayed the highest titers, with average values ranging from 4.65 to 5.79 (bE: 5.01–5.79; C2-bE: 4.89–5.61; C2-qE: 4.65–5.19) (Fig. 7.36a). In contrast, IgG3 titers were the lowest observed, with means between 2.36 and 4.11 (bE: 3.26–3.81; C2-bE: 2.36–2.72; C2-qE: 2.48–4.11) (Fig. 7.36b). Further analysis of IgG2a and IgG2b revealed differences across groups and serotypes: in the bE group, IgG2a titers surpassed IgG2b for serotypes 1 and 2, but this trend reversed for serotypes 3 and 4 (Fig. 7.36c, d). In the C2-bE group, IgG2b titers were consistently higher than IgG2a for every serotype. Conversely, in the C2-qE group, IgG2a generally exceeded IgG2b, except for serotype 3. Notably, 1EDIII showed strong IgG1 and IgG2a responses, while 3EDIII was associated with higher IgG2b and IgG3 responses. To evaluate Th1/Th2 immune skewing induced by C2-bE and C2-qE, IgG2a/IgG1 and

IgG2b/IgG1 ratios were determined for each serotype. Ratios above 1 indicate a Th1-dominant response, whereas ratios below 0.5 suggest a Th2 orientation. The results indicated that both ratios fell within the range consistent with a Th2-type response. Specifically, IgG2a/IgG1 ratios ranged from 0.28 to 0.6 for bE, 0.07 to 0.14 for C2-bE, and 0.13 to 0.33 for C2-qE (Fig. 7.36e). Similarly, IgG2b/IgG1 ratios were between 0.11 and 0.55 for bE, 0.07 and 0.23 for C2-bE, and 0.07 and 0.19 for C2-qE (Fig. 7.36f).

In summary, a predominance of IgG1 antibodies, along with low IgG2a/IgG1 and IgG2b/IgG1 ratios, suggests a Th2-skewed immune response. While IgG1 plays a key role in humoral immunity, an imbalance favoring IgG1 over IgG2a raises concerns. These results are consistent with typical responses to recombinant protein vaccines, especially those lacking robust Th1-stimulating components, which often result in a Th2 bias and increased IgG1 production in BALB/c mice. Both C2-bE and C2-qE vaccine candidates can induce neutralizing antibodies against the dengue virus. The C2-bE formulation showed superior neutralization across all four dengue serotypes, especially against DENV2-SRIP, while C2-qE also demonstrated strong neutralizing activity against DENV-1, DENV-3, and DENV-4, with performance similar to C2-bE for most strains. Although C2-bE holds a slight advantage in broad neutralization, both approaches effectively present tetravalent EDIII antigens on dengue capsid virus-like particles.

7.4 Summary

This section compares the morphology, immune response, and neutralization activity of tetravalent VLPs targeting four dengue serotypes. CPV-LPs displayed EDIII antigens more efficiently than DENV C-LPs, likely due to favorable structural features, resulting in superior antibody binding and neutralization. NoV-LPs were used to display either separate or quartet EDIIIs; both forms induced similar IgG responses, with NoV::quartetEDIII effectively generating neutralizing antibodies against all serotypes and simplifying vaccine composition for large-scale production. DENV C-LPs displaying bifurcated or quartet EDIIIs maintained structural integrity and elicited comparable IgG responses, though bifurcated EDIII produced higher neutralization titers. All platforms predominantly induced a Th2-type immune response, which may limit effectiveness and warrants further investigation. Ultimately, CPV-LP showed better display efficiency, DENV C-LP had higher expected affinity, and bifurcated EDIII outperformed quartet EDIII in neutralization. However, optimal scaffold-EDIII combinations remain undetermined, necessitating additional studies to identify the most effective VLP platform for dengue vaccines.

References

Amelia F, Iyori M, Emran TB, Yamamoto DS, Genshi K, Otsuka H, Onoue Y, Yusuf Y, Islam A, Yoshida S (2019) Down-selecting circumsporozoite protein-based malaria vaccine: a comparison of malaria sporozoite challenge model. Parasite Immunol 41:e12624. https://doi.org/10.1111/pim.12624

Amelung S, Nerlich A, Rohde M, Spellerberg B, Cole JN, Nizet V, Chhatwal GS, Talay SR (2011) The FbaB-type fibronectin-binding protein of Streptococcus pyogenes promotes specific invasion into endothelial cells. Cell Microbiol 13:1200–1211. https://doi.org/10.1111/j.1462-5822.2011.01610.x

Boonyakida J, Utomo DIS, Soma FN, Park EY (2022) Two-step purification of tag-free norovirus-like particles from silkworm larvae (Bombyx mori). Protein Expr Purif 190:106010. https://doi.org/10.1016/j.pep.2021.106010

Boonyakida J, Khoris IM, Nasrin F, Park EY (2023) Improvement of modular protein display efficiency in SpyTag-implemented norovirus-like particles. Biomacromolecules 24:308–318. https://doi.org/10.1021/acs.biomac.2c01150

Boonyakida J, Nakayama K, Kusakisako K, Ikadai H, Park EY (2024) Modular display of Plasmodium yoelii circumsporozoite surface protein and Merozoite Surface Protein-1 on norovirus-like particles. Bioconjug Chem 35:1933–1943. https://doi.org/10.1021/acs.bioconjchem.4c00460

Boonyakida J, Matsuda M, Suzuki R, Muthuraman KR, Park EY (2025) Virus-like particle-based multiserotype quartet vaccine of dengue envelope protein domain III elicited potent anti-dengue responses. Biomacromolecules 26:4449–4463. https://doi.org/10.1021/acs.biomac.5c00459

Brown KE, Hibbs JR, Gallinella G, Anderson SM, Lehman ED, Mccarthy P, Young NS (1994) Resistance to parvovirus B19 infection due to lack of virus receptor (erythrocyte P antigen). N Engl J Med 330:1192–1196. https://doi.org/10.1056/NEJM199404283301704

Brune KD, Leneghan DB, Brian IJ, Ishizuka AS, Bachmann MF, Draper SJ, Biswas S, Howarth M (2016) Plug-and-display: decoration of virus-like particles via isopeptide bonds for modular immunization. Sci Rep 6:19234. https://doi.org/10.1038/srep19234

Chang D, Liu Y, Chen Y, Hu X, Burov A, Puzyr A, Bondar V, Yao L (2020) Study of the immunogenicity of the VP2 protein of canine parvovirus produced using an improved Baculovirus expression system. BMC Vet Res 16:202. https://doi.org/10.1186/s12917-020-02422-3

Chawla B, Mahajan B, Oakley M, Majam VF, Belmonte A, Sedegah M, Shimp RL Jr, Kaslow DC, Kumar S (2019) Antibody-dependent, gamma interferon-independent sterilizing immunity induced by a subunit malaria vaccine. Infect Immun 87. https://doi.org/10.1128/IAI.00236-19

Chen YL, Huang CT (2020) Establishment of a two-step purification scheme for tag-free recombinant Taiwan native norovirus P and VP1 proteins. J Chromatogr B 1159:122357. https://doi.org/10.1016/j.jchromb.2020.122357

Chen X, Zaro JL, Shen WC (2013) Fusion protein linkers: property, design and functionality. Adv Drug Deliv Rev 65:1357–1369. https://doi.org/10.1016/j.addr.2012.09.039

Child MA, Epp C, Bujard H, Blackman MJ (2010) Regulated maturation of malaria merozoite surface protein-1 is essential for parasite growth. Mol Microbiol 78:187–202. https://doi.org/10.1111/j.1365-2958.2010.07324.x

Choi JY, Woo SD, Lee HK, Hong HK, Je YH, Park JH, Song JY, An SH, Kang SK (2000) High-level expression of canine parvovirus VP2 using Bombyx mori nucleopolyhedrovirus vector. Arch Virol 145:171–177. https://doi.org/10.1007/s007050050014

Comas-Garcia M, Colunga-Saucedo M, Rosales-Mendoza S (2020) The role of virus-like particles in medical biotechnology. Mol Pharm 17:4407–4420. https://doi.org/10.1021/acs.molpharmaceut.0c00828

Cunha MG, Rodrigues MM, Soares IS (2001) Comparison of the immunogenic properties of recombinant proteins representing the Plasmodium vivax vaccine candidate MSP1(19) expressed in distinct bacterial vectors. Vaccine 20:385–396. https://doi.org/10.1016/s0264-410x(01)00359-0

Das S, Hertrich N, Perrin AJ, Withers-Martinez C, Collins CR, Jones ML, Watermeyer JM, Fobes ET, Martin SR, Saibil HR, Wright GJ, Treeck M, Epp C, Blackman MJ (2015) Processing of Plasmodium falciparum Merozoite Surface Protein MSP1 activates a spectrin-binding function enabling parasite egress from RBCs. Cell Host Microbe 18:433–444. https://doi.org/10.1016/j.chom.2015.09.007

Deo VK, Yoshimatsu K, Otsuki T, Dong J, Kato T, Park EY (2013) Display of *Neospora caninum* surface protein related sequence 2 on Rous sarcoma virus-derived gag protein virus-like particles. J Biotechnol 165:69–75. https://doi.org/10.1016/j.jbiotec.2013.02.013

Deo VK, Yui M, Alam MJ, Yamazaki M, Kato T, Park EY (2014) A model for targeting colon carcinoma cells using single-chain variable fragments anchored on virus-like particles via glycosyl phosphatidylinositol anchor. Pharm Res 31:2166–2177. https://doi.org/10.1007/s11095-014-1316-4

Deo VK, Kato T, Park EY (2015) Chimeric virus-like particles made using GAG and M1 capsid proteins providing dual drug delivery and vaccination platform. Mol Pharm 12:839–845. https://doi.org/10.1021/mp500860x

Deo VK, Kato T, Park EY (2016) Virus-like particles displaying recombinant short-chain fragment region and interleukin 2 for targeting colon cancer tumors and attracting macrophages. J Pharm Sci 105:1614–1622. https://doi.org/10.1016/j.xphs.2016.02.011

Devant JM, Hansman GS (2021) Structural heterogeneity of a human norovirus vaccine candidate. Virology 553:23–34. https://doi.org/10.1016/j.virol.2020.10.005

Dijkman PM, Marzluf T, Zhang Y, Chang SS, Helm D, Lanzer M, Bujard H, Kudryashev M (2021) Structure of the merozoite surface protein 1 from Plasmodium falciparum. Sci Adv 7. https://doi.org/10.1126/sciadv.abg0465

Ding X, Liu D, Booth G, Gao W, Lu Y (2018) Virus-like particle engineering: from rational design to versatile applications. Biotechnol J 13:e1700324. https://doi.org/10.1002/biot.201700324

Duffy PE, Patrick Gorres J (2020) Malaria vaccines since 2000: progress, priorities, products. NPJ Vaccines 5:48. https://doi.org/10.1038/s41541-020-0196-3

Elia G, Desario C, Pezzoni G, Camero M, Brocchi E, Decaro N, Martella V, Buonavoglia C (2012) Recombinant ELISA using baculovirus-expressed VP2 for detection of antibodies against canine parvovirus. J Virol Methods 184:98–102. https://doi.org/10.1016/j.jviromet.2012.05.025

Feng H, Liang M, Wang HL, Zhang T, Zhao PS, Shen XJ, Zhang RZ, Hu GQ, Gao YQ, Wang CY, Wang TC, Zhang W, Yang ST, Xia XZ (2011) Recombinant canine parvovirus-like particles express foreign epitopes in silkworm pupae. Vet Microbiol 154:49–57. https://doi.org/10.1016/j.vetmic.2011.06.022

Feng H, Hu GQ, Wang HL, Liang M, Liang H, Guo H, Zhao P, Yang YJ, Zheng XX, Zhang ZF, Zhao YK, Gao YW, Yang ST, Xia XZ (2014) Canine parvovirus VP2 protein expressed in silkworm pupae self-assembles into virus-like particles with high immunogenicity. PLoS One 9:e79575. https://doi.org/10.1371/journal.pone.0079575

Fernandez S, Thomas SJ, De La Barrera R, Im-Erbsin R, Jarman RG, Baras B, Toussaint JF, Mossman S, Innis BL, Schmidt A, Malice MP, Festraets P, Warter L, Putnak JR, Eckels KH (2015) An adjuvanted, tetravalent dengue virus purified inactivated vaccine candidate induces long-lasting and protective antibody responses against dengue challenge in rhesus macaques. Am J Trop Med Hyg 92:698–708. https://doi.org/10.4269/ajtmh.14-0268

Geng X, Doitsh G, Yang Z, Galloway NL, Greene WC (2014) Efficient delivery of lentiviral vectors into resting human CD4 T cells. Gene Ther 21:444–449. https://doi.org/10.1038/gt.2014.5

Georgiev GI, Malonis RJ, Wirchnianski AS, Wessel AW, Jung HS, Cahill SM, Nyakatura EK, Vergnolle O, Dowd KA, Cowburn D, Pierson TC, Diamond MS, Lai JR (2022) Resurfaced ZIKV EDIII nanoparticle immunogens elicit neutralizing and protective responses in vivo. Cell Chem Biol 29:811–823 e817. https://doi.org/10.1016/j.chembiol.2022.02.004

Gilbert L, Toivola J, Valilehto O, Saloniemi T, Cunningham C, White D, Makela AR, Korhonen E, Vuento M, Oker-Blom C (2006) Truncated forms of viral VP2 proteins fused to EGFP assemble into fluorescent parvovirus-like particles. J Nanobiotechnol 4:13. https://doi.org/10.1186/1477-3155-4-13

Goetschius DJ, Hartmann SR, Organtini LJ, Callaway H, Huang K, Bator CM, Ashley RE, Makhov AM, Conway JF, Parrish CR, Hafenstein SL (2021) High-resolution asymmetric structure of a Fab-virus complex reveals overlap with the receptor binding site. Proc Natl Acad Sci USA 118. https://doi.org/10.1073/pnas.2025452118

Goffar MG, Deo VK, Kato T, Park EY (2022) Dual display hemagglutinin 1 and 5 on the surface of enveloped virus-like particles in silkworm expression system. Protein Expr Purif 197:106106. https://doi.org/10.1016/j.pep.2022.106106

Goldmann C, Stolte N, Nisslein T, Hunsmann G, Luke W, Petry H (2000) Packaging of small molecules into VP1-virus-like particles of the human polyomavirus JC virus. J Virol Methods 90:85–90. https://doi.org/10.1016/s0166-0934(00)00226-3

Gromowski GD, Barrett ND, Barrett AD (2008) Characterization of dengue virus complex-specific neutralizing epitopes on envelope protein domain III of dengue 2 virus. J Virol 82:8828–8837. https://doi.org/10.1128/JVI.00606-08

Haider N, Hasan MN, Onyango J, Asaduzzaman M (2024) Global landmark: 2023 marks the worst year for dengue cases with millions infected and thousands of deaths reported. IJID Reg 13:100459. https://doi.org/10.1016/j.ijregi.2024.100459

Haldar K, Bhattacharjee S, Safeukui I (2018) Drug resistance in plasmodium. Nat Rev Microbiol 16:156–170. https://doi.org/10.1038/nrmicro.2017.161

Halstead SB, Nimmannitya S, Cohen SN (1970) Observations related to pathogenesis of dengue hemorrhagic fever. IV. Relation of disease severity to antibody response and virus recovered. Yale J Biol Med 42:311–328

Hansman GS, Taylor DW, Mclellan JS, Smith TJ, Georgiev I, Tame JR, Park SY, Yamazaki M, Gondaira F, Miki M, Katayama K, Murata K, Kwong PD (2012) Structural basis for broad detection of genogroup II noroviruses by a monoclonal antibody that binds to a site occluded in the viral particle. J Virol 86:3635–3646. https://doi.org/10.1128/JVI.06868-11

Huhti L, Blazevic V, Nurminen K, Koho T, Hytonen VP, Vesikari T (2010) A comparison of methods for purification and concentration of norovirus GII-4 capsid virus-like particles. Arch Virol 155:1855–1858. https://doi.org/10.1007/s00705-010-0768-z

Hui G, Hashimoto C (2007) Plasmodium falciparum anti-MSP1-19 antibodies induced by MSP1-42 and MSP1-19 based vaccines differed in specificity and parasite growth inhibition in terms of recognition of conserved versus variant epitopes. Vaccine 25:948–956. https://doi.org/10.1016/j.vaccine.2006.08.041

Hurtado A, Rueda P, Nowicky J, Sarraseca J, Casal JI (1996) Identification of domains in canine parvovirus VP2 essential for the assembly of virus-like particles. J Virol 70:5422–5429. https://doi.org/10.1128/JVI.70.8.5422-5429.1996

Ikwuagwu B, Tullman-Ercek D (2022) Virus-like particles for drug delivery: a review of methods and applications. Curr Opin Biotechnol 78:102785. https://doi.org/10.1016/j.copbio.2022.102785

Izac JR, Kwee EJ, Gaigalas A, Wang L (2024) Quantitative and standardized Pseudovirus neutralization assay for COVID-19. Methods Mol Biol 2779:259–271. https://doi.org/10.1007/978-1-0716-3738-8_11

Janitzek CM, Matondo S, Thrane S, Nielsen MA, Kavishe R, Mwakalinga SB, Theander TG, Salanti A, Sander AF (2016) Bacterial superglue generates a full-length circumsporozoite protein virus-like particle vaccine capable of inducing high and durable antibody responses. Malar J 15:545. https://doi.org/10.1186/s12936-016-1574-1

Jegerlehner A, Tissot A, Lechner F, Sebbel P, Erdmann I, Kundig T, Bachi T, Storni T, Jennings G, Pumpens P, Renner WA, Bachmann MF (2002) A molecular assembly system that renders antigens of choice highly repetitive for induction of protective B cell responses. Vaccine 20:3104–3112. https://doi.org/10.1016/s0264-410x(02)00266-9

Jin H, Xia X, Liu B, Fu Y, Chen X, Wang H, Xia Z (2016) High-yield production of canine parvovirus virus-like particles in a baculovirus expression system. Arch Virol 161:705–710. https://doi.org/10.1007/s00705-015-2719-1

Jones CT, Ma L, Burgner JW, Groesch TD, Post CB, Kuhn RJ (2003) Flavivirus capsid is a dimeric alpha-helical protein. J Virol 77:7143–7149. https://doi.org/10.1128/jvi.77.12.7143-7149.2003

Kang HJ, Coulibaly F, Clow F, Proft T, Baker EN (2007) Stabilizing isopeptide bonds revealed in gram-positive bacterial pilus structure. Science 318:1625–1628. https://doi.org/10.1126/science.1145806

Kato T, Yui M, Deo VK, Park EY (2015) Development of Rous sarcoma virus-like particles displaying hCC49 scFv for specific targeted drug delivery to human colon carcinoma cells. Pharm Res 32:3699–3707. https://doi.org/10.1007/s11095-015-1730-2

Kato T, Arai S, Ichikawa H, Park EY (2016) Versatility of chitosan/BmNPV bacmid DNA nanocomplex as transfection reagent of recombinant protein expression in silkworm larvae. Biotechnol Lett 38:1449–1457. https://doi.org/10.1007/s10529-016-2144-x

Kato T, Kakuta T, Yonezuka A, Sekiguchi T, Machida Y, Xu J, Suzuki T, Park EY (2023) Expression and purification of porcine rotavirus structural proteins in silkworm larvae as a vaccine candidate. Mol Biotechnol 65:401–409. https://doi.org/10.1007/s12033-022-00548-3

Keeble AH, Banerjee A, Ferla MP, Reddington SC, Anuar I, Howarth M (2017) Evolving accelerated Amidation by SpyTag/SpyCatcher to analyze membrane dynamics. Angew Chem Int Ed Engl 56:16521–16525. https://doi.org/10.1002/anie.201707623

Kleinschmidt I, Bradley J, Knox TB, Mnzava AP, Kafy HT, Mbogo C, Ismail BA, Bigoga JD, Adechoubou A, Raghavendra K, Cook J, Malik EM, Nkuni ZJ, Macdonald M, Bayoh N, Ochomo E, Fondjo E, Awono-Ambene HP, Etang J, Akogbeto M, Bhatt RM, Chourasia MK, Swain DK, Kinyari T, Subramaniam K, Massougbodji A, Oke-Sopoh M, Ogouyemi-Hounto A, Kouambeng C, Abdin MS, West P, Elmardi K, Cornelie S, Corbel V, Valecha N, Mathenge E, Kamau L, Lines J, Donnelly MJ (2018) Implications of insecticide resistance for malaria vector control with long-lasting insecticidal nets: a WHO-coordinated, prospective, international, observational cohort study. Lancet Infect Dis 18:640–649. https://doi.org/10.1016/S1473-3099(18)30172-5

Kolodny N, Kitov S, Vassell MA, Miller VL, Ware LA, Fegeding K, De La Vega P, Sacci JB Jr, Lanar DE (2001) Two-step chromatographic purification of recombinant Plasmodium falciparum circumsporozoite protein from Escherichia coli. J Chromatogr B Biomed Sci Appl 762:77–86. https://doi.org/10.1016/s0378-4347(01)00340-1

Ladenstein R, Morgunova E (2020) Second career of a biosynthetic enzyme: Lumazine synthase as a virus-like nanoparticle in vaccine development. Biotechnol Rep (Amst) 27:e00494. https://doi.org/10.1016/j.btre.2020.e00494

Lal AA, De La Cruz VF, Welsh JA, Charoenvit Y, Maloy WL, Mccutchan TF (1987) Structure of the gene encoding the circumsporozoite protein of Plasmodium yoelii. A rodent model for examining antimalarial sporozoite vaccines. J Biol Chem 262:2937–2940

Lee MF, Long CM, Poh CL (2025) Current status of the development of dengue vaccines. Vaccine X 22:100604. https://doi.org/10.1016/j.jvacx.2024.100604

Li L, Fierer JO, Rapoport TA, Howarth M (2014) Structural analysis and optimization of the covalent association between SpyCatcher and a peptide tag. J Mol Biol 426:309–317. https://doi.org/10.1016/j.jmb.2013.10.021

Li J, Zhang Y, Soubias O, Khago D, Chao FA, Li Y, Shaw K, Byrd RA (2020) Optimization of sortase a ligation for flexible engineering of complex protein systems. J Biol Chem 295:2664–2675. https://doi.org/10.1074/jbc.RA119.012039

Liu Y, Zhou J, Yu Z, Fang D, Fu C, Zhu X, He Z, Yan H, Jiang L (2014) Tetravalent recombinant dengue virus-like particles as potential vaccine candidates: immunological properties. BMC Microbiol 14:233. https://doi.org/10.1186/s12866-014-0233-3

Liu ZH, Deng ZF, Lu Y, Fang WH, He F (2022) A modular and self-adjuvanted multivalent vaccine platform based on porcine circovirus virus-like nanoparticles. J Nanobiotechnol 20:493. https://doi.org/10.1186/s12951-022-01710-4

Lopez-Sagaseta J, Malito E, Rappuoli R, Bottomley MJ (2016) Self-assembling protein nanoparticles in the design of vaccines. Comput Struct Biotechnol J 14:58–68. https://doi.org/10.1016/j.csbj.2015.11.001

Makela AR, Oker-Blom C (2006) Baculovirus display: a multifunctional technology for gene delivery and eukaryotic library development. Adv Virus Res 68:91–112. https://doi.org/10.1016/S0065-3527(06)68003-2

Malonis RJ, Georgiev GI, Haslwanter D, VanBlargan LA, Fallon G, Vergnolle O, Cahill SM, Harris R, Cowburn D, Chandran K, Diamond MS, Lai JR (2022) A Powassan virus domain III nanoparticle immunogen elicits neutralizing and protective antibodies in mice. PLoS Pathog 18:e1010573. https://doi.org/10.1371/journal.ppat.1010573

Matsuda M, Yamanaka A, Yato K, Yoshii K, Watashi K, Aizaki H, Konishi E, Takasaki T, Kato T, Muramatsu M, Wakita T, Suzuki R (2018) High-throughput neutralization assay for multiple flaviviruses based on single-round infectious particles using dengue virus type 1 reporter replicon. Sci Rep 8:16624. https://doi.org/10.1038/s41598-018-34865-y

Mellid-Carballal R, Gutierrez-Gutierrez S, Rivas C, Garcia-Fuentes M (2023) Viral protein-based nanoparticles (part 2): pharmaceutical applications. Eur J Pharm Sci 189:106558. https://doi.org/10.1016/j.ejps.2023.106558

Modis Y, Ogata S, Clements D, Harrison SC (2004) Structure of the dengue virus envelope protein after membrane fusion. Nature 427:313–319. https://doi.org/10.1038/nature02165

Moss DK, Remarque EJ, Faber BW, Cavanagh DR, Arnot DE, Thomas AW, Holder AA (2012) Plasmodium falciparum 19-kilodalton merozoite surface protein 1 (MSP1)-specific antibodies that interfere with parasite growth in vitro can inhibit MSP1 processing, merozoite invasion, and intracellular parasite development. Infect Immun 80:1280–1287. https://doi.org/10.1128/IAI.05887-11

Muthuraman KR, Utomo DIS, Matsuda M, Suzuki R, Park EY (2024) Expression of dengue capsid-like particles in silkworm and display of envelope domain III of dengue virus serotype 2. Protein Expr Purif 222:106543. https://doi.org/10.1016/j.pep.2024.106543

Muthuraman KR, Boonyakida J, Matsuda M, Suzuki R, Kato T, Park EY (2025a) Comparative immunogenicity study of quartet and bifurcated tetravalent dengue virus envelope domain III displayed virus-like particle vaccine candidates in BALB/c mice. Vaccine 63:127670. https://doi.org/10.1016/j.vaccine.2025.127670

Muthuraman KR, Boonyakida J, Matsuda M, Suzuki R, Kato T, Park EY (2025b) Tetravalent virus-like particles engineered to display envelope domain IIIs of four dengue serotypes in silkworm as vaccine candidates. Biomacromolecules 26:2003–2013. https://doi.org/10.1021/acs.biomac.4c01831

Nerome K, Imagawa T, Sugita S, Arasaki Y, Maegawa K, Kawasaki K, Tanaka T, Watanabe S, Nishimura H, Suzuki T, Kuroda K, Kosugi I, Kajiura Z (2023) The potential of a universal influenza virus-like particle vaccine expressing a chimeric cytokine. Life Sci Alliance 6:e202201548. https://doi.org/10.26508/lsa.202201548

Neves-Martins TC, Mebus-Antunes NC, Neto CHG, Barbosa GM, Almeida FCL, Caruso IP, Da Poian AT (2023) Self-assembly of dengue virus empty capsid-like particles in solution. iScience 26:106197. https://doi.org/10.1016/j.isci.2023.106197

Nilchan N, Kraivong R, Luangaram P, Phungsom A, Tantiwatcharakunthon M, Traewachiwiphak S, Prommool T, Punyadee N, Avirutnan P, Duangchinda T, Malasit P, Puttikhunt C (2024) An engineered N-glycosylated dengue envelope protein domain III facilitates epitope-directed selection of potently neutralizing and minimally enhancing antibodies. ACS Infect Dis 10:2690–2704. https://doi.org/10.1021/acsinfecdis.4c00058

Oyen D, Torres JL, Cottrell CA, Richter King C, Wilson IA, Ward AB (2018) Cryo-EM structure of P. falciparum circumsporozoite protein with a vaccine-elicited antibody is stabilized by somatically mutated inter-Fab contacts. Sci Adv 4:eaau8529. https://doi.org/10.1126/sciadv.aau8529

Park EY (2025) From silkroad to bioroad: silkworm biotechnology. Protein Expr Purif 234:106762. https://doi.org/10.1016/j.pep.2025.106762

Park J, Kim J, Jang YS (2022) Current status and perspectives on vaccine development against dengue virus infection. J Microbiol 60:247–254. https://doi.org/10.1007/s12275-022-1625-y

Perera KL, Handunnetti SM, Holm I, Longacre S, Mendis K (1998) Baculovirus merozoite surface protein 1 C-terminal recombinant antigens are highly protective in a natural primate model for human Plasmodium vivax malaria. Infect Immun 66:1500–1506. https://doi.org/10.1128/IAI.66.4.1500-1506.1998

Peretti S, Schiavoni I, Pugliese K, Federico M (2006) Selective elimination of HIV-1-infected cells by Env-directed, HIV-1-based virus-like particles. Virology 345:115–126. https://doi.org/10.1016/j.virol.2005.09.054

Phan TTN, Hvasta MG, Kudlacek ST, Thiono DJ, Tripathy A, Nicely NI, De Silva AM, Kuhlman B (2022) A conserved set of mutations for stabilizing soluble envelope protein dimers from dengue and Zika viruses to advance the development of subunit vaccines. J Biol Chem 298:102079. https://doi.org/10.1016/j.jbc.2022.102079

Plassmeyer ML, Reiter K, Shimp RL Jr, Kotova S, Smith PD, Hurt DE, House B, Zou X, Zhang Y, Hickman M, Uchime O, Herrera R, Nguyen V, Glen J, Lebowitz J, Jin AJ, Miller LH, Macdonald NJ, Wu Y, Narum DL (2009) Structure of the Plasmodium falciparum circumsporozoite protein, a leading malaria vaccine candidate. J Biol Chem 284:26951–26963. https://doi.org/10.1074/jbc.M109.013706

Rabinovich RN, Drakeley C, Djimde AA, Hall BF, Hay SI, Hemingway J, Kaslow DC, Noor A, Okumu F, Steketee R, Tanner M, Wells TNC, Whittaker MA, Winzeler EA, Wirth DF, Whitfield K, Alonso PL (2017) malERA: An updated research agenda for malaria elimination and eradication. PLoS Med 14:e1002456. https://doi.org/10.1371/journal.pmed.1002456

Ramamoorthy S, Sriranganathan N, Lindsay DS (2005) Gerbil model of acute neosporosis. Vet Parasitol 127:111–114. https://doi.org/10.1016/j.vetpar.2004.09.016

Reddington SC, Howarth M (2015) Secrets of a covalent interaction for biomaterials and biotechnology: SpyTag and SpyCatcher. Curr Opin Chem Biol 29:94–99. https://doi.org/10.1016/j.cbpa.2015.10.002

Rueda P, Martinez-Torrecuadrada JL, Sarraseca J, Sedlik C, Del Barrio M, Hurtado A, Leclerc C, Casal JI (1999) Engineering parvovirus-like particles for the induction of B-cell, CD4(+) and CTL responses. Vaccine 18:325–332. https://doi.org/10.1016/s0264-410x(99)00202-9

Samandoulgou I, Hammami R, Morales Rayas R, Fliss I, Jean J (2015) Stability of secondary and tertiary structures of virus-like particles representing noroviruses: effects of pH, ionic strength, and temperature and implications for adhesion to surfaces. Appl Environ Microbiol 81:7680–7686. https://doi.org/10.1128/AEM.01278-15

Schneider IC, Hartmann J, Braun G, Stitz J, Klamp T, Bihi M, Sahin U, Buchholz CJ (2018) Displaying tetra-membrane spanning Claudins on enveloped virus-like particles for cancer immunotherapy. Biotechnol J 13:e1700345. https://doi.org/10.1002/biot.201700345

Seesen M, Jearanaiwitayakul T, Limthongkul J, Sunintaboon P, Ubol S (2022) Mice immunized with trimethyl chitosan nanoparticles containing DENV-2 envelope domain III elicit neutralizing antibodies with undetectable antibody-dependent enhancement activity. J Gen Virol 103. https://doi.org/10.1099/jgv.0.001768

Sharma J, Shepardson K, Johns LL, Wellham J, Avera J, Schwarz B, Rynda-Apple A, Douglas T (2020) A self-Adjuvanted, modular, antigenic VLP for rapid response to influenza virus variability. ACS Appl Mater Interfaces 12:18211–18224. https://doi.org/10.1021/acsami.9b21776

Shukla R, Rajpoot RK, Arora U, Poddar A, Swaminathan S, Khanna N (2017) Pichia pastoris-expressed bivalent virus-like particulate vaccine induces domain III-focused bivalent neutralizing antibodies without antibody-dependent enhancement in vivo. Front Microbiol 8:2644. https://doi.org/10.3389/fmicb.2017.02644

Stander J, Chabeda A, Rybicki EP, Meyers AE (2021) A plant-produced virus-like particle displaying envelope protein domain III elicits an immune response against West Nile virus in mice. Front Plant Sci 12:738619. https://doi.org/10.3389/fpls.2021.738619

Thoresen D, Matsuda K, Urakami A, Ngwe Tun MM, Nomura T, Moi ML, Watanabe Y, Ishikawa M, Hau TTT, Yamamoto H, Suzaki Y, Ami Y, Smith JF, Matano T, Morita K, Akahata W (2024) A tetravalent dengue virus-like particle vaccine induces high levels of neutralizing antibodies and reduces dengue replication in non-human primates. J Virol 98:e0023924. https://doi.org/10.1128/jvi.00239-24

Tsao J, Chapman MS, Agbandje M, Keller W, Smith K, Wu H, Luo M, Smith TJ, Rossmann MG, Compans RW et al (1991) The three-dimensional structure of canine parvovirus and its functional implications. Science 251:1456–1464. https://doi.org/10.1126/science.2006420

Tsuji Y, Deo VK, Kato T, Park EY (2011) Production of Rous sarcoma virus like particles displaying human transmembrane protein in silkworm larvae and its application to ligand-receptor binding assay. J Biotechnol 155:185–192. https://doi.org/10.1016/j.jbiotec.2011.07.008
Urakami A, Ngwe Tun MM, Moi ML, Sakurai A, Ishikawa M, Kuno S, Ueno R, Morita K, Akahata W (2017) An envelope-modified tetravalent dengue virus-like-particle vaccine has implications for Flavivirus vaccine design. J Virol 91:e01181–e01117. https://doi.org/10.1128/JVI.01181-17
Utomo DIS, Hirono I, Kato T, Park EY (2019) Formation of virus-like particles of the dengue virus serotype 2 expressed in silkworm larvae. Mol Biotechnol 61:852–859. https://doi.org/10.1007/s12033-019-00210-5
Utomo DIS, Pambudi S, Sjatha F, Kato T, Park EY (2020) Production of dengue virus-like particles serotype-3 in silkworm larvae and their ability to elicit a humoral immune response in mice. AMB Express 10:147. https://doi.org/10.1186/s13568-020-01087-3
Utomo DIS, Pambudi S, Park EY (2022) Humoral immune response induced with dengue virus-like particles serotypes 1 and 4 produced in silkworm. AMB Express 12:8. https://doi.org/10.1186/s13568-022-01353-6
Valdes I, Marcos E, Suzarte E, Perez Y, Brown E, Lazo L, Cobas K, Yaugel M, Rodriguez Y, Gil L, Guillen G, Hermida L (2017) A dose-response study in mice of a tetravalent vaccine candidate composed of domain III-capsid proteins from dengue viruses. Arch Virol 162:2247–2256. https://doi.org/10.1007/s00705-017-3360-y
Veggiani G, Nakamura T, Brenner MD, Gayet RV, Yan J, Robinson CV, Howarth M (2016) Programmable polyproteams built using twin peptide superglues. Proc Natl Acad Sci USA 113:1202–1207. https://doi.org/10.1073/pnas.1519214113
Vuitika L, Cortes N, Malaquias VB, Silva JDQ, Lira A, Prates-Syed WA, Schimke LF, Luz D, Duraes-Carvalho R, Balan A, Camara NOS, Cabral-Marques O, Krieger JE, Hirata MH, Cabral-Miranda G (2024) A self-adjuvanted VLPs-based Covid-19 vaccine proven versatile, safe, and highly protective. Sci Rep 14:24228. https://doi.org/10.1038/s41598-024-76163-w
Wicht KJ, Mok S, Fidock DA (2020) Molecular mechanisms of drug resistance in Plasmodium falciparum malaria. Ann Rev Microbiol 74:431–454. https://doi.org/10.1146/annurev-micro-020518-115546
Wilson DW, Fowkes FJ, Gilson PR, Elliott SR, Tavul L, Michon P, Dabod E, Siba PM, Mueller I, Crabb BS, Beeson JG (2011) Quantifying the importance of MSP1-19 as a target of growth-inhibitory and protective antibodies against Plasmodium falciparum in humans. PLoS One 6:e27705. https://doi.org/10.1371/journal.pone.0027705
Win SJ, Ward VK, Dunbar PR, Young SL, Baird MA (2011) Cross-presentation of epitopes on virus-like particles via the MHC I receptor recycling pathway. Immunol Cell Biol 89:681–688. https://doi.org/10.1038/icb.2010.161
Wriggers W, Chakravarty S, Jennings PA (2005) Control of protein functional dynamics by peptide linkers. Biopolymers 80:736–746. https://doi.org/10.1002/bip.20291
Xie Q, Chapman MS (1996) Canine parvovirus capsid structure, analyzed at 2.9 A resolution. J Mol Biol 264:497–520. https://doi.org/10.1006/jmbi.1996.0657
Xu J, Guo HC, Wei YQ, Dong H, Han SC, Ao D, Sun DH, Wang HM, Cao SZ, Sun SQ (2014) Self-assembly of virus-like particles of canine parvovirus capsid protein expressed from Escherichia coli and application as virus-like particle vaccine. Appl Microbiol Biotechnol 98:3529–3538. https://doi.org/10.1007/s00253-013-5485-6
Xu J, Hiramatsu R, Suhaimi H, Kato T, Fujimoto A, Tokiwa T, Ike K, Park EY (2019a) Rous sarcoma virus antigens displaying virus-like particles as a bivalent vaccine candidate against neosporosis. Vaccine 37:6426–6434. https://doi.org/10.1016/j.vaccine.2019.09.002
Xu J, Kato T, Park EY (2019b) Development of SpyTag/SpyCatcher-Bacmid expression vector system (SpyBEVS) for protein bioconjugations inside of silkworms. Int J Mol Sci 20:4228. https://doi.org/10.3390/ijms20174228
Xu J, Sekiguchi T, Boonyakida J, Kato T, Park EY (2023) Display of multiple proteins on engineered canine parvovirus-like particles expressed in cultured silkworm cells and silkworm larvae. Front Bioeng Biotechnol 11:1096363. https://doi.org/10.3389/fbioe.2023.1096363

Yamanaka A, Suzuki R, Konishi E (2014) Evaluation of single-round infectious, chimeric dengue type 1 virus as an antigen for dengue functional antibody assays. Vaccine 32:4289–4295. https://doi.org/10.1016/j.vaccine.2014.06.017

Yamanaka A, Matsuda M, Okabayashi T, Pitaksajjakul P, Ramasoota P, Saito K, Fukasawa M, Hanada K, Matsuura T, Muramatsu M, Shioda T, Suzuki R (2021) Seroprevalence of Flavivirus neutralizing antibodies in Thailand by high-throughput neutralization assay: endemic circulation of Zika virus before 2012. mSphere 6:e0033921. https://doi.org/10.1128/mSphere.00339-21

Yuan W, Parrish CR (2001) Canine parvovirus capsid assembly and differences in mammalian and insect cells. Virology 279:546–557. https://doi.org/10.1006/viro.2000.0734

Zabel F, Kundig TM, Bachmann MF (2013) Virus-induced humoral immunity: on how B cell responses are initiated. Curr Opin Virol 3:357–362. https://doi.org/10.1016/j.coviro.2013.05.004

Zakeria B, Fierera JO, Celikb E, Chittocka EC, Schwarz-Linekc U, Moyb VT, Howartha M (2012) Peptide tag forming a rapid covalent bond to a protein, through engineering a bacterial adhesin. Proc Natl Acad Sci USA 109:E690–E697. https://doi.org/10.1073/pnas.1115485109

Chapter 8
Perspectives on the Industrial Application of the Silkworm Biofactory and Outlook

Abstract This chapter discusses the industrial applications and future outlook of the silkworm (*Bombyx mori*) biofactory for recombinant protein production. It traces the history of commercialization, starting with the first successful veterinary interferon in the 1990s, and highlights recent advancements in scalable production and downstream processing. The chapter reviews various commercialized silkworm-derived products, emphasizing the platform's advantages, such as cost-effectiveness and appropriate post-translational modifications. Looking ahead, it explores the potential of utilizing transgenic silkworms and the integration of artificial intelligence to optimize expression systems, paving the way for broader adoption in biopharmaceuticals and materials science.

Keywords Silkworm biofactory • Interferon • Silkworm-derived products • Transgenic silkworm

Growing interest among researchers in using silkworms for protein production has led to the emergence of several new business ventures in this field. This trend is driven by the unique advantages silkworms offer as biofactories, such as their high efficiency in expressing recombinant proteins, relatively low rearing costs, and the ability to perform complex post-translational modifications. As a result, the silkworm system has attracted significant attention as an alternative to traditional protein production platforms such as bacteria, yeast, and mammalian cells. Despite this progress, only a handful of products have reached the market so far, as shown in Table 8.1. The earliest example of a silkworm-derived protein product was an interferon for cats and dogs, introduced in the mid-1990s and approved for use against animal diseases in 45 different countries. This milestone demonstrated the commercial viability and global regulatory acceptance of silkworm-produced proteins, opening the door for further exploration in the field. In the 2000s, new start up companies were launched, and industry activity surged, but only a limited number of products were ultimately commercialized. This was due in part to technical challenges, such as scaling up production, ensuring consistent product quality, and meeting rigorous safety standards. Additionally, market competition from

E. Y. Park, *Silkworm Biofactory*, https://doi.org/10.1007/978-981-92-0056-6_8

Table 8.1 Commercialized products using the silkworm expression system (Park 2025)

Product	Application	Commercialization	Manufacturer	Reference
Feline interferon (FeIFN)	Treatment of viral infections in cats, especially feline calicivirus infections and parvovirus infections in dogs	1994	Toray Ind. Inc. (Ehime, Japan)	Ueda et al. (1993)
Canine interferon-γ	Treatment of canine atopic dermatitis in dogs	1997	Toray Ind. Inc.	Okano et al. (2000)
Neosilk collagen I	Research reagent or cosmetic raw material to improve skin elasticity and firmness	2008	Immuno-Biological Laboratories (IBL), Inc. (Fujioka-shi, Gunma, Japan)	Adachi et al. (2010)
Neosilk collagen III	Cosmetic raw material to improve skin elasticity and firmness	2021	IBL, Inc.	Adachi et al. (2010)
Human Amyloidβ (1–40) Assay Kit-IBL	Amyloid β measuring ELISA kit	2012	IBL, Inc.	Iizuka et al. (2009)
ELISA kit	Measurement of SARS-CoV-2 antibody	2022	KAICO Ltd. (Fukuoka, Japan)	Goto et al. (2023)
NoV-LP GI.2–4, 17	Research reagent	~2022	KAICO Ltd.	Masuda et al. (2023)
Fibronectin Neosilk, Cellular	Scaffold material for cell culture	2023	IBL, Inc.	Yanagisawa et al. (2007)
Fibronectin Neosilk, Plasma	Scaffold material for cell culture	2023	IBL, Inc.	Yanagisawa et al. (2007)

established protein production systems and the complexity of regulatory approval processes have posed significant barriers to rapid commercialization.

Today, the primary applications are antibodies for diagnostic purposes, ingredients for cosmetics, and proteins used in veterinary medicine; notably, no pharmaceuticals for human use have yet been commercialized. Diagnostic antibodies produced in silkworms are valued for their specificity and cost-effectiveness, making them attractive for use in medical diagnostics and research. In the cosmetics industry, proteins such as silk fibroin and recombinant growth factors derived from silkworms are incorporated into skincare products for their purported benefits for skin regeneration and hydration. Veterinary medicine has also benefited from silkworm-derived proteins, with several therapeutic proteins and vaccines for animal health developed using this platform. However, the transition to human therapeutics has been slow, largely due to the need for extensive clinical testing and strict regulatory oversight.

In Japan, the production of genetically engineered proteins using silkworms must comply with the Cartagena Protocol on Biosafety (Guideline 2023). This process requires filing applications under both Type 1 and Type 2 regulatory frameworks and implementing various containment measures for genetically modified silkworms. Type 1 regulations mandate close oversight of breeding practices and the identification of moths to avoid unintentional cross-breeding with wild populations. These measures are crucial to prevent the uncontrolled spread of genetically modified traits in the environment. Type 2 regulations are aimed at preventing genetically modified organisms (GMOs) from spreading into the environment and require strict containment facilities and protocols. Successfully bringing silkworm-produced proteins to market involves adhering to stringent regulatory procedures, including, in some cases, an application and review under the Law for Ensuring the Safety of GMO. This comprehensive regulatory landscape ensures the safety and traceability of silkworm-derived products but can also lengthen development timelines and increase costs. Looking forward, continued innovation in genetic engineering, process optimization, and regulatory compliance will be essential to unlocking the full potential of silkworm biofactories for industrial protein production. As more stakeholders recognize the benefits and reliability of this system, the range of commercial products—and their impact on medicine, diagnostics, and other industries—will likely continue to expand.

In the pharmaceutical manufacturing, stringent measures must be taken to minimize the risk of contamination. The primary issue is not simply the presence of DNA but rather the presence of specific DNA sequences—most notably those that encode active oncogenes—or other contaminants originating from the host cells used in production. Regulatory agencies such as the Food and Drug Administration (FDA), the European Medicines Agency (EMA), and the International Council for Harmonisation (ICH), specifically through guideline Q5A(R2), mandate that each product undergo a detailed, case-by-case risk assessment regarding residual DNA before it can receive market authorization (Guideline 2023). These evaluations consider the nature and potential hazards posed by any DNA present in the final product. Generally, it is accepted that when the amount of residual DNA from continuous cell lines is kept at or below 100 pg per parental dose (Grachev et al. 1998), the associated risk is extremely low. Nevertheless, certain vaccine types are subject to even more rigorous standards. For example, the European Pharmacopeia has established explicit maximum limits for residual host cell DNA in vaccines: for the human papillomavirus-like particle (HPV-LP) vaccine, the threshold is 10 ng per dose, while for the hepatitis B vaccine, the limit is set at 10 pg of DNA per dose of purified HPV-LPs (Park and Minkner 2025). These strict regulatory benchmarks highlight the high level of safety oversight required for biopharmaceutical products. At present, there are numerous legal and regulatory challenges that must be addressed before recombinant proteins produced in silkworms can be used in pharmaceuticals. The pathway to approval is complex and can vary depending on the jurisdiction and intended application. However, the adoption of silkworms as a global platform for protein production could potentially streamline regulatory processes and facilitate the broader commercialization of silkworm expressed proteins.

By harmonizing standards and building international acceptance, it may become easier to bring silkworm-derived therapeutics to the market worldwide.

From an engineering perspective, the silkworm-based protein expression platform has matured into a reliable and adaptable system for producing recombinant proteins at scale. This approach serves as a promising alternative to conventional mass-production methods, particularly those relying on *E. coli* and yeast, which, while widely used, do not always meet the sophisticated post-translational modifications necessary for certain therapeutic and diagnostic proteins. In contrast, the silkworm system can perform complex modifications similar to those in mammalian cells, making it highly advantageous for producing proteins with specific functional properties required in advanced biotechnological applications. One notable application of the silkworm expression system is the production of VLPs, which are increasingly important in the development of modern vaccines. Silkworms can generate VLPs efficiently and economically, with the added benefit of avoiding competition for resources used in food production, such as eggs. This efficiency, combined with the ability to rapidly scale up production, positions silkworms as a valuable resource in responding to emerging infectious diseases. However, before these VLPs can be widely commercialized, comprehensive safety assessments and regulatory clearances are essential to ensure their suitability for use in human and animal health. Advances in chemical conjugation have further expanded the capabilities of silkworm-derived VLPs, enabling researchers to attach multiple antigens to their surfaces. This technological progress opens the door to the creation of multivalent vaccines, potentially targeting complex diseases such as dengue and malaria. Such innovations could significantly enhance global health efforts, especially in regions where these diseases are prevalent. Currently, silkworm expression technology is most prevalent in Asia, but it has substantial potential for worldwide adoption. By developing new methods for rearing silkworms and establishing robust distribution channels, the reach of this technology could be extended far beyond its traditional markets. Historically, silkworms have played an important role in cultural and economic exchanges from east to west, along the Silk Roads, by providing valuable silk. Today, they stand at the forefront of a new era in biotechnology, potentially forging a "Bioroad" that connects innovative scientific advancements "From Silk Road to Bioroad" with practical applications across the globe.

References

Adachi T, Wang X, Murata T, Obara M, Akutsu H, Machida M, Umezawa A, Tomita M (2010) Production of a non-triple helical collagen alpha chain in transgenic silkworms and its evaluation as a gelatin substitute for cell culture. Biotechnol Bioeng 106:860–870. https://doi.org/10.1002/bit.22752

Goto T, Sasaki T, Chong Y, Taniguchi M, Lee JM, Masuda A, Ebihara T, Shiraishi K, Tani N, Yonekawa A, Gondo K, Kuwano H, Shimono N, Ikematsu H, Akashi K, Kusakabe T (2023) SARS-CoV-2 strain-specific anti-spike IgG ELISA utilizing spike protein produced by silkworms. Hum Antibodies 31:27–33. https://doi.org/10.3233/HAB-230006

Grachev V, Magrath D, Griffiths E (1998) WHO requirements for the use of animal cells as in vitro substrates for the production of biologicals (Requirements for biological susbstances no. 50). Biologicals 26:175–193. https://doi.org/10.1006/biol.1998.0153

Guideline ICH (2023) Viral safety evaluation of biotechnology products derived from cell lines of human or animal origin Q5A (R1). ICH Expert Working Group

Iizuka M, Ogawa S, Takeuchi A, Nakakita S, Kubo Y, Miyawaki Y, Hirabayashi J, Tomita M (2009) Production of a recombinant mouse monoclonal antibody in transgenic silkworm cocoons. FEBS J 276:5806–5820. https://doi.org/10.1111/j.1742-4658.2009.07262.x

Masuda A, Man Lee J, Miyata T, Sato S, Masuda A, Taniguchi M, Fujita R, Ushijima H, Morimoto K, Ebihara T, Hino M, Kakino K, Mon H, Kusakabe T (2023) High yield production of norovirus GII.4 virus-like particles using silkworm pupae and evaluation of their protective immunogenicity. Vaccine 41:766–777. https://doi.org/10.1016/j.vaccine.2022.12.015

Okano F, Satoh M, Ido T, Okamoto N, Yamada K (2000) Production of canine IFN-gamma in silkworm by recombinant baculovirus and characterization of the product. J Interf Cytokine Res 20:1015–1022. https://doi.org/10.1089/10799900050198462

Park EY (2025) From silkroad to bioroad-silkworm biotechnology. Protein Expr Purif 234:106762. https://doi.org/10.1016/j.pep.2025.106762

Park EY, Minkner R (2025) A systematic approach for scalable purification of virus-like particles. Protein Expr Purif 228:106664. https://doi.org/10.1016/j.pep.2025.106664

Ueda Y, Sakurai T, Yanai A (1993) Homogeneous production of feline interferon in silkworm by replacing single amino acid code in signal peptide region in recombinant baculovirus and characterization of the product. J Vet Med Sci 55:251–258. https://doi.org/10.1292/jvms.55.251

Yanagisawa S, Zhu Z, Kobayashi I, Uchino K, Tamada Y, Tamura T, Asakura T (2007) Improving cell-adhesive properties of recombinant *Bombyx mori* silk by incorporation of collagen or fibronectin derived peptides produced by transgenic silkworms. Biomacromolecules 8:3487–3492. https://doi.org/10.1021/bm700646f

www.ingramcontent.com/pod-product-compliance
Ingram Content Group UK Ltd.
Pitfield, Milton Keynes, MK11 3LW, UK
UKHW020632260926
475829UK00003BA/45

* 9 7 8 9 8 1 9 2 0 0 5 5 9 *